Aerial Plant Surface Microbiology

Aerial Plant Surface Microbiology

Edited by

Cindy E. Morris and
Philippe C. Nicot
Institut National de la Recherche Agronomique (INRA)
Station de Pathologie Végétale
Avignon, France

and

Christophe Nguyen-The
Institut National de la Recherche Agronomique (INRA)
Station de Technologie des Produits Végétaux
Avignon, France

Plenum Press • New York and London

Library of Congress Cataloging-in-Publication Data

Aerial plant surface microbiology / edited by Cindy E. Morris and
   Philippe C. Nicot and Christophe Nguyen-The.
       p.   cm.
     "Proceedings of the Sixth International Symposium on the
   Microbiology of Aerial Plant Surfaces, held September 11-15, 1995,
   in Bandol, France"--T.p. verso.
     Includes bibliographical references and index.
     ISBN 0-306-45382-7
     1. Plant surfaces--Microbiology--Congresses.   I. Morris, Cindy E.
   II. Nicot, Philippe C.   III. Nguyen-The, Christophe.
   IV. International Symposium on the Microbiology of Aerial Plant
   Surfaces (6th : 1995 : Bandol, France)
   QR351.A393   1996
   576'.15--dc20   96-31150
                                                                  CIP

Front cover: Naturally occurring microbial biofilms on leaves of broad-leaved endive (*Cichorium endivia* var. *latifolia*) (magnification 764×; courtesy of Jean-Michel Monier).

Proceedings of the Sixth International Symposium on the Microbiology of Aerial Plant Surfaces, held September 11–15, 1995, in Bandol, France

ISBN 0-306-45382-7

© 1996 Plenum Press, New York
A Division of Plenum Publishing Corporation
233 Spring Street, New York, N. Y. 10013

Printed in the United States of America

To the young researchers who will find their inspiration
in the beauty and enigma of life on the leaf surface.

PREFACE

Since 1970, scientists from diverse disciplines have met every five years to discuss issues related to the biology and ecology of micro-organisms associated with aerial surfaces of plants. These symposia have helped advance our understanding of a tremendously complex habitat that harbours plant pathogens, as well as micro-organisms that are beneficial for plant health, that may influence global weather, that have an impact on food technology, and that may be noxious to animals. Each of the six symposia since 1970 has culminated in a volume; this series of books is a valuable testament to the status and evolution of this field of research. The references for the five volumes preceding this present volume are given in Chapter 19 by J. H. Andrews.

This present volume issues from the 6th International Symposium on the Microbiology of Aerial Plant Surfaces, which convened from 11–15 September 1995 on the Island of Bendor, Bandol, France. This symposium was attended by 105 scientists from 22 countries. The chapters in this volume represent the major presentations given at this symposium, although here they have been expanded and adapted to a more general audience. These chapters consider the microbiology not only of surfaces of leaves but also of bark, stems, and fruits at different stages of maturity from emergence until senescence and death.

This volume is organised into five principal sections representing the major themes of the symposium as established by its scientific committee (S. Droby, S. S. Hirano, L. L. Kinkel, J. L. Luisetti, C. E. Morris, and T. G. Villa). These five themes were chosen because they represent both the areas of this field where much progress has been made since the last symposium and domains where we need to be asking provocative questions to ensure that this progress continues. For example, the first two chapters, by J. Schönherr and P. Baur and by S. Derridj, describe the major advances that have been made in the techniques and models available for evaluating the nature of the chemical environment on leaf surfaces, but they also illustrate the immeasurable task that lies ahead of researchers aspiring to characterise nutritional resource pools in space and time from the perspective of a micro-organism. The chapters in the part on quantification and modelling of microbial population dynamics are meant to stimulate thought about experimental strategies for measuring microbial population size or assessing the competitiveness of micro-organisms. Once the data from such experiments are collected and analysed, we are reminded in the last chapter of this part by C. D. Upper and S. S. Hirano to consider if the results of our laboratory tests reflect what is really happening in natural settings.

This book has a lengthy part dedicated to the importance of plant surface micro-organisms for agricultural and food quality. This section considers how agricultural practices in greenhouse and field settings have been adapted to the biology and ecology of pathogens of aerial plant parts. This part is complemented by Chapter 5 by Droby *et al.* (a previous section) on the use of yeasts for biocontrol of postharvest pathogens of fruits. The previous volume in this series, edited by J. H. Andrews and S. S. Hirano, was the first to include chapters about the biology and ecology of micro-organisms of medical importance associated with aerial plant parts. In this present volume, we have expanded on this approach by including chapters addressing the importance for food quality of organisms such as *Listeria monocytogenes* and toxinogenic fungi that are adapted to life on aerial plant parts. The parts on the interactions of epiphytic micro-organisms with the plant and with other micro-organisms clearly illustrate the tremendous contribution of microbial and plant genetics and molecular biology to our understanding of phenomena such as attachment, pathogenicity, induced resistance, antagonism, *in situ* gene transfer, and coevolution of plants and parasites. The last chapter in the volume is an overview of the progress made in this field in recent years and an outlook for the future. It includes a synthesis of the views expressed by symposium participants during a highly animated discussion organised to motivate collective reflection on the status of our field and where we should be going.

We would like to thank all of the authors for their efforts to contribute very stimulating chapters to this volume. Some of these authors also participated in reviewing manuscripts contributed to this text, for which we are grateful. We would also like to thank the external reviewers for the time and care they have taken to help us ensure the quality of the information presented here: C. Boucher (INRA, Castanet-Tolosan, France); J. J. Burdon (CSIRO, Canberra City, Australia); T. J. Burr (Cornell University, USA); M. E. di Menna (Ruakura Agricultural Research Center, New Zealand); C. Jeffree (University of Edinburgh, UK); C. Kenerley (Texas A & M University, USA); M. A. Lachance (University of Western Ontario, Canada); J. W. Mansfield (Wye College, University of London, UK); M. O. Moss (University of Surrey, UK); J. E. Smith (University of Strathclyde, UK); N. R. Towers (Ruakura Agricultural Research Center, New Zealand); J. D. van Elsas (IPO-DLO, The Netherlands); G. Yuen (University of Nebraska, USA); J. M. Whipps (Horticulture Research International, UK); and M. Wilson (Auburn University, USA).

Financial support from the following sponsors is gratefully acknowledged: Campbell Scientific; Conseil Régional de Provence-Alpes Côtes d'Azur; Elsevier Science and Elsevier-France; French Embassy in Uzbekistan; French Ministry of Foreign Affairs; Groupe LVMH; Institut National de la Recherche Agronomique (Bureau of Conferences, Direction of International Relations, and the Plant Pathology Station at Avignon); Jouan, S.A.; Lesaffre Développement; Sandoz Agro, AG and Société Ricard. We also thank the Federation of European Microbiological Societies for a generous grant that provided financial assistance for a dozen young researchers to attend the symposium. Successful organisation of the symposium was possible due to the hard work and enthusiasm of many volunteers; we thank M. Bardin, M.-F. Cornic, V. Decognet, C. Glaux, P. Gros, S. Menassieu, J.-M. Monier, R. Olivier from the Plant Pathology Station of INRA-Avignon, and J.-P. Prunier from INRA-Paris.

We hope that this volume can convey part of the passion that those of us dedicated to this field have for life on plant surfaces and the importance of these regular interdisciplinary symposia. This sentiment was best summarised by one of the participants of the 6th symposium, Barrie Seddon of the University of Aberdeen, in his report to the British Society for Applied Bacteriology. Dr. Seddon was grateful for the "opportunity to be involved with and enjoy this symposium in such idyllic and romantic surroundings. This was indeed a place to remind oneself that science can have a magnetic attraction— almost like falling in love—something that is becoming increasingly rare in our modern day society where time seems not to be given over to such focused discussions. 'Put away your PC's, stop surfing the Internet and get back to your laboratories, research work and discussions' is the call!'' (*Soc. Appl. Bacteriol. Notices and News*, Dec. 1995, pp. 17–19).

Cindy E. Morris, Philippe C. Nicot, and Christophe Nguyen-The

CONTENTS

Part IV: Plant Surface Micro-Organisms, Agricultural Practices, and Food Quality

Part V: Quantification and Modelling of Population Dynamics of Micro-Organisms Associated with Aerial Plant Surfaces

Part VI: Research Outlook

Aerial Plant Surface
Microbiology

CUTICLE PERMEABILITY STUDIES

A Model for Estimating Leaching of Plant Metabolites to Leaf Surfaces

Jörg Schönherr and Peter Baur

The Phytodermatology Group
Institute of Fruit and Nursery Sciences
University of Hannover
Am Steinberg 3
D-31157 Sarstedt
Germany

INTRODUCTION

Plant leaf surfaces constitute an important habitat for micro-organisms in many vegetation types. Bacteria, yeasts, fungi and lichens can be found and the bacterial population alone ranges from 10^4 to 10^6 cells/cm^2 (Libbert and Manteuffel, 1970; Hirano and Upper, 1983). These micro-organisms are essentially immobile on leaves and for those which are not pathogenic and do not penetrate the interior of the leaves the origins of nutrients essential for their growth is still an unresolved problem. Can we regard these organisms as epiphytes in the classical sense, which would imply that leaves are used only as base for attachment or do they serve as sources for substrates at least temporarily? Heterotrophic epiphytes could obtain organic carbon from volatiles contained in ambient air or emitted from the host plant through open stomata, or they may utilise metabolites which diffuse through the cuticle. Some reports have demonstrated various interactions between host plants and micro-organisms on their leaves. Volatile compounds such as C_6 saturated and unsaturated alcohols and aldehydes, which are responsible for the green odour of leaves (Hatanaka, 1993), can inhibit proliferation of micro-organisms in a bioassay system if applied via the vapour phase in concentrations of some 100 µg/litre (Deng et al., 1993). These compounds also inhibited pollen germination (Hamilton et al., 1991) and development of some arthropods (Deng et al., 1993). Many other volatile compounds are emitted from leaves, flowers and fruits such as the typical aroma constituents small alcohols, aldehydes, and esters (e.g. Mattheis et al., 1991) and monoterpenes which play a role in allelopathy (e.g. Vaughn and Spencer, 1993). Some of these compounds are released in large quantities, e.g. conifers are reported to emit under extreme temperature and light conditions up to 10 µg terpenes per g leaf and minute (after Hock and Elstner, 1988), but if they can serve as substrates for micro-organisms appears not to have been investigated.

Aerial Plant Surface Microbiology, edited by Morris et al.
Plenum Press, New York, 1996

Transfer of compounds from epiphytes to the plant host via the cuticle has been demonstrated. Examples are the various pre-infection events of pathogens (Harborne, 1982) and diffusion of indoleacetic acid (IAA) from different epiphytes to corn coleoptiles (Libbert and Silhengst, 1970). It has been suggested by Hirano and Upper (1990) that some stimulants may be produced by epiphytes which could increase the flux of nutrients from plant to micro-organism. Biosurfactants are possible candidates for such compounds and they are produced by many micro-organisms during growth on water insoluble substrates (Cooper and Zajic, 1980). They have been shown to enhance wetting of leaf surfaces (Bunster et al., 1989) and as emulsifiers they might solubilise constituents of plant cuticles such as waxes, which could serve as a source of organic carbon. Amino acid-containing surfactants and the group of glycolipid surfactants (Fiechter, 1992) are typical representatives of biosurfactants. However, there is not much known about their occurrence in the surroundings of epiphytes and it is unknown if they can increase availability of plant metabolites other than cuticular waxes to epiphytes.

The mass of cuticles and cuticular waxes is large relative to the mass of epiphytic micro-organisms. Typical cuticles from mesophytic plants have masses ranging from about 100 to 250 $\mu g/cm^2$ (corresponding to a thickness of 0.9 to 2.3 μm.) and the quantities of cuticular waxes range from about 5 to 60 $\mu g/cm^2$ (Baker, 1982; Watanabe and Yamaguchi, 1991). Fruit cuticles may even have up to 10 times more wax than leaf cuticles. In contrast, a bacterial population of 10^6 cells/cm² (at an average "fresh" cell mass of about 10^{-6} μg) has a total dry mass in the μg range. This shows that the amount of cuticular wax may be larger by an order of magnitude than the mass of the epiphytic population. Thus, epicuticular waxes are a potential source of organic carbon, especially since epicuticular waxes can be regenerated if removed (Schieferstein and Loomis, 1959). Polypeptides and polysaccharides do not occur on the surface of cuticles but they are important constituents of the polymer matrix of cuticles (Schönherr and Bukovac, 1973; Riederer and Schönherr, 1984) and are potential nutrient sources only for fungi which can synthesise cutinase (Kollatukudy, 1985; Pascholati et al., 1992). Amino acids and carbohydrates occur in the apoplast and their concentrations in barley and spinach leaves range from 0.01 to 10 millimolar (Lohaus et al., 1995). However, they are not readily available to epiphytic micro-organisms unless these compounds diffuse through the cuticle. Intimate contact between epiphytic micro-organisms and plant cuticles facilitates exchange of dissolved substances. Some fungi can produce exudates which make the leaf surface more hydrophilic and establish close contact (Freytag and Mendgen, 1991; Pascholati et al., 1992) and the production of extracellular polysaccharides by bacterial pathogens on the surface of mushroom caps (Goodman et al., 1986) and plant leaves (Lindow, 1990) is known.

Plant cuticles are lipid membranes and their permeabilities for lipophilic compounds can be substantial (Schönherr and Riederer, 1989). The best example is secretion of waxes, which are synthesised in epidermal cells and deposited within cutin (intracuticular waxes) and on the surface of the cuticle (epicuticular waxes). But even fairly polar solutes can penetrate cuticles as has been shown with many plant protection agents (Hull, 1970). Since foliar uptake is possible, the reverse process is also possible and Bauer and Schönherr (1992) have shown that permeability of Citrus leaf cuticles to 2,4-dichlorophenoxy acetic acid (2,4-D) in the steady state is practically the same in both directions. Many different compounds can be recovered when leaves are washed with water. This process is called leaching (washing-off and washing-out are usually not distinguished). The last comprehensive review on leaching was prepared by Tukey 26 years ago (Tukey, 1970). He defined the term leaching as the loss of substances from above ground parts of plants by the action of water (rain, dew, fog) and many different inorganic and organic compounds have been identified in leachates, for example K^+, Ca^{2+}, Mg^{2+}, Mn^{2+}, sugars, sugar alcohols, amino acids, pectic substances, organic acids, growth substances, vitamins, phenolic compounds and alkaloids. Some of the examples quoted are astounding. For instance, 50% of Ca^{2+} and 80 % of K^+ were reported to have been leached from apple leaves in 24 hours. Total losses of carbohydrates from apple leaves were estimated to amount to 800 kg per hectare and year. Furthermore, 52% of the Ca^{2+} in mature barley leaves were leached by soaking them in water for a few minutes. These figures suggest that epiphytes should be amply supplied with all kinds of nutrients, but they are inconsistent with current knowledge of permeabilities of cuticles. Tukey considered contributions by nectaries, glandular trichomes, damaged leaves, but there is the uncertainty that many organic and inorganic compounds can be deposited on leaf surfaces as wet and dry deposition from the atmosphere. The possibility of leaching of nutrients from epiphytic populations was not considered either. More materials could be leached from mature and older leaves than from young developing ones. Since mature and older leaves usually contain more epiphytic organisms than young ones (Hirano and Upper, 1990) the increased leaching with age may have been due at least in part, to leaching from micro-organisms. Schreiber and Schönherr (1992) demonstrated that epiphytic organisms living on spruce needles take up methylglucose very rapidly such that equilibrium with the bathing medium was established within 2 h. Leaching from epiphytes was very rapid too, as 90% of the methylglucose previously taken up could be washed out with water within 2 h. These observations and our current knowledge about cuticular permeability (Schönherr and Riederer, 1989; Schönherr and Baur, 1994) indicate that some of the older data on foliar leaching should be considered with caution and a new look at this phenomenon appears necessary.

A compartmental analysis as used by Schreiber and Schönherr (1992) could reveal which compounds can be leached from various compartments, such as epiphytes, cuticles, apoplast and symplast. However, such comprehensive data are not available and for this reason we shall use a different approach. Starting with a quantitative and mechanistic analysis of diffusion across cuticles as affected by properties of cuticles, penetrants and temperature we shall outline the basic laws of cuticular penetration and point out those plant metabolites for which permeabilities of cuticles are sufficiently high to permit significant diffusional flows across cuticles, provided the necessary driving forces prevail.

In our modelling approach it is implicit that cuticles are solubility membranes, that means all compounds are dissolved in cutin and cuticular waxes in which they diffuse. Since diffusion in aqueous pores traversing the cuticle has never been demonstrated in our isolated cuticular membranes (Schönherr and Riederer, 1989) we feel that this approach is justified. This does not preclude that aqueous pores might exist in other types of cuticles, even though evidence in favour of the existence of water filled pores large enough to accommodate amino acids, sugars or proteins is lacking.

COMPOSITION AND STRUCTURE OF PLANT CUTICLES AS RELATED TO SOLUTE DIFFUSION

Plant cuticles are lipid membranes composed of cuticular waxes and the lipophilic cutin polymer which always contains some polar polysaccharides (pectins, celluloses, polypeptides). Lateral heterogeneity of leaf surfaces and cuticles is due to the presence of stomata, hydathodes and various types of trichomes. Thickness and permeability of cuticles over these structures usually differs. This has been reviewed before (c.f. Franke, 1967; Hull, 1970; Martin and Juniper, 1970; Bukovac, 1976) and we shall not deal with these aspects here. Cuticle thickness ranges from about 0.1 to 20 µm, but thickness is not related to permeability (Schönherr, 1982; Schönherr and Riederer, 1989; Kerstiens, 1995). Cutin is a characteristic high molecular weight polymer of mainly esterified hydroxy fatty acids (Holloway, 1982b). The outer layer of cuticles is composed of cutin and waxes but towards the epidermal cell wall cutin is linked to the cell wall constituents such as celluloses, hemicelluloses and pectin. Long chain aliphatic waxes with melting points ranging form 60 to 120 °C occur inside the cuticle and on its surface. Flavanoids and phenols also occur covalently bound to cuticles (Baker, 1982). This non-uniform distribution of polar polymers imparts to the cuticle a pronounced transversal heterogeneity, which is evident as a gradient of uptake of stains on all cuticle cross sections (Wattendorff and Holloway, 1980; Merida *et al.*, 1981; Schmidt *et al.*, 1981; Holloway 1982a). We call isolated cuticles cuticular membranes (CM). If waxes are extracted from CM intact membranes are obtained which contain cutin and the polysaccharides mentioned above, since the polymer matrix (MX) is insoluble in all known solvents.

Cuticular waxes constitute the main transport barrier in plant cuticles, even though they amount to only a few percent of the mass of cuticles. This is readily shown by comparing permeabilities of cuticles prior to and after extraction of waxes, which increases water permeability by factors up to 1500 (Schönherr, 1982) and permeability to organic compounds by up to 9200 fold (Riederer and Schönherr, 1985). It is not really clear to what extent epicuticular, intracuticular waxes and the outermost layers of cutin co-operate in the formation of the barrier, but it is has been shown, that a limiting skin (LS) is located at the outer surface (Fig. 1) and that it amounts to only a small fraction (about 10 %) of the total thickness of the cuticle (Schönherr and Riederer, 1988). The nature of this waxy limiting skin has been the subject of considerable speculation (*c.f.* Schönherr and Baur 1994; Riederer and Schreiber 1995).

LEAF COMPARTMENTS AND PATHWAYS FOR SOLUTE DIFFUSION

Leaching of plant metabolites to the leaf surfaces is a passive process and takes place by diffusion. Passive diffusional transport across membranes and interfaces is usually described by equations similar to Ohm's law. Interfaces and membranes separate the

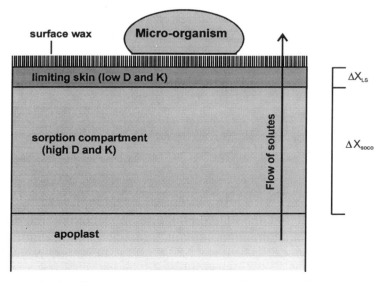

Figure 1. Schematic drawing of a cross-section of a cuticular membrane as interface between the epidermal apoplast and an epiphytic micro-organism. Permeability of cuticles is determined by the properties of the thin limiting skin of thickness Δx_{ls}, while in the relatively thick sorption compartment (Δx_{soco}) both diffusion (D) and partition coefficients (K) are much higher resulting in considerably higher permeability.

compartments and they are considered resistances which might be arranged in series or in parallel. By measuring or estimating the magnitudes of the individual resistances it is possible to evaluate their contributions to the total resistance of the diffusion path.

The overall process we are concerned with here is diffusion of metabolites from the symplast of the leaves to micro-organisms on their surfaces and as a first step we shall identify the compartments involved and the resistances encountered along the diffusion path. Later on, we shall characterise cuticular resistance in more detail and show how it depends on properties of leaves, solutes and temperature.

Metabolites are synthesised in leaf cells and some are released into the aqueous cell walls which constitute the apoplast (Lohaus et al., 1995). In our treatment the apoplast is the donor compartment (neglecting its potential transport resistance) while micro-organisms on the leaf surfaces serve as the receiver (Fig. 1). Donor and receiver are connected by two parallel pathways, which will be called the cuticular and the vapour pathways (Fig. 2). The cuticular pathway consists of two resistances in series, the cuticle (r_c) and the interfacial resistance between cuticular waxes and cell walls of the micro-organisms (r_i), where they are in direct contact. Along the vapour path three resistances in series must be overcome which are the intercellular air space, (r_a), stomata (r_s) and the boundary layer (r_b). Due to their small size micro-organisms are always inside the boundary layer. The surfaces of cuticles are never completely covered with epiphytic micro-organisms and at vacant sites volatile diffusates could evaporate and diffuse through the boundary layer (r_b) towards the ambient air. Alternatively, diffusates may form liquid or solid deposits which might dissolve in fog, dew or rain. Morphology of leaves, the presence of open stomata and properties of metabolites all determine the relative importance of these parallel pathways and the contributions of the individual resistances to the total resistance of the diffusion path.

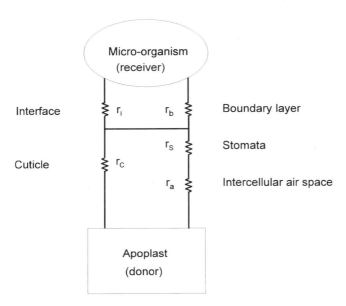

Figure 2. Pathways and the resistance (r) network for the transport of organic compounds from the leaf apoplast to epiphytes.

EQUATIONS DESCRIBING SOLUTE TRANSPORT ACROSS INTERFACES

In the steady state the net flow (F, in mol s^{-1}) is proportional to the area (A, in m^2) of the interface and the driving force. We restrict our attention to neutral organic solutes, that is, to non-electrolytes and to non-ionised weak acids and bases at relatively low concentrations. Under these conditions donor and receiver are in the liquid state and the difference in their concentrations (C) may be taken as the driving force:

$$F = PA(C^{don} - C^{rec}) \tag{1}$$

The proportionality coefficient (P) is called permeance or permeability coefficient and it characterises the permeability of a given interface (or membrane) for a given solute. P usually increases with temperature and it may depend on solute concentration. It has the dimension of velocity (ms^{-1}) and its reciprocal value is equivalent to the resistances (r) shown in Fig. 2, that is, $1/P = r$. P is a popular parameter as it easy to determine by simply measuring steady state fluxes resulting from a concentration difference between donor and receiver. If diffusion across an interface proceeds in the same phase (for instance an unstirred layer of air or water) P is simply the ratio of the diffusion coefficient (D) over the thickness of the interface (Δx):

$$P = \frac{D}{\Delta x} \tag{2}$$

However, if a change in phase is involved as in diffusion across a lipid membrane separated by aqueous solutions, then P also depends on the solvent properties of donor, receiver and membrane. A phase change requires the introduction of a partition coefficient (K) which accounts for differential solubility in the two phases. This is necessary because concentration differences are used as driving forces, instead of chemical potentials (Schönherr and Baur, 1994). Eq. 2 therefore assumes the form

$$P = \frac{DK}{\Delta x} \tag{3}$$

where, for instance, the partition coefficient membrane/water (K_{mw}) would be equal to

$$K_{mw} = \frac{C_{membrane}}{C_{water}} \tag{4}$$

and the concentrations are equilibrium concentrations. The permeability of a barrier increases with increasing solubility of a solute in the barrier. (This is the reason why permeabilities of lipid membranes increase with increasing lipophilicity of solutes.) Substituting eq. 3 in 1 gives

$$F = ADK \frac{C^{don} - C^{rec}}{\Delta x} \tag{5}$$

which is the steady state solution of Fick's first law (Cussler, 1984). Diffusion across all resistances involved can be described by equation 5 if the appropriate partition coefficients and concentrations are used.

The Cuticular Pathway

Solutes diffuse from the aqueous epidermal walls as part of the apoplast into the lipophilic cuticles made up of cutin and waxes and on the outer surface of the cuticle they reach the interfacial resistance between cuticular waxes and cell walls of the micro-organisms. It is assumed here that this interface is aqueous due to mucilage and water in cell walls. Thus, both entry into and exit out of cuticles involves a phase transition characterised by the cuticle/water partition coefficient (K_{cw}) defined by eq. 4, Δx_c is the thickness of the cuticle and the driving force is the difference in aqueous concentrations in the apoplast (apo) and the interface (i), respectively. Rearranging and simplifying equation 5 by assuming that any metabolite appearing in the interface will be taken up rapidly by the micro-organisms, which would make the interfacial concentration zero, leads to

$$\frac{F_c}{A_c} = \frac{D_c}{\Delta x_c}\left(K_{cw}C_w^{apo}\right)$$

(6)

and the flow of metabolites per unit area of cuticle can be calculated if their apparent diffusion coefficients in the cuticles (D_c), their cuticle/water partition coefficients and their concentrations in the apoplast are known. K_{cw} values do not differ much between species and they are very similar to octanol/water partition coefficients (K_{ow}) which can be taken from the literature. Some metabolite concentrations in cell walls are also known, and if not they must be measured or estimated. Diffusion coefficients in cuticles are not available and cannot be measured directly due to heterogeneity of cuticles (see below). Only a few estimates of D in reconstituted cuticular waxes have been published (Schreiber and Schönherr 1993) and the method of determination is limited to lipophilic compounds. However, solute mobilities in the limiting skins of cuticles can be estimated by a new method (Bauer and Schönherr, 1992) called UDOS (unilateral desorption from the outer surface). The rate constants of desorption (k*) are related to solute mobility in the limiting skin (D_{ls}):

$$D_{ls} = k * \cdot \Delta x_{ls} \cdot \Delta x_{soco}$$

(7)

The thickness of the limiting skin of cuticles (Δx_{ls}) is not known precisely, but it is estimated to be about one tenth the thickness of the cuticle, the remainder being the sorption compartment (Δx_{soco}) as defined in Fig. 1 (see also Schönherr and Riederer, 1988; Schönherr and Baur, 1994). Any deviation of the length of the diffusional pathway from the thickness of the limiting skin is included in k* as a tortuosity factor. These uncertainties are not a serious drawback if mobilities for different compounds in the same type of cuticle are to be compared. Since:

$$\Delta x_c = \Delta x_{ls} + \Delta x_{soco}$$

(8)

we can rewrite eq. 6 if it is assumed that $\Delta x_c \approx \Delta x_{soco}$:

$$\frac{F_c}{A_c} = k * \Delta x_{ls}\left(K_{cw}C_w^{apo}\right)$$

(9)

and diffusional flows can be predicted from the solute concentration in the apoplast, K_{cw} and k*. The latter parameter depends only on molar volumes of solutes and can be estimated for a given type of cuticle. Thus molar volumes and lipophilicity of metabolites are the only solute properties needed for predicting flows of metabolites across cuticles if their concentrations in the apoplast are known.

The Vapour Pathway

Similar equations can be used to characterise the parallel pathway in which diffusion takes place in the vapour phase. At the interface between cell walls and intercellular air spaces, solutes must vapourise. The driving force of vapour diffusion is the concentration difference between the vapour concentration at the immediate surface of the cell walls and the vapour concentration in the stagnant layer on the leaves or at the cell walls of the epiphytic micro-organisms. The solute concentrations in the apoplast (liquid phase) and in the air of the intercellular air space in contact with the surface of the cell walls (vapour phase) are in equilibrium, but they are not numerically identical. This equilibrium at the liquid/vapour interface is characterised by the air/water partition coefficient ($K_{aw} = C_a/C_w$). Written in terms of the concentration in the aqueous apoplast (C_w) eq. 6 for the vapour pathway thus assumes the form:

$$\frac{F_v}{A_v} = \frac{D_v}{\Delta x_v}(K_{aw}C_w^{apo}) \tag{10}$$

The air/water partition coefficient is related to saturation vapour pressure of the solute (p_s) and to water solubility (S_w) (or to the Henry constant (H)) and to the gas constant (R) and temperature (T):

$$K_{aw} = \frac{p_s}{S_w RT} = \frac{H}{RT} \tag{11}$$

Substituting eq. 11 into eq. 10 results in an expression which relates vapour flux (per unit area) to vapour pressure and water solubility:

$$\frac{F_v}{A_v} = \frac{D_v}{\Delta x_v}\left(\frac{p_s C_w^{apo}}{S_w RT}\right) = \frac{D_v}{\Delta x_v}\left(\frac{HC_v^{apo}}{RT}\right) \tag{12}$$

According to equations 2 to 5 a diffusion and a partition coefficient as well as the cross section and the lengths of the vapour diffusion paths are needed to quantitatively characterise each resistance.

We shall restrict our attention to the cuticular pathway. Only a few comments concerning the vapour path will be made to identify solutes for which the vapour path makes a significant contribution to the total flow. In the vapour pathway stomatal pores are generally the limiting resistances and for this reason we shall compare the cuticular with the stomatal pathway. The cuticular path is open to all compounds which are sufficiently soluble in water and in the lipids of the cuticle, irrespective of their vapour pressures. The vapour path is available only for compounds which are volatile. Unit area of the cuticular path is 1 m^2 of adaxial or abaxial leaf surfaces. Unit area for the vapour path is zero with astomatous surfaces and when stomata are closed. When stomata are open about 2% of the leaf surface consists of open stomata. If stomata are open only 12 h per day 100 m^2 of stomatous leaf surface is needed to give 1 m^2 of pores open to vapour diffusion. Diffusion coefficients in the vapour phase are about 10^{-5} m^2/s while apparent diffusion coefficients in the limiting skins of cuticles range from 10^{-18} to 10^{-20} m^2/s (Schönherr and Baur, 1994). Thickness of the limiting skin is about 0.1 μm, while for stomatal pores 10μm may be taken as path length. The terms D/Δx therefore range from 10^{-11} to 10^{-13} m/s for cuticles and 1 m/s for stomatal pores. This means that the term D/Δx for the stomatal path is 10^{11} to 10^{13} times larger than for the cuticular path. These figures must be divided by 100 to correct for the ratio $A_{cuticle}/A_{stomata}$. It then

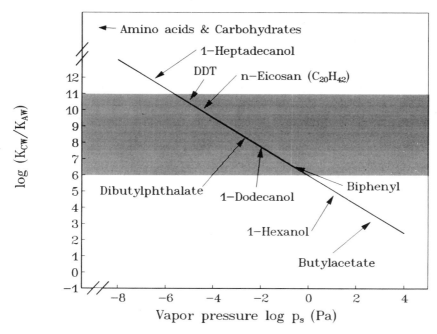

Figure 3. Dependence of the ratio of K_{CW}/K_{AW} on the saturation vapour pressure p_S of organic compounds (data taken from Reid *et al.*, 1977; Hansch and Leo, 1979; Suntio *et al.*, 1988; Sangster, 1989 and our own data).

follows that equal flows across cuticles and stomata can be expected if the ratio K_{cw}/K_{aw}, which is equal to the cuticle/air partition coefficient K_{CA}, ranges from 10^9 to 10^{11}. These are estimates for cuticles having a very low permeability. Below we shall argue that permeabilities of cuticles may be higher by 3 to 4 orders of magnitude and this means that equal flows across cuticles and stomata can be expected in the K_{cw}/K_{aw} -range of 10^6 to 10^{11}. In Fig. 3 log K_{cw}/K_{aw} is plotted against the saturation vapour pressure (p_s). The range in which flows across cuticles equal flows across open stomata is indicated. For compounds located below the hatched area significant vapour flows across open stomata can be expected to occur. For compounds above the hatched area flows in the vapour phase are insignificant. It should be pointed out that vapour pressure and water solubility are not independent and the cuticle/water partition coefficient is also inversely related to water solubility (Schönherr and Riederer, 1989). Still, Fig. 3 shows that for compounds with high vapour pressure (> 1 Pa) significant diffusion across open stomata takes place, while for compounds with p < 10^{-6} Pa only cuticular diffusion needs to be considered.

Solutes in the vapour phase escaping from open stomata are released into the stagnant air layer from which they can diffuse into turbulent air. This means that epiphytic micro-organisms will be able to secure only a small fraction of the vapour molecules emanating from stomatal pores. This fraction will depend on many biotic and environmental factors but it should increase with proximity to stomata.

The resistances of the aqueous compartments indicated in Fig. 2 are in the majority of the cases very low and not rate limiting (Schönherr and Riederer, 1989; Schönherr and Baur, 1994; Riederer and Schreiber, 1995). For this reason we can restrict our attention to mobilities and solubilities of penetrants in cuticles.

SOLUTE MOBILITIES IN CUTICLES AS AFFECTED BY PROPERTIES OF CUTICLES, SOLUTES, AND TEMPERATURE

Solute diffusion across cuticles is the limiting step in the resistance network depicted in Fig. 2. The main reason for this is the fact that diffusion coefficients in cutin and cuticular waxes are smaller by many orders of magnitude than in liquid phases. Additional details and sample calculations can be found in recent reviews (Schönherr and Riederer, 1989; Schönherr and Baur, 1994; Riederer and Schreiber, 1995) and they will not be repeated here. Instead, we restrict our attention to effects of solute and cuticle properties on solute mobility and solubility.

Solute Mobilities in Cuticular and in Polymer Matrix Membranes from Different Plant Species

Solute mobilities in the waxy limiting skin of the cuticle are determined using isolated cuticles and unilateral desorption from the outer surface (Bauer and Schönherr, 1992). In this type of experiment the sorption compartment of the cuticle (*c.f.* Fig. 1) is loaded with a test compound which is subsequently desorbed from the outer surface of the cuticle. Velocity of desorption is limited by diffusion across the waxy limiting skin. This is a first order process which can be quantitatively expressed using a first order rate constant k*. If the natural logarithm of the relative amounts of solutes (remaining in the sorption compartment at time

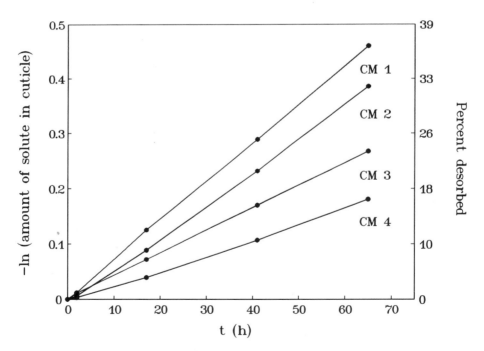

Figure 4. UDOS desorption plot showing desorption of the model compound bifenox from 4 individual pear leaf cuticular membranes (CM). If the logarithm (-ln) of the amount of bifenox in the cuticles is plotted vs. time straight lines are obtained. Their slopes are the first order rate constants of desorption (k*). For reference purposes percent desorbed is plotted on the right ordinate. Note that its scale is not linear and plotting data vs. percent desorbed would not give straight lines. This presentation was also adopted for the following SOFU graphs.

t) are plotted against time, straight lines are obtained and their slopes are equivalent to k*. An example is shown in Fig. 4 where the radio-labelled herbicide bifenox was desorbed from 4 individual pear cuticles. Bifenox is used in our laboratory routinely to test if adjuvants have accelerator properties.

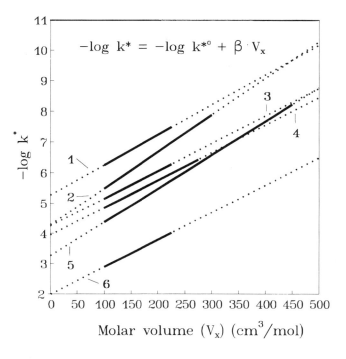

No.	Species	-log k*⁰	β	r²
1	*Ilex* CM	5.27 ± 0.80	0.010 ± 0.004	0.95
2	*Citrus* CM /cyclic compounds	4.28 ± 0.53	0.012 ± 0.003	0.86
3	*Pyrus* CM	4.25 ± 0.36	0.009 ± 0.002	0.96
4	*Capsicum* CM	3.95 ± 0.38	0.009 ± 0.002	0.87
5	*Citrus* CM /aliphatic compounds	3.27 ± 0.97	0.011 ± 0.003	0.88
6	*Pyrus* MX	1.99 ± 0.57	0.009 ± 0.003	0.91

Figure 5. The effect of molar volumes (V_x) of solutes on rate constants of desorption (k*) at 25° C measured with leaf cuticular membranes (CM) of different species and polymer matrix membranes (MX) from pear leaves (see text). Solid bars mark the range of V_x covered by the model compounds used to estimate the parameters of the linear equations (-log k* = - log k*⁰ + β·V_x) given on the right side, where -log k*⁰ is the y intercept and β the slope of the lines, respectively (calculated from data published by Baur *et al.*, 1996b, and from unpublished *Ilex* data).

Such experiments have been conducted using isolated leaf cuticles from bitter orange (*Citrus aurantium* L.), pear (*Pyrus communis* L.), chaplet flower (*Stephanotis floribunda* Brogn.), mate (*Ilex paraguariensis* St.-Hil.) and from pepper (*Capsicum annuum* L.) fruit cuticles. Up to 27 different solutes were tested and good correlations between k* and molar volumes were observed for all CM and for pear leaf polymer matrix (cuticles from which waxes had been extracted), provided cyclic and aliphatic compounds were plotted separately (Fig. 5). There was no significant correlation between k* and lipophilicity (K_{ow}) or vapour pressure (p_s) of the compounds.

The slopes of these lines are a quantitative measure of the size selectivity (β) of the cuticles and they were very similar, while the y-intercepts (k^{*o}) differ considerably (Baur *et al.*, 1996b). From Fig. 5 it can be concluded that size of solutes as expressed by molar volume and the shape of the molecules (cyclic vs. aliphatic) are the only solute properties which determine mobility in a given type of cuticle. Differences among plant species and between cuticle and polymer matrix are due mainly to differences in the y-intercepts. The y-intercept represents the mobility of a hypothetical molecule having zero molar volume. Comparing the results for pear leaf cuticles (CM) and pear leaf polymer matrix (MX) shows that extracting the waxes had no effect of size selectivity (β) but greatly increased the y-intercept. Similar results have been obtained with bitter orange leaf CM and MX. If this is a general feature of leaf cuticles then the increase in k^{*o} on extraction of waxes could be used as a quantitative measure of effects of waxes on solute mobility in cuticles. Likewise, differences in k^{*o} among different plant species would characterise the barrier properties of different cuticular waxes. Mobilities (k*) are highest in the polymer matrix and impregnating cutin with waxes can reduce solute mobility by more than 100 fold. Increasing the size of the solutes from 100 to 500 cm^3/mol reduces mobility by a factor of 3162 if the slope is 0.009. Thus, the presence of cuticular waxes in cuticles and size selectivities affect solute mobilities independently. Once the regression equations shown in Fig. 5 have been established using a limited number of model compounds as a training set, they can be used to calculate and predict k* for any other solute.

Effects of Accelerators on Solute Mobilities in Cuticular Waxes

During the course of our experiments it was observed that mobility of the ethoxylated alcohol octaethyleneglycol dodecylether ($C_{12}E_8$) was much higher than predicted by the data summarised in Fig. 5. This indicated that this compound may have changed barrier properties of the cuticles. In a systematic study using alcohol ethoxylates differing in length of the carbon chains and in degree of ethoxylation their effects on solute mobility in cuticles was studied. 2,4-D was used as a model compound, but we have also observed that mobilities of all solutes including wax constituents are increased. In fact, the effect increases with increasing molar volume of the test compounds (unpublished results).

UDOS was again used and results from a typical experiment with radio labelled chlorfenvinphos and *Citrus* CM is depicted in Fig. 6A. Initially, the rate constants of desorption (slopes) were determined for each cuticle using an inert desorption medium for which it is known that it does not change barrier properties of cuticles. For instance, an aqueous 1% phospholipid suspension (PLS) may be used. At the time marked by the arrow, desorption was continued using 25 mM C_8E_4. The sudden increase in slopes shows that solute (chlorfenvinphos) mobility increased, and the ratio of the two slopes ($k^*_{surfactant}/k^*_{PLS}$) is a quantitative measure of the effect of C_8E_4 on solute mobility in the limiting skin. Compounds which increase rate constants of desorption (mobility) have been termed accelerators or accelerator adjuvants. If the effect is plotted vs. $1/k^*_{pls}$ a good linear relationship is often observed as seen in Fig. 6B. This shows that the effect of an accelerator is greater with cuticles having a low initial mobility.

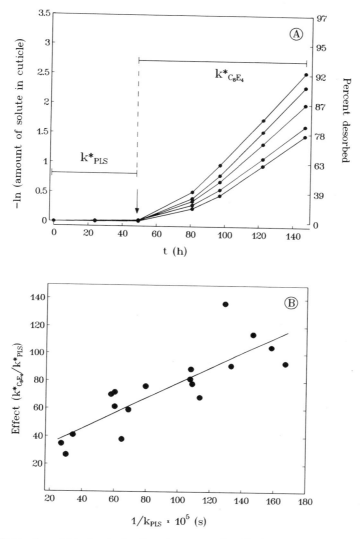

Figure 6. (A) Desorption of chlorfenvinphos from the outer surface (T = 25° C) of 5 selected *Citrus* CM using a phospholipid suspension (1% PLS, = control) or (beginning at the time marked by the arrow) aqueous solutions of tetraethylenglycol monooctylether (25 mM). (B) Dependence of the maximum effect on initial rate constants measured using PLS as a desorption medium (k*$_{pls}$). The maximum effect is the ratio of maximum rate constants of desorption measured with C_8E_4 (k*$_{C8E4}$) over k*$_{pls}$. This ratio was calculated for each CM separately.

Solute mobilities in cuticles can be increased by up to two orders of magnitude, depending on the type of cuticle, solute, accelerator.

Accelerators must be sorbed in cuticular waxes in order to be effective. They increase fluidity of cuticular waxes in the limiting layer of the cuticle (see below and L. Schreiber, unpublished) and possibly reduce tortuosity of the diffusion path (Baur *et al.*, 1996a, b). This effect is proportional to the amount of accelerator sorbed in the wax (Riederer *et al.*, 1995; Baur *et al.*, 1996a). That means, accelerators must be lipophilic and soluble in cuticular waxes. Some accelerators have been shown to increase water permeability of plant cuticles (Geyer and Schönherr, 1988). Later it was found that many compounds can have accelerator activity and ethoxylation or surface activity is not necessary. For instance, dialkylesters of aliphatic dicarboxylic acids (diethylsuberate and diethylsebacate) and trialkylesters of phosphoric acid (tributylphosphate) are very powerful accelerators, as are heptanol, octanol, nonanol and decanol (Schönherr, 1993), as well as octane, octanoic acid and octylamine (unpublished results). These aliphatic compounds are synthesised by plants and we believe that their presence in cuticular waxes can lead to a modulation of solute permeability *in vivo*. Since mobilities of solutes in cuticular waxes increase with accelerator concentration their effects will depend on their rates of synthesis and rates of evaporation from the cuticles. Both rates depend on environmental conditions which therefore could modify permeability of cuticles.

These relatively short chain alcohols are volatile. If they are present in the cuticle prior to isolation then they are lost during processing and storage of isolated cuticles. This could mean that solute mobilities measured using isolated cuticles mark the lowest values that can be obtained with a given type of cuticle. While many volatile compounds can be found in the atmosphere over canopies, they are generally not found as constituents of cuticular waxes because they are lost during work up prior to GC-MS analysis. These aspects are important and need to be investigated.

Accelerators soften cuticular waxes and they make them more fluid by a plasticising action. This offers the possibility for epiphytic micro-organisms to actively increase permeability of cuticles by secreting compounds having accelerator activity. In this context it is notable that epiphytic micro-organisms are in close contact with the most sensitive, that is the rate limiting part of the cuticle (see Fig. 1). Possible candidates for accelerators are

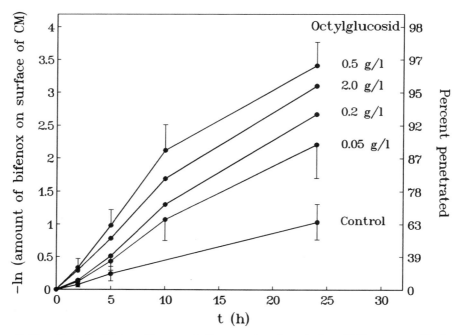

Figure 7. The effect of octylglucoside concentrations on rates of penetration of bifenox across *Stephanotis* cuticular membranes (means of 12 to 18 CM; bars represent 95% confidence intervals).

lipophilic biosurfactants which are not volatile. We have tested octylglucoside as a representative of the group of glycolipids.

The test procedure used was SOFU (*simulation of foliar uptake*) where we studied the effect of this surfactant on rates of penetration from the outer to the inner surfaces of isolated cuticles. In this test procedure small droplets (5μL) containing the test compound (^{14}C-bifenox having a K_{cw} = 27500, molar volume = 216 cm^3/mol) and octylglucoside dissolved in water are applied to the outer surface of cuticular membranes. As soon as the water had evaporated desorption was started by adding 1% PLS into the desorption chamber and the test compound was desorbed from the inner surface of the cuticle (for further details see Schönherr and Baur, 1994). Both compounds penetrate the cuticle, but only the flux of the radio labelled bifenox was measured. Bifenox was used at a concentration of 0.05 g/l and octylglucoside at concentrations given in Fig. 7. The control was bifenox in water/acetone. A straight line was observed which shows, that penetration from the surface to the PLS receiver was a first order process. The slope of the line is the first order rate constant. In the presence of octylglucoside initial rates of penetration were much higher and highest rates were observed with an octylglucoside concentration of 0.5 g/l, which corresponds to 2.5 μg per droplet covering an area of approximately 0.2 cm^2. However, octylglucoside was effective even at a much lower concentration (0.05 g/l). *Stephanotis* leaf cuticles have about 20 to 30 μg wax per cm^2 and it appears that small amounts of accelerator (considerably smaller than the amounts of wax) can significantly increase permeability of cuticles. Curves level off after 10 h because octylglucoside penetrated the cuticle such that its concentration in cuticular wax - and therefore the effect on mobility of bifenox in wax - decreased with time. Increasing the concentration of octylglucoside even slowed penetration since it serves as a solvent and a diluent on the surface of the cuticle and therefore reduces the driving force for penetration. This dilution effect would not be operative if octylglucoside were applied to the outer surface and bifenox to the inner surface of the cuticles.

From Fig. 7 it is clear that biosurfactants can increase permeabilities of cuticles. This effect is not limited to *Stephanotis* CM and bifenox. Mobilities of plant metabolites will be increased as well, but the magnitude of the effect depends on the type of biosurfactant, on molar volume of metabolites and on properties of the waxy limiting skin. It remains to be seen which epiphytic bacteria produce sufficient amounts of biosurfactants having accelerator properties.

Effect of Temperature on Solute Mobility in Cuticles

Temperature affects both sorption and solute mobility in membranes. In addition, temperature may change membrane structure, by inducing phase transitions. While studying water permeability it was found that isolated cuticles underwent a phase transition in the range of 40 to 50°C, which caused a sudden and steep increase in water permeability (Schönherr et al.. 1979). Using differential thermal analysis of wet isolated cuticles Eckl and Gruler (1980) observed exotherms at 38 and 44°C, the first of which was attributed to cuticular waxes, while the second was associated with the polymer matrix. Schreiber and Schönherr (1990) investigated temperature induced volumetric expansion of cuticles from 12 plant species. They found for all species second order phase transitions in the temperature range of 38 to 55°C, depending on species. *Citrus aurantium* and *Ficus elastica* cuticles exhibited an additional phase transition between 15 and 25°C. At the phase transition temperature, volume expansion coefficients showed a sudden increase leading to strain between the various components of the cuticles (cellulose, cutin, wax) having different thermal expansion coefficients. This probably caused disorder and defects between polymer matrix and cuticular waxes, and water permeability increased as a consequence.

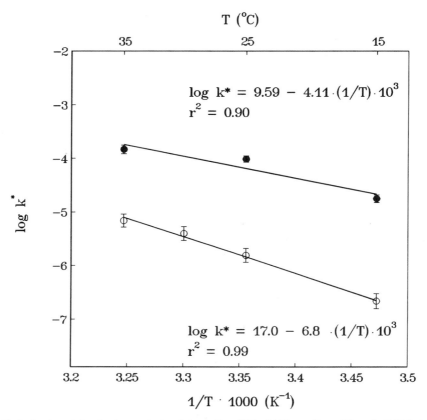

Figure 8. Arrhenius plots of rate constants of desorption (k*) of bifenox from *Citrus* leaf CM at various temperatures (T). Desorption was carried out either with a phospholipid suspension (control, empty symbols) or with 25 mM diethylsuberate dissolved in 1% PLS (filled symbols). (Bars represent 95% confidence intervals).

Phase transitions at temperatures above 45 °C probably are of little consequence as far as availability of nutrients to epiphytic organisms is concerned, so in studying temperature effects on solute mobility we restricted the temperature range to 15 to 40 °C. *Stephanotis* CM and bifenox were used and desorption from the outer surfaces was carried out using an inert phospholipid suspension. Arrhenius plots were linear in this temperature range where phase transitions must not be expected (Fig. 8, empty symbols). Activation energies of diffusion can be calculated from the slopes of the Arrhenius graphs using the equation $E_D = (-2.3R) \times$ slope (Baur and Schönherr, 1995), where R is the gas constant (8.3143 J/mol K). This calculation results in an activation energy of 130 kJ/mol. Activation energies have been measured for cuticles of other species and various solutes and they ranged from 110 to 170 kJ/mol, depending on type of cuticle and molar volume of solutes, such that E_D was greater with cuticles having low solute mobility and compounds of large molar volumes (Baur and Schönherr, 1995). From these data Q_{10} values ranging from 5 to 10 can be calculated. This shows that solute mobilities in cuticles increase very rapidly with increasing temperatures such that mobilities and permeabilities (see eq. 5) increase by factors of 5 to 10 for every 10 degree temperature interval. Temperature effects are larger with cuticles having a low intrinsic mobility and/or for solutes having a low mobility due to their large molar volumes.

If activation energies were plotted vs. $1/k^*$ an excellent linear relationship was obtained which can be used to predict activation energies from a single determination of k^* for any cuticle/compound combination.

If the accelerator adjuvant diethylsuberate (0.025 mol/L) was added to the PLS used for desorption two effects were observed: (1) Rate constants (k^*), that is solute mobilities, were higher by more than an order of magnitude and (2) the slope of the Arrhenius graph was not as steep, such that the activation energy of diffusion amounted to only 78.6 kJ/mol (Fig. 8, filled symbols) and the Q_{10} was reduced to 2.2. The decrease in the activation energy is good evidence for the plasticising effect of accelerators on cuticular waxes (Baur *et al.*, 1996a). The accelerator effect was greatest at low temperatures, which means that the production of biosurfactants or other compounds with accelerator properties would be most advantageous at low temperatures, when permeabilities of cuticles are low and accelerator effects are large.

EFFECT OF TEMPERATURE ON SOLUBILITY OF SOLUTES IN CUTICLES

Permeability of membranes is proportional to solute mobility and solubility in the membrane (eq. 3 and 10). In the above section we have shown that solute mobility increases rapidly with rising temperatures, both in absence and presence of accelerators. Along the cuticular pathway there is a transition from the aqueous (cell wall) to a lipophilic phase (cuticle) which requires the introduction of the cuticle/water partition coefficient (eqs. 3 and 4). Two questions therefore arise: (1) To what extent do partition coefficients vary among cuticles from different plant species? (2) Are partition coefficients affected by temperature?

Riederer and Schönherr (1984) studied sorption of 2,4-D in CM and polymer matrix (MX) from 11 different species. They found that fruit cuticles tended to have somewhat higher partition coefficients (K_{CW}= 424-479) than leaf cuticles (340-470). The ranges given show that variability between species amounted to not more than a factor of two. Extracting cuticular waxes increased partition coefficients by about 30 to 160 % as sorption sites previously occupied by waxes become available for 2,4-D molecules when the waxes are removed. Kerler and Schönherr (1988) compared cuticle/water and octanol/water partition coefficients for 8 solutes differing in K_{ow} by about 5 orders of magnitude and they observed excellent linear correlations for both CM and MX with all 4 plant species, such that K_{cw} and K_{mxw} can be predicted from octanol water partition coefficients (see also Sabljic *et al.*, 1990 and Schönherr and Baur, 1994 for additional data).

Above we had pointed out that cuticular waxes associated with cutin form the limiting skin of cuticles and permeability of these limiting skins is proportional to the partition coefficients in this region. These partition coefficients cannot be measured experimentally, but a few wax/water partition coefficients have been measured using reconstituted barley leaf wax (Schreiber and Schönherr, 1992) which indicate that wax/water partition coefficients are smaller than cuticle/water partition coefficients, since waxes have fewer sorption sites than cutin. The ratios K_{cw}/K_{wxw} reported range from 8.5 to 23, depending on solutes.

Temperature effects on sorption in CM and MX were investigated by Riederer and Schönherr (1986) using 4-nitrophenol (4-NP). They found that partition coefficients decreased slightly with increasing temperature and with increasing 4-NP concentration in the external aqueous phase. The concentration effect should be of little consequence for metabolite sorption in cuticles, since concentrations in the apoplast are not likely to be higher than 1 to 10 millimolar. Up to these concentration sorption isotherms where linear, that is K_{CW} was constant and independent of concentration.

Temperature dependence of partition coefficients (K) can be estimated from the Gibbs free energy ($\Delta G°$) which ranged (depending on type of cuticle and solute concentration) from approximately -10 to -11.8 kJ/mol. Using the equation

$$\Delta G° = -RT\ln K \qquad (13)$$

it can be calculated that partition coefficients decrease by factors of up to 2.8 when temperature increases from 5 to 40 °C. Thus, neither the interspecific variation of K_{cw} nor the effect of temperature on K_{cw} are very large. It should be recognised, however, that increasing temperatures decrease partition coefficients somewhat, while they increase solute mobility greatly (Baur and Schönherr, 1995), such that permeability (eq. 3) and flow rates at constant driving forces (eq. 5) increase with increasing temperatures.

PREDICTION OF FLOWS OF METABOLITES ACROSS CUTICLES IN RELATION TO NUTRITIONAL DEMANDS OF MICRO-ORGANISMS

Solute flows across cuticles are proportional to permeability of cuticles and to the concentration in the apoplast (eq. 1). Permeability of cuticles is proportional to solute mobility (D) and lipophilicity as expressed by the cuticle/water partition coefficient. Solute mobilities can be estimated from molar volumes of solutes (Fig. 5) and partition coefficients may be derived from tabulated octanol/water partition coefficients or from water solubilities (Schönherr and Riederer, 1989). Hence, we have all the information needed for estimating metabolite permeabilities of cuticles and to quantitatively correct for temperature effects on sorption as well as mobilities. Complications arise because solute mobilities in cuticles from different species vary by orders of magnitude, and accelerator affects can be large, but are difficult to predict. Currently, the state of the art permits predictions only for specific types of cuticles and accelerator concentrations, which is probably not what microbiologists are most interested in. Therefore, we have decided to first estimate maximum permeabilities and then calculate the concentration in the apoplast needed to result in metabolite flows which are needed to maintain a minimum level of growth of epiphytic micro-organisms.

Estimating Maximum Permeabilities of Cuticles

Permeabilities were estimated in three steps. First we calculated molar volumes of selected metabolites or groups of metabolites shown in Fig. 9 using the formalism of Abraham and McGowan (1987). Mobilities of these compounds (k*) were calculated based on the data for pear leaf polymer matrix (*Pyrus* MX) given in Fig. 5. These k*-values were multiplied with octanol/water partition coefficients taken from the literature and the products k*K_{ow} for each compound or groups of compounds are plotted (Fig. 9). The range of permeabilities covers 10 orders of magnitude. Polar metabolites such as amino acids and carbohydrates were lowest. Five amino acids including the smallest (glycine) and biggest (tryptophan), very hydrophilic (glycine, histidine) and most lipophilic ones (tryptophan, phenylalanine) and five carbohydrates (monosaccharides including glucose-1-phosphate and glucose-6-phosphate) and other compounds were included. Adenosine permeability is very low too, but phosphorylation tends to increase permeability somewhat. Highest values were calculated for fatty acids and fatty alcohols the latter selected as typical wax constituents

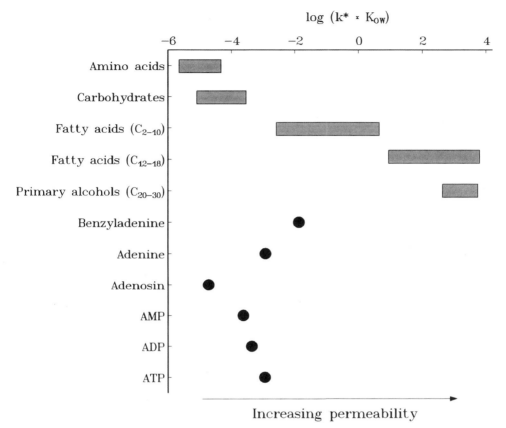

Figure 9. Estimated values of the permeability coefficients calculated as the product $k^* \times K_{OW}$ for some common plant metabolites. Partition coefficients K_{OW} (T = 20 or 25° C) were taken from Hansch and Leo (1979) or Sangster (1989).

representative of other common aliphatic and very lipophilic compounds like aldehydes and alkylesters. The product $k^* K_{ow}$ increased with increasing numbers of carbon atoms because this increases K_{ow} much faster than it decreases mobility (k^*). Estimates are maximum permeabilities as they refer to polymer matrix membranes which are free of cuticular waxes. It was pointed out above, that waxes in cuticles reduce permeabilities by up to 3 orders of magnitude. Accelerators and high temperatures increase permeability but probably not by more than 3 orders of magnitude. We have adopted this approach, because we want to point out for which types of metabolites significant cuticular flows can be expected and which compounds will practically not penetrate cuticles.

Estimating Nutrient Demands of Micro-Organisms

As an example we shall estimate the demand for organic carbon, since organic carbon is an essential nutrient for probably all epiphytic micro-organisms. It was assumed that unicellular microbes of 10^{-6} µg fresh weight are in contact with the cuticle at a concentration of 10^9 cells/m² cuticle. The microbes shall have a dry weight of 30% with a carbon content of 50% (Schlegel, 1985). Thus, the microbes contain $1.5 \cdot 10^{-4}$ g/m² carbon. If this population doubles in 24 h the carbon demand for this requires a carbon flux of $1.74 \cdot 10^{-9}$ g carbon/s m^2. If the microbes are regarded as cubes with a volume of 1 μm^3 they occupy only a small fraction of the total area of the cuticle, namely 10^{-3} m^2 and the flux of non-volatile metabolites through the cuticle under the microbes must be $1.74 \cdot 10^{-6}$ g/s m^2. This calculation neglects carbon demand for generation of ATP.

Estimating Required Metabolite Concentrations in the Apoplast

We again assume that the receiver concentration (*e.g.* the bacteria/leaf interphase) remains practically zero such that the solute concentration in the apoplast may be taken as the driving force of penetration (eq. 9) and the flows are assumed to be steady. Solving eq. 9 for the concentration in the apoplast (C_w^{apo}) the concentrations needed to obtain a carbon flux of $1.74 \cdot 10^{-6}$ g/s m^2 can be calculated when Δx_{ls} amounts to $1 \cdot 10^{-7}$ m and corrections for the carbon content of the different compounds are made (e.g. glycine has a carbon content of 32% w/w). These calculations resulted in the following molar concentrations required in the apoplast: Amino acids, > 2.6; carbohydrates, > 0.2; ATP, 0.1; benzyladenine, 10^{-2}; fatty acids (C_2 to C_{10}), 0.25 to $3 \cdot 10^{-5}$; fatty acids (C_{12} to C_{18}) and alcohols, 10^{-5} to 10^{-8}.

Lohaus *et al.* (1995) measured amino acids, glucose, fructose and sucrose concentrations in the apoplast of barley and spinach leaves to be in the range of 0.1 to 5 millimolar. These concentrations are more than 3 orders of magnitude smaller than those required for a moderate carbon flow across cuticles of $1.74 \cdot 10^{-6}$ g/s m^2. Since we started our calculations using very high permeabilities of cuticles we must conclude, that amino acids and monosaccharides will not diffuse across intact cuticles in rates required to sustain growth of epiphytic micro-organisms. This is not too surprising since cuticles are made to protect the contents of the apoplast against loss in order to allow apoplastic transport of inorganic ions and phloem loading from the apoplast (Lohaus *et al.*, 1995).

We found no data about apoplastic concentrations of fatty acids and fatty alcohols. But with these compounds required concentrations are only in the milli- and micromolar range which is also the range of their aqueous solubilities. Thus, significant cuticular transport is possible. Fatty acids are much less volatile than alcohols of equal carbon numbers. As shown in Fig. 3 short chain alcohols can take both the cuticular and the vapour pathway, while fatty acids will be confined to the cuticular pathway. Non-volatile fatty acids and alcohols are constituents of cuticular waxes. If they are utilised by epiphytic micro-organisms, they can be replaced quickly, because they are synthesised in the epidermis and permeability in cuticles for these compounds can be high. The carbon content of microbes per leaf surface area was estimated above to be $1.5 \cdot 10^{-4}$ g/m^2 which is only 0.015 $\mu g/cm^2$. This is orders of magnitude less than the lowest wax coverages commonly found on plants; only the amount of fatty acids will reach this value.

CONCLUSIONS

Permeabilities of cuticles are relatively high for lipophilic molecules and they are low for polar compounds, even if they are relatively small. Inorganic ions and polar metabolites are not likely to diffuse across cuticles because both permeabilities and apoplastic concentrations are too low. Only relatively lipophilic molecules can be expected to penetrate cuticles in significant amounts. Rates of penetration can be increased by biological accelerators and by increasing temperatures. Lipophilic amines could provide microbes with nitrogen. However, there is ample deposition of nitrogenous compounds from the atmosphere. It is unlikely that ATP or other polar P-containing compounds will penetrate cuticles in significant amounts. Alkylesters of phosphoric acid are lipophilic, they are accelerators of permeability and their own permeability of cuticles is very high (unpublished results). But we do not know if they are synthesised by higher plants. If no other lipophilic phosphorus compounds are made by higher plants, availability of P may well be growth-limiting for epiphytic micro-organisms which can utilise fatty acids or fatty alcohols as carbon sources.

Our modelling approach can be used for any compound to test its potential nutritional value, if the molar volume and the octanol/water partition coefficient are known. This approach can also be used to test if the older data on foliar leaching are erroneous or if they represent loss from damaged leaves, excretion from nectaries and glandular trichomes, washing off of deposits from leaves or leaching of micro-organisms.

REFERENCES

Abraham, M.H. and McGowan, J.C. 1987, The use of characteristic volumes to measure cavity terms in reversed phase liquid chromatography. *Chromatographia* 23:243-246.

Baker, E.A. 1982, Chemistry and morphology of plant epicuticular waxes. pp. 45-95 In: Cutler, D.F., Alvin, K.L. and Price, C.E. (eds.), *The Plant Cuticle*, Academic Press, London.

Bauer, H. and Schönherr, J. 1992, Determination of mobilities of organic compounds in plant cuticles and correlation with molar volumes. *Pestic. Sci.* 35:1-11.

Baur, P. and Schönherr, J. 1995, Temperature dependence of the diffusion of organic compounds across plant cuticles. *Chemosphere* 30:1331-1340.

Baur, P., Grayson, B.T. and Schönherr, J. 1996a, Concentration dependent mobility of chlorphenvinophos in isolated plant cuticles. *Pestic. Sci.* 47: 171-180.

Baur, P., Marzouk, H., Schönherr, J. and Bauer, H. 1996b, Mobilities of organic compounds in plant cuticles as affected by structure and molar volumes of chemicals and plant species. *Planta* 199: 404-412.

Bukovac, M.J. 1976, Herbicide entry into plants, pp. 335-364, In: Audus L.J. (ed) Herbicides: Physiology, biochemisty and ecology, Academic Press, New York, San Francisco.

Bunster, L., Fokkema, N.J. and Schippers, B. 1989, Effect of surface-active *Pseudomonas* spp. on leaf wettability. *Appl. Environ. Microbiol.* 55:1340-1345.

Cooper, D.G. and Zajic, J.E. 1980, Surface-active compounds from microorganisms. *Adv. Appl. Microbiol.* 26:229-253.

Cussler, E.L. 1984, *Diffusion, Mass Transfer in Fluid Systems*, Cambridge University Press, Cambridge.

Deng, W., Hamilton-Kemp, T.R., Nielsen, M.T. andersen, R.A., Collins, G.B. and Hildebrand, D.F. 1993, Effects of six-carbon aldehydes and alcohols on bacterial proliferation. *J. Agric. Food Chem.* 41:506-510 .

Eckl, K. and Gruler, H. 1980, Phase transitions in plant cuticles. *Planta* 150:102-113.

Fiechter, A. 1992, Biosurfactants: moving towards industrial application. *Tibtech* 10:208-217.

Franke, W. 1967, Mechanisms of foliar penetration of solutions. *Annu. Rev. Plant Physiol.* 18:281-300.

Freytag, S. and Mendgen, K. 1991, Carbohydrates on the surface of urediniospore- and basidiospore-derived infection structures of heteroecious and autoecious rust fungi. *New Phytol.* 119:527-534.

Goodman, R.N., Kivaly, Z. and Wood, K.R. 1986, *The Biochemistry and Physiology of Plant Disease*, University of Missouri Press, Columbia.

Geyer, U. and Schönherr, J. 1988, *In vitro* test for effects of surfactants and formulations on permeabilty of plant cuticles, pp. 22-23 In: Cross, B. and Scher, H.B. (eds.) *Pesticide Formulations: Innovations and Developments*, ACS Symposium Series 371, Washington D.C.

Hamilton-Kemp, T.R., Loughrin, J.H., Archbold, D.D. andersen, R.A. and Hildebrand, D.F. 1991, Inhibition of pollen germination by volatile compounds including 2-hexenal and 3-hexenal. *J. Agric. Food Chem.* 32:952-956.

Hansch, C. and Leo, A. 1979, *Substituent Constants for Correlation Analysis in Chemistry and Biology*, John Wiley and Sons, New York, Chister, Brisbane, Toronto.

Harborne, J.B. 1982, *Introduction to Ecological Biochemistry*, 2nd. ed, Academic Press, London, New York, Toronto, Sydney, San Francisco.

Hatanaka, A. 1993, The biogeneration of green odour by green leaves. *Phytochemistry* 34:1201-1218.

Hirano, S.S. and Upper, C.D. 1983, Ecology and epidemiology of foliar bacterial plant pathogens. *Annu. Rev. Phytopatol.* 21:243-269.

Hirano, S.S. and Upper, C.D. 1990, Population biology and epidemiology of *Pseudomonas syringae*. *Annu. Rev. Phytopathol.* 28:155-77 .

Hock, B. and Elstner, E.F. 1988, *Schadwirkungen auf Pflanzen*. Bibliographisches Institut und F.A. Brockhaus AG, Zürich.

Holloway, P.J. 1982a, Structure and histochemistry of plant cuticular membranes: an overview, pp. 1-32 In: Cutler, D.F., Alvin, K.L. and Price, C.E. (eds.), *The Plant Cuticle*, Academic Press, London.

Holloway, P.J. 1982b, The chemical constitution of plant cuticles, pp. 45-85 In: Cutler, D.F., Alvin, K.L. and Price, C.E. (eds.), *The Plant Cuticle*, , Academic Press, London.

Hull, H.M. 1970, Leaf structure as related to absorption of pesticides and other compounds. *Resid. Rev.* 31:1-150.

Kerler, F. and Schönherr, J. 1988, Permeation of lipophilic chemicals across plant cuticles: Prediction from octanol/water partition coefficients and molecular volumes. *Arch. Environ. Contam. Toxicol.* 17:7-12.

Kerstiens, G. 1995, Cuticular water permeability of European trees and shrubs grown in polluted and unpolluted atmospheres and its relation to stomatal response to humidity in beech (*Fagus sylvatica* L.). *New Phytol.* 129:495-503.

Kollatukudy, P.E. 1985, Enzymatic penetration of the plant cuticle by fungal pathogens. *Annu. Rev. Phytopathol.* 23:223-250.

Libbert, E. and Manteuffel, R. 1970, Interactions between plants and epiphytic bacteria regarding their auxine metabolism. VII. The influence of the epiphytic bacteria on the amount of diffusible auxin. *Physiol. Plant.* 23:93-98.

Libbert, E. and Silhengst, P. 1970, Interactions between plants and epiphytic bacteria regarding their auxine metabolism. VIII. Transfer of ^{14}C-indoleacetic acid from epiphytic bacteria to corn coleoptiles. *Physiol. Plant.* 23:480-487.

Lindow, S.E. 1990, Determinants of epiphytic fitness in bacteria, pp. 295-314 In: Andrews, J.H. and Hirano, S.S. (eds.), *Microbial Ecology of Leaves*, Springer-Verlag, New York.

Lohaus, G., Winter, H., Riens, B. and Heldt, H.W. 1995, Further studies of the phloem loading process in leaves of barley and spinach. The comparison of metabolite concentrations in the apoplastic compartment with those in the cytosolic compartment and in the sieve tubes. *Bot. Acta* 108:270-275.

Martin, J.T. and Juniper, B.E. 1970, *The Cuticle of Plants*. Arnolds, London.

Mattheis, J.P., Fellman, J.K., Chen, P.M. and Patterson, M.E. 1991, Changes in headspace volatiles during physiological development of Bisbee Delicious apple fruit. *J. Agric. Food Chem.* 39:1902-1906.

Merida, T., Schönherr, J. and Schmidt, H.W. 1981, Fine structure of cuticles in relation to water permeability: the fine structure of the cuticle of *Clivia miniata* Reg. leaves. *Planta* 152:259.

Pascholati, S.F., Yoshioka, H., Kunoh, H. and Nicholson, R.L. 1992, Preparation of the infection court of *Erysiphe graminis* f. sp. *hordei*: cutinase is a component of the conidial exudate. *Physiol. Mol. Plant Pathol.* 41:53-59.

Reid, C.R., Prausnitz, J.M. and Sherwood, T.K. 1977, *The Properties of Gases and Liquids*. McGraw-Hill, New York.

Riederer, M. and Schönherr, J. 1984, Accumulation and transport of (2,4-dichlorophenoxy) acetic acid in plant cuticles: I. Sorption in the cuticular membrane and its components. *Ecotoxicol. Environ. Safety* 8: 236-247.

Riederer, M., Schönherr, J. 1985, Accumulation and transport of (2,4-dichlorophenoxy) acetic acid in plant cuticles: II. Permeability of the cuticular membrane. *Ecotoxicol. Environ. Safety* 9:196-208.

Riederer, M., Burghardt, M., Mayer, S. and Obermeier, H. 1995, Sorption of monodisperse alcohol ethoxylates and their effects on mobility of 2,4-D in isolated plant cuticles. *J. Agric. Food Chem.* 43:1067-1075.

Riederer, M. and Schreiber, L. 1995, Waxes: The transport barriers of plant cuticles, pp. 131-156 In: Hamilton, R.J. (ed.) *Waxes: Chemistry, Molecular Biology and Functions*, The Oily Press, Dundee, Scotland.

Sablic, A., Güsten, H., Riederer, M. and Schönherr, J. 1990, Modeling plant uptake of airborne organic chemicals. I. Plant cuticle/water partitioning and molecular connectivity. *Environ. Sci. Technol.* 24:1321-1326.

Sangster, J. 1989, Octanol-water partition coefficients of simple organic compounds. *J. Phys. Chem. Ref. Data* 18:1111-1229.

Schieferstein, R.H and Loomis, W.E. 1959, Development of the cuticular layer in angiosperm leaves. *Am. J. Bot.* 46:625-635.

Schlegel, H.G. 1985, *Allgemeine Mikrobiologie*, Georg Thieme Verlag, Stuttgart, New York.

Schmidt, H.W., Merida, T. and Schönherr, J. 1981, Water permeability and fine structure of cuticular membranes isolated enzymatically from leaves of *Clivia miniata* Reg. *Z. Pflanzenphysiol.* 105:41-51.

Schönherr J. 1982, Resistance of plant surfaces to water loss: Transport properties of cutin, suberin and associated lipids, pp. 153-179 IN: Lange, O.L., Nobel, P.S., Osmond, C.B. and Ziegler, H. (eds.) *Encyclopedia of Plant Physiology*, New Series, Vol. 12B, Physiological Plant Ecology. Springer , Berlin, Heidelberg.

Schönherr, J. 1993, Effects of alcohols, glycols and monodisperse ethoxylated alcohols on mobility of 2,4-D in isolated plant cuticles. *Pestic. Sci.* 39:213-223.

Schönherr, J. and Baur, P. 1994, Modelling penetration of plant cuticles by crop protection agents and effects of adjuvants on their rates of penetration. *Pestic. Sci.* 42:185-208.

Schönherr, J. and Bukovac, M.J. 1973, Ion exchange of isolated tomato fruit cuticles: Exchange capacity, nature of fixed charges and cation selectivity. *Planta* 109:73-93.

Schönherr, J., Eckl, K. and Gruler, H. 1979, Water permeability of plant cuticles: the effect of temperature on diffusion of water. *Planta* 147:21-26.

Schönherr, J. and Riederer, M. 1988, Desorption of chemicals from plant cuticles: evidence for asymmetry. *Arch. Environ. Contam. Toxicol.* 17:13-19.

Schönherr, J. and, Riederer, M. 1989, Foliar penetration and accumulation of organic chemicals in plant cuticles. *Rev. Environ. Contam. Toxicol.* 108:1-70.

Schreiber, L. and Schönherr, J. 1990, Phase transitions and thermal expansion coefficients of plant cuticles: the effect of temperature on structure and function. *Planta* 182:186-193.

Schreiber, L. and Schönherr, J. 1992, Leaf surface microflora may significantly affect studies on foliar uptake of chemicals. *Bot. Acta* 105:345-347.

Suntio , L.R., Shiu, W.Y., Mackay, D., Seiber, J.N. and Glotfelty, D. 1988, Critical review of Henry's law constants for pesticides. *Rev. Environ. Contam. Toxicol.* 103:1-59.

Tukey, H.B. 1970, The leaching of substances from plants. *Ann. Rev. of Plant Physiol.* 21:305-324.

Vaughn, S.F. and Spencer, G.F. 1993, Volatile monoterpenes as potential parent structures for new herbicides. *Weed Sci.* 41:114-119.

Watanabe, T. and Yamaguchi, I. 1991, Wettability characteristics of crop leaf surfaces. *J. Pestic. Sci.* 16:651-663.

Wattendorf, J. and Holloway, P.J. 1984, Periclinal penetration of potassium permanganate into mature cuticular membranes of *Agave* and *Clivia* leaves: new implications for plant cuticle development. *Planta* 161, 1-11.

2

NUTRIENTS ON THE LEAF SURFACE

Sylvie Derridj

INRA, Unité de Phytopharmacie et des Médiateurs Chimiques
Route de Saint Cyr
78026 Versailles cedex, France

INTRODUCTION

The leaf surface constitutes the interface between the external environment and the plant tissues. It is limited by a thin layer (about 1.0 µm), the cuticle, composed of non soluble lipid polymers (the cutin) in which soluble lipid waxes are embedded. The function of the cuticle is to protect the leaf from external (biotic and abiotic) aggressions and to regulate the passage of water from the plant to its environment (Schönherr, 1976), of inorganic ions (Ferrandon and Chamel, 1989), and organic solutes (Tukey, 1970; Mengel *et al.*, 1990) coming from tissues and particularly from the extracellular matrix and intercellular air spaces (apoplast). Under natural conditions many factors may cause injury to leaf surfaces: indirectly such as physiological disorders, nutrient and moisture deficiencies, adverse temperatures, poor aeration, deficient light and leaching and directly such as invasion of micro-organisms, attack by insects, wind and air pollution, (Tukey, 1963).

Protection against external factors is linked to cuticle resistance and mostly to the epicuticular waxes (wax quantities, composition, structure and arrangement). In many plant species water soluble organic compounds such as terpenoids, flavonoid aglycones and simple phenolics may be imbedded in the epicuticular material (Wollen and Dietz, 1981). Some of these substances protect leaves against ultraviolet radiation damage (Stephanou and Manetas, 1995).

Contact of aqueous solutions (rain, dew, mist and fog) with the leaf surface may induce losses of substances from plants (leaching). A great diversity of substances can be leached and there are differences of ease and degree according to the nature of the molecules, the plant species, and biotic and abiotic factors as described in the previous chapter by Schönherr and Baur. For example, leaves from healthy plants are much less susceptible to leaching than are injured leaves. Annual throughfall deposition from injured conifer spruce showed an increase in organic and inorganic chemicals. Foliar leaching seems to be an overall stress reaction in injured trees in Europe and the US (Alanas and Skärby, 1988). In many situations the loss of inorganic nutrients and primary metabolites was estimated in kg per ha per year. For certain plant species these substances may also originate from guttation at night through hydathodes (Von Scheffer *et al.*, 1965; Goatley and Lewis, 1966).

Aerial Plant Surface Microbiology, edited by Morris et al.
Plenum Press, New York, 1996

The permeability of plant cuticles to solutes is much more closely linked to the chemical composition arrangements and physicochemical properties of the intracuticular waxes and is usually measured using enzymatically isolated membranes which are intact. The previous chapter deals with the state of results obtained in this field. The objective of this chapter is to review the literature describing the chemical environment at the leaf surface, with particular reference to chemicals that could be sources of nutrients for epiphytic micro-organisms. Whereas Schönherr and Baur in the previous chapter have used modelling as an approach to understanding the chemical nature of the leaf surface, this chapter will explore data based on various extraction procedures of field- and greenhouse-grown plants.

The nature of aerial plant surfaces has been intensively studied with regard to insect feeding and oviposition behaviour. Numerous insect species are known to explore the leaf surface before feeding (Woodhead and Chapman, 1986; Chapman and Bernays, 1989; Bernays and Chapman, 1994) or ovipositing (Städler, 1986; Städler and Roessingh, 1991; Renwik and Chew, 1994) on a suitable host plant. Combinations of physical and chemical stimuli influence the evaluation process of the leaf surface by the insect. Plant glandular structures and secretions, leaf epicuticular waxes (fatty acids, alkanes, primary alcohols, esters of the alcohols, aldehydes, ketones) as reported by Eigenbrode and Espelie (1995), nonpolar substances (generally secondary compounds extractable by organic solvents from the waxes) (Städler, 1992), water-soluble secondary metabolites embedded in epicuticular waxes (such as flavonoid glycosides) and, more recently, primary metabolites such as sugars, amino acids, leaking out of waxes (Derridj et al., 1989; Fiala et al., 1990), are perceived by insects via the leaf surface.

In spite of a great number of plant species specific stimulatory chemicals that have been identified for insects, no real consistent evidence has been given to explain host plant recognition due to a single secondary chemical (Jermy, 1993) and a fortiori from the leaf surface. Explanations for this situation may be that the effect of chemicals on insect behaviour is that of (i) a complex chemical mixture of secondary compounds (ii) a mixture of primary metabolites (iii) a mixture of both chemical groups and inorganic ions. It is also possible that extraction techniques do not reveal substances which are in real contact with the insect.

This chapter will focus on water soluble substances and particularly primary metabolites of leaf surfaces. They have still been neglected for explaining host-plant selection by insects in spite of the fact that they could constitute basic information on plant health, physiology and nutrition. They are known to be detected by insects and, if they are present on the leaf surface, they could influence their behaviour. Three points will be emphasised: their collection and identification, their sources of variability, and the information they can provide for insects about the plant.

EXTRACTION TECHNIQUES AND IDENTIFICATION OF NUTRIENTS

The techniques to employ for extracting water soluble substances present on leaf surfaces depend on our objectives. The first step in defining a suitable extraction procedure for putative recognition substances is to define what the insects are doing when they interact with the plant surface during their exploratory routines. If they only touch the surface, then there is scarcely any point in looking for compounds in the cuticular layer. If they excavate a patch of the cuticle, e.g. by scraping, then a physical abrasion mechanism designed to stimulate this might be an appropriate procedure. For studies of the role of nutrients in the recognition of host plants by insects, there are the following difficulties to consider: (i) The leaf surface must be put in contact with water in spite of the presence of epicuticular waxes which limit wetting. The wetability is dependent on the properties of the chemical groups and wax fine structure at the leaf surface (Taylor *et al.*, 1981) (ii) Substances should be collected without inducing any leaching from the inside of the leaf (iii) Only the quantities of substances that the insect could encounter during the different behaviour events on the leaf surface should be extracted iv) We should also verify that only water-soluble compounds are collected. For studies of the nature of the nutritional environment of micro-organisms on leaves, similar difficulties would be encountered.

Fiala *et al.* (1990) have described two extraction techniques to collect soluble carbohydrates: (i) dipping of corn leaves in water for different durations and (ii) spraying the leaf surface with 25 ml of ultrapure water for 30s with a pulveriser held 30 cm from the leaf surface having a flow of nitrogen gas of 17 L.min^{-1} (Figure 1). Comparison of these two techniques showed that the latter one makes it possible to collect larger amounts of substances. A similarity in the sugar and amino acid composition collected from mutant corn leaf surfaces without epicuticular waxes (Glossy 1) (Lorenzoni and Salamini, 1975) and from the normal isogenic hybrid with epicuticular waxes indicated that the spraying technique is probably adequate for collecting these substances from corn leaves. The quantities of total amino acids on the ear leaf at the late tassel stage were 2617 and 2347 nmol cm^{-2} on normal and Glossy 1, respectively. The quantities of sugars on the two plant lines were 470 and 512 nmol cm^{-2} for glucose, 479 and 462 nmol cm^{-2} for fructose, and 1751 and 1190 nmol cm^{-2} for sucrose. The majority of substances (75% of the sugars, and 52% of the amino acids) was collected after the first spraying. There was a risk of leaching amino acids with a second

Figure 1. Collection of leaf surface nutrients on corn (*Zea mays* L.) by water spraying. A sprayer with a flow of nitrogen gas of 17 L min^{-1} was used to obtain a very low water pressure. The drop size varied from 0.1 to 0.3 mm diameter and the leaf surface was covered uniformly.

or third spraying when the duration between each spraying was 30 mn as opposed to 1 mn. For durations between spraying of 1 mn the quantities of amino acids collected were rather constant between each spraying.

Wounding of leaves during sampling and extraction probably also has important consequences for the types and amounts of substances detected. When corn leaf trichomes were broken with a razor there were modifications in the proportions of amino acids collected on the leaf surface; aspartic acid, glutamic acid and alanine constituted 14.7%, 14.1% and 30.8%, respectively of the total amino acids on wounded plants whereas these same amino acids constituted only 6.8%, 5.5% and 13.2% of the total amino acids on control plants (J.P Boutin, unpublished). In all our experiments we chose leaves with no apparent wounds and tried not to produce any lesions before collecting substances. Nevertheless, we suspected that, in certain cases, lesions were present because quantities of some substances detected were 100 or 1000 times that of the usual values. When evaluating the quantities of substances on the leaf surface, it is better to collect substances on several single leaves so as to identify quantities that may be due to wounds.

Concerning electrolyte substances such as amino acids, the acidity of the water used for spraying could influence the chemical composition of the extract. But use of water at pH 3.7 compared to 6.4 did not result in different amino acid profiles. Nevertheless, identifying polar compounds could require the use of water at the pH of the leaf surface. The pH of the corn ear leaf surface, measured with a micro-pH meter on the upper leaf side, in the green house and in field conditions was about 5 to 6 which corresponds to the pH of the ultrapure water used for spraying (Elgastat UHQ II, international standard ASTM, CAP, NCCLS, resistance 18 Megaohm-cm at 25°C, organics at less than 0.0001 OD at 254 nm, bacteria < 1 colony per ml, absolute filtration 0.005 microns). (S. Derridj, unpublished). Under natural conditions corn leaf pH did not vary over a day but varied with location of fields. On plants grown in fields near a road, the pH may be rather low (2 to 4).

All substances which leak from leaf tissues may be present on the leaf surface more or less embedded in the cuticular waxes. Their extraction depends on where they are located and on the information we need. The extraction of carbohydrates and organic acids embedded in corn leaf waxes can be accomplished with chloroform; their separation from the waxes can be achieved by a mixture of chloroform and water (1/1, v/v). The extraction dynamics of carbohydrates from corn waxes (from 2s to 200s) showed that their distribution in waxes was variable. Two sorts of distributions were observed: carbohydrates could be concentrated and extracted after dipping 50s or they could be more dispersed and be extracted gradually over 200s. The effect of time on the extraction of sugars and organic acids is illustrated in Figure 2. In the case of corn leaves, organic acids were extracted much earlier than were carbohydrates (Figure 2).

Extraction from waxes of substances on the leaf surface is dependent on the physical and chemical properties of the cuticle. The chemical composition of epicuticular wax deposits alters as leaves expand (Baker and Hunt, 1986) and also by weathering. The spraying technique we adopted, which is very gentle, mechanically collected lipid substances such as alkanes (C18-C35), alkanoic acids (C16-C32), alkanols (C16-C35), and polar lipids (C16-C32). We compared the profiles of the carbon chain lengths of these different groups collected from the corn leaf surface by spraying, to those extracted from epicuticular wax with chloroform for 2s and to those extracted from the apoplast (Figure 3). Those obtained by the spraying technique were most similar to those obtained from the apoplast (except for alkanoic acids). This result suggests that lipids on the corn leaf surface give biochemical information about the apoplastic composition and could be very easily available for the insect.

The identification methods used for the extracted chemicals need to be adapted to the properties of the molecules and to their quantities. Quantities of primary metabolites collected were very low (about 10^{-6} to 10^{-5} M.m$^{-2)}$. Gas chromatography was used to analyse

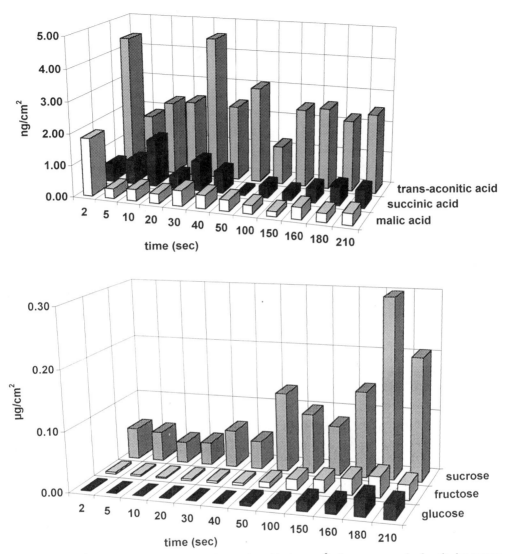

Figure 2. Extraction dynamics of sugars and organic acids (ng cm^{-2}) from corn ear leaf cuticular waxes removed with chloroform. The data are means of seven replicates of one or two ear leaves.

water soluble carbohydrates after silylation and made it possible to analyse single corn leaf segments of about 250 cm^2. Progress in detection by HPLC should mean that, in the near future, substances can be analysed more rapidly.

The extensive list of substances reported to leach from plants to the surface includes a great majority of primary metabolites and inorganic ions (Tukey, 1970; Leonardi and Flückiger, 1987; Turner and Broekhuizen, 1992). We collected them by spraying water on the leaf surfaces of different plant species (*Allium porum, Cichorium endivia, Galinsoga sp., Helianthus annuus L., Nicotiana plumbaginifolia, Prunus laurocerasus, Senecio sp., Zea mays L.*). With this technique we found soluble carbohydrates (fructose, glucose, sucrose),

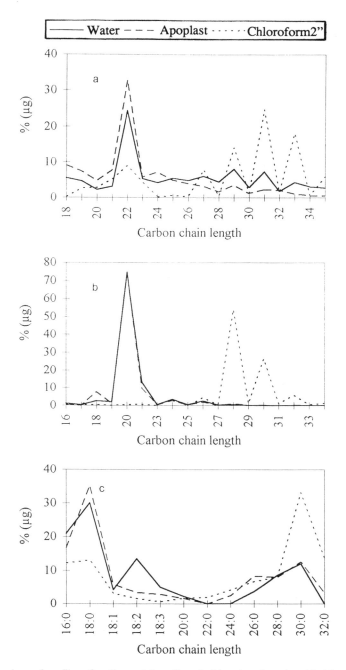

Figure 3. Comparison of profiles of n-alkanes (a), n-alkanols (b) and n-alkanoic acids (c) collected from the corn ear leaf surface by spraying water (——), dipping the leaf in chloroform for 2 sec (– – –) and from the apoplast collected by infiltration of water and centrifugation of the leaf (— –). Carbon chain length and numbers of double bonds are indicated.

the twenty free amino acids, urea, ammonia, other undetermined but stable ninhydrin positive compounds, organic acids and inorganic ions (Na, K, Mg, Ca). These substances will be described later in this chapter.

The quantities of substances which are collected are in proportion to the leaf area unit. In reality, if substances are not uniformly distributed on the leaf surface, their local concentrations could be underestimated. This can be further complicated by the heterogeneous topography of the leaf surface and by abiotic and exogenous factors.

SOURCES OF VARIATIONS

Leaves

There are numerous endogenous and exogenous factors which can effect the presence of substances on the leaf surface. Variations among individual leaves have been observed for leaching (Tukey, 1970), for cuticle waxes (Blaker and Greyson, 1987), and for pH-values of the leaf surface (Harr et al., 1980). Surface waxes also show variations according to plant species, health, physiology, leaf sides (Baker, 1981) and abiotic factors. Under windy conditions abrasions of the upper surface of leaves can be caused by impact between leaves and buds and the impact of wind-borne particles (Jeffree, 1994). This damage can increase the magnitude and heterogeneity of solute uptake. The fine structure of wax, the epicuticular wax level, chemistry and wetability, cuticular permeability and spray retention all change during leaf expansion (Bukovac et al., 1979). In our experiments, under controlled conditions, although care was taken to maintain homogenous leaf populations, differences in quantities of substances collected among individual leaves at the same position and at the same stage of development were recorded (Fiala et al., 1990).

Leaf Sides. On corn plants grown in greenhouses, the upper and under leaf surfaces can show differences in sugar quantities. This could be due in part to the fact that more exogenous substances deposit on the upper than on the lower side of leaves. It is less likely that a difference in the number of stomata or cuticle permeability between upper and lower sides contributes to differences in sugar quantities (Fiala et al., 1990). Under natural conditions cherry laurel leaves were richer in carbohydrates on the upper leaf side which was without stomata than on the under side with stomata (V. Fiala, unpublished).

Leaf Positions. As described in the chapter by Jacques (this volume) more sugars were generally observed on older leaf surfaces than on younger ones for corn and tobacco. The proportions of each of the three sugars (sucrose, fructose, glucose) varied with leaf position for corn and were different from those found in leaf tissues. Concentrations of sugars in leaf tissues varied little with leaf position (Fiala et al., 1990). This difference between leaf ages was also observed for inorganic ions which were more concentrated on older leaves (Figure 4). Similar observations were made for sugars and malic acid on tobacco leaves at the early flowering stage (Figure 5). Nevertheless, these observations can differ according to the plant growth stages.

Micro-Localisation of Carbohydrates

We have conducted a study of the micro-localisation of monosaccharides (glucose, galactose, mannose, fructose, arabinose and xylose) on the leaf surface (i) to confirm that the substances collected are not the result of induced leaching and that sugars are really in contact with insects on leaf surfaces, and (ii) to observe the micro-distribution of carbohy-

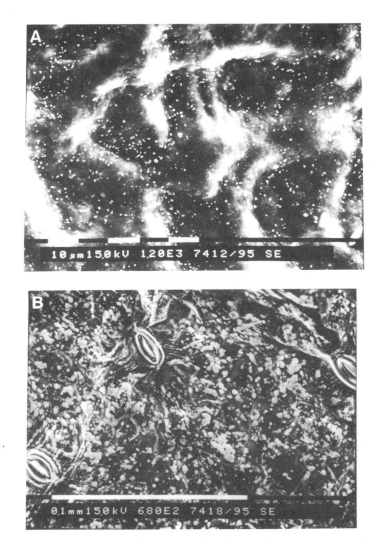

Figure 4. Characterisation of monosaccharides by scanning electron microscopy associated with X-ray microanalysis on leaf surfaces of (A) cherry laurel (*Prunus laurocerasus* L.) and (B) broad-leaved endive (*Cichorium endivia* var. *latifolia*).

drates and evaluate if local concentrations are underestimated by spraying and (iii) to have an idea of their major paths through the cuticle. The study followed different steps which were tissue fixation (with nitrogen paste and lyophylisation) to avoid any leaching, and precipitation, characterisation, and insolubilisation of monosaccharides by baryum-oxide and then silver-oxide. The silver-monosaccharide complexes were visualised by electronic microscopy associated with X-ray spectroscopy microanalysis. On corn leaves in greenhouses sampled in the morning (10 a.m.) no monosaccharides were visible (P. Barry, unpublished). Later in the day at two hours before sunset, monosaccharides were visualised in small scattered areas. The majority of monosaccharides were localised along the anticlinal

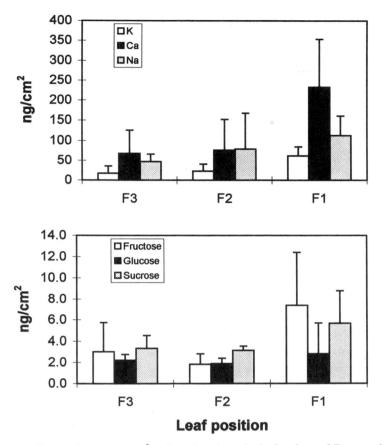

Figure 5. Sugars and inorganic ions (ng cm^{-2}) collected on the under leaf surfaces of *Zea mays* L in relation to the leaf position (or age), F1 being the oldest and F3 the youngest leaves on a plant. Plants were grown in greenhouses and used at the late tassel stage. Error bars represent standard deviation.

walls of the epidermal cells, and scattered in small granulations over epidermal cells (Fiala *et al.*, 1993). They could be found more rarely in the shape of "craters" and over some cells around stomata. This was different from the homogenous distribution observed on broad-leaved endive and cherry laurel leaves (Figure 6). The monosaccharide micro-distribution was much more dense and homogenous on endive leaves. On young cherry laurel leaves, monosaccharides were also homogeneously distributed but were less dense than on endive. Spectroscopy associated with X-ray microanalysis revealed the presence of several inorganic ions, including K, Ca, Cl, and Mg, within the monosaccharide granulations.

Plant Species and Varieties

The quantity of carbohydrates at the leaf surface (Fiala *et al.*, 1993) and in the leaf tissues (Derridj *et al.*, 1990) two hours before sunset can vary among corn hybrids and may be a heritable trait. But in a comparison of corn and sunflower at different growth stages, the quantities of sugars collected on leaf surfaces depended more on the growth stage of development of plants than on the plant species (Derridj *et al.*, 1989).

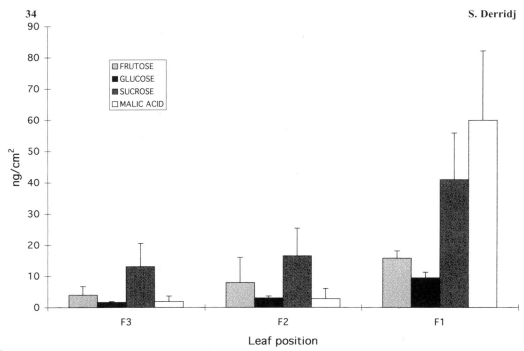

Figure 6. Sugars and malic acid (ng cm^{-2}) collected on the under leaf surfaces of *Nicotiana plumbaginifolia* in relation to leaf positions or age. They were divided in three groups of about 3 leaves (F1, F2, F3). F1 being the oldest leaves. Plants were grown in greenhouses and used at the early flowering stage. Error bars represent standard deviation.

This was very different for free amino acids which showed plant species specificity in their quantities and proportions on leaf surfaces whatever the plant hybrids (corn, sunflower), the growth stage of the plant, or the growth conditions. Surprisingly, 8 to 10 out of 22 amino acids were sufficient to discriminate different members of the genus *Senecio: S. jacobaea, S. sylvaticus, S. vulgaris, S. viscosus* (Soldaat *et al.*, 1996) (Figure 7), or different plant species such as corn, sunflower, leek and tansy ragwort (Derridj *et al.*, 1996) (Figure 8). Among ninhydrin positive substances, some specific unknown chemicals (three peaks) were collected from leek leaf surfaces. They were each in much higher concentrations than free amino acids (6.6, 5.6 and 6.1 µmol m^{-2} as opposed to a total of free amino acids of about 6 µmol m^{-2}) (Derridj *et al.*, 1996). This shows that widespread substances like free amino acids can have a plant species specificity due to the combination in which they are found. This is probably the case for other substances like organic acids (Morgan and Tukey, 1963) and cations. The difficulty that entomologists have in explaining the host specificity of a specialist insect by a single secondary chemical could be overcome by considering the specificity resulting from a pattern of primary metabolites. Added to the more or less limited number of chemicals detected by insects and interactions between substances at the sensorial perception and central nervous system levels, this could explain a very selective recognition of plants by insects (Van der Meijden *et al.*, 1989).

Time of Day

On broad-leaved endive (*Cichorium endivia* var. *latifolia*) grown in May in green-houses, quantities of fructose, glucose and sucrose collected from the leaf surface showed a progressive rise from 9 a.m. until 1 p.m. and then a progressive fall until reaching the low

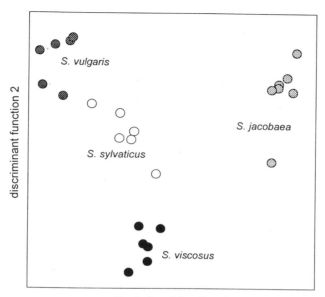

Figure 7. Discrimination of four *Senecio* species on the basis of their proportions of amino acids on the leaf surface. The first and second discriminant functions contain 77 and 16.9% of the variation, respectively.

values observed 24 hours before. Leaking, and then accumulation of carbohydrates on the leaf surface were followed by a decrease in sugars which is most probably due to the re-entry of molecules into the plant tissues (Figure 9). Results of the same experiment conducted on corn which showed less variations throughout 24 hours. It was rather difficult to attribute the curve obtained throughout 24 hours to micro-organisms. But in other situations epiphytic micro-organisms may interfere in the penetration of the molecules into cuticles (Schreiber and Schönherr, 1993).

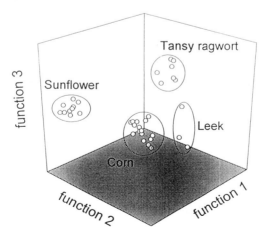

Figure 8. Discrimination of four plant species. The first and second discriminant functions contain 81 and 17% of the variation, respectively.

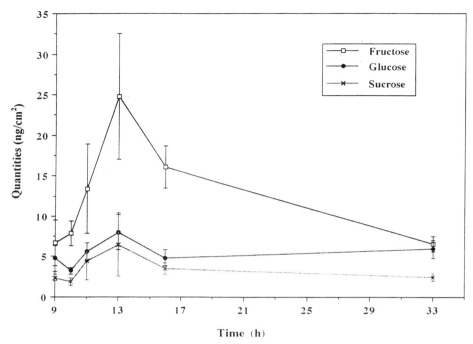

Figure 9. Quantities of fructose, glucose and sucrose (ng.cm^{-2}) collected by spraying water on the upper side of leaves of *Cichorium endivia* var. *latifolia* during 24 hours. Error bars represent standard errors for 3 replicates of two leaf samples.

The kinetics of penetration of different molecules in the cuticle has been examined (Schönherr and Riederer, 1989; Chamel *et al.*, 1991). Likewise, it would be interesting to study the kinetics of emergence of different types of molecules at the leaf surface. In addition to the physico-chemical properties of the molecules, types and states of cuticles, abiotic factors like relative humidity and temperature, and the duration that dew or rain droplets remain on the leaf surface should have a non negligible influence (Price, 1982). The time of day is also a factor which has to be considered when extracting substances from leaf surfaces. For example, because of the crepuscular behaviour of the insects we have studied (*Ostrinia nubilalis Hbn., Lepidoptera pyralidae*), all sampling was carried out two hours before sunset.

Epiphytic Micro-Organisms

Transported by wind, rain, air-borne pollen, dust, plant debris and insects to the leaf surface, micro-organisms need moisture and nutrients to grow and multiply. Nutrients may come from the plant as well as from foreign materials. Von Kunert and Libbert (1972) showed the importance of corn shoot exudates as a nutritional resource for epiphytic bacteria and as precursors for bacterial auxin production (21 amino acids including low amounts of tryptophan, free sugars: fructose, glucose, sucrose, and galactose). It is plausible that epiphytic micro-organisms can influence the chemical composition of the leaf surface by metabolising nutrients, secreting products of catabolism, and by uptake and desorption of chemicals. Thus they can modify the biochemical information given by the plant. They might also limit the contact of the insect with the leaf surface and with chemical stimuli by occupying sites where nutrients have leaked.

Leaf surfaces of greenhouse-grown corn were artificially inoculated with the bacterium *Erwinia herbicola*. The quantity of nutrients on the leaf surface was influenced by the presence of bacteria. Concentrations of arginine, fructose, sucrose and glucose increased after inoculation and were highest on the upper sides of leaves where the bacterial densities were the highest (Fiala *et al.*, 1993).

CHEMICAL INFORMATION GIVEN BY NUTRIENTS ON LEAF SURFACES

Photosynthetic Origin of Primary Metabolites on Leaf Surfaces

Experiments with radioisotopes reported by Tukey (1970) showed movements of molecules along cell walls and through the cuticle above anticlinal walls. These results can be compared to the microdistribution of monosaccharides we observed by electron microscopy on corn leaf surfaces and confirm the endogenous origin of leaf surface molecules. With the advent of radioisotopes, the phenomenon of the leaching action of rain, mist and dew was conclusively demonstrated (Long *et al.*, 1956; Tukey *et al.*, 1958). There was, nevertheless, no direct evidence of the photosynthetic origin of the primary metabolites present at the leaf surface. This evidence has been provided by experiments at the late tassel stage of corn grown in a phytotron chamber (Derridj *et al.*, 1996). The aim was to follow the evolution of ^{13}C in carbohydrates present at the leaf surface from $^{13}CO_2$. As early as 0.5 h after the 30 min $^{13}CO_2$ pulse, ^{13}C-labelled sugars (fructose, glucose, sucrose) appeared in the cells, the apoplast, and at the leaf surface. The maximum of ^{13}C-labelled sugars on the leaf surface was reached between 3 and 6.5 hours after the pulse. During the dark period, the proportions of ^{13}C-labelled molecules decreased. The decrease could be explained by continuous diffusion of unlabelled sugars and also re-entry of labelled sugars into the leaf. The very short time needed for substances to appear on the leaf surface after their biosynthesis and the fact that they do not necessarily accumulate on the surface suggest that further studies are needed about their kinetics and about factors which modulate them during a day and throughout leaf development. The movement of molecules through the cuticle seems rather rapid and reflects plant metabolism and cuticle properties. The coincidence in time and localisation between events of insect behaviour and molecules at the leaf surface have to be taken into consideration when explaining plant-insect coevolution.

Relationship between Plant Tissues and Leaf Surface Nutrients

Compounds which are collected at the leaf surface result from movements of variable duration of these compounds through the cuticle. For practical reasons it was interesting to evaluate the relationship between concentrations and ratios of compounds in leaf tissues and those found at the leaf surface. For example, it would be useful for plant breeders to know if screening based on leaf tissue chemical contents would be sufficient to evaluate chemicals present on the leaf surface. This would be useful for selecting plant lines with modified nutritional environments on their leaf surfaces that were repulsive to insect pests or that gave a competitive advantage to micro-organisms antagonistic to plant pathogens. An attempt at this latter strategy is described in the last chapter on future research trends (Andrews, this volume). The comparison of tissue and leaf surface nutrient contents was done on several plant species grown in greenhouses for samples collected two hours before sunset. We analysed quantitative

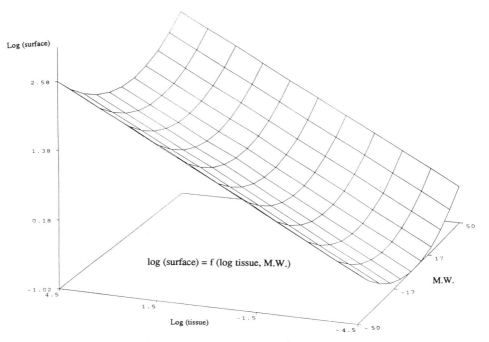

Figure 10. Partial graphic representation of the model : Log (surface) = 0.360 + b_1 "plant specific factors" + *0.31 log (tissue concentration) + 0.06 molecular weight+pK_2+0.0003 (molecular weight)^2+0.54 pK_2.pH* isoelectric point, which explains quantities of free amino acids collected on the leaf surface of three plant species (*Zea mays* L., *Helianthus. annuus* L. and *Senecio jacobaea*). Underlined parameters are represented in the figure.

and qualitative parameters related to the composition of amino acids at the leaf surface. The parameters were: plant species, different stages of their development, 20 free amino acids, several physico-chemical properties of molecules such as molecular weight, pK_1(COOH), pK_2 (NH_3), pH at the isoelectric point, polarity, hydrophobicity, proportions of C, N, O, H in the molecules and carbon chain length, and tissue amino acid content. For a general logarithmic-linear model a few parameters explained 78% of the variation of amino acid composition at the leaf surface. Three factors were equally important in explaining quantities and proportions of amino acids at the leaf surface: unknown "specific factors" linked to the plant species, tissue content, and physico-chemical properties of the molecules (Molecular weight/pK_2, Molecular weight^2, pK_2·pH at isoelectric point) (Figure 10). The proportions of the 20 amino acids considered as a whole were explained by the plant species parameter, which corroborates the previously mentioned evidence for the plant-specific character of the amino acid pattern. Our hypothesis was that the cuticle and some of the properties regulating how substances cross it, are probably involved in the "plant-specific factor". Focusing on a single plant species, a particular stage of development and a single substance may allow us to more quickly reveal the relationship between the chemical composition of tissue and the leaf surface.

Selectivity of the Cuticle to Carbohydrates

On isolated cuticles of cherry laurel (without stomata), the rates of permeability of fructose, glucose and sucrose have been observed during 30 min to 24 h periods (Stammitti *et al.*, 1995). Glucose (21.7 x 10^{-9} m s^{-1}) crossed more quickly than sucrose (4.13 x10^{-9} m

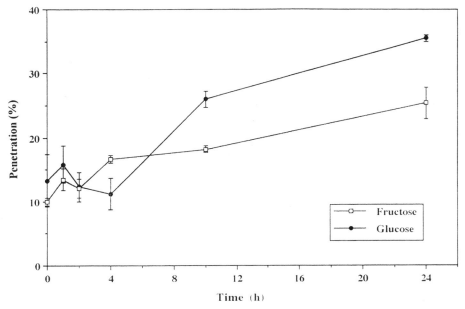

Figure 11. Penetration of ^{14}C-labelled sugars (fructose and glucose) into the cuticle. A total of 80 ng cm^{-2} was deposited in water droplets on the upper leaf side of young *Cichorium endivia* var. *latifolia* two hours before dark. The data at time 0 represent proportions of penetration 15 min after deposition.

s^{-1}) and fructose (2.34 x 10^{-9} m s^{-1}). Similar results were observed in controlled conditions on the penetration of glucose and fructose labelled with ^{14}C deposited in small droplets on endive leaf surfaces during 24 hours (Figure 11). These preliminary results would mean that the natural cuticle is able to exert a similar selectivity on sugars in both directions from inside and outside the cuticle.

We have tried to evaluate the role of chemical composition of cuticular waxes in the natural selectivity of the cuticle to sugars. When comparing the cuticular wax composition of corn to the quantities of sugars (fructose, glucose, sucrose) embedded in it, we observed a relationship with alkanoic acids, C16-C18 waxes having a major role (Wu *et al.*, 1995). Furthermore there was a temporal relationship between the extraction of alkanoic acids (C16-18) with chloroform and the release of sugars embedded in the waxes. Observations on isolated cuticles of cherry laurel showed that the rate of sugars passing through the cuticle increased when wax extraction by chloroform was long enough (100s) to extract the major part of alkanoic acids. But the selectivity to sugars was not changed by this treatment. This would probably mean that the natural cuticular selectivity to sugars is linked to several factors including alkanoic acids, their localisation in the cuticle, and to other more stable characters not destroyed by 100s of extraction in chloroform (*i.e.* cutin) as well as to the physico-chemical properties of the sugars.

CONCLUSION

The loss of substances from leaf surfaces has been known for a long time. More recently, the phenomenon of cuticular absorption has been studied. However two characteristics of the chemical environment of the leaf surface have been overlooked: the kinetics of accumulation of molecules on the leaf surface and their localisation. An understanding of this variability may give clues about spatial distribution patterns of micro-organisms on leaf surfaces or about preferred feeding and oviposition sites of insects.

Nutrients are present on the leaf surface, usually in very small quantities (ng cm^{-2}). They are outside of the cuticle more or less hidden by the epicuticular waxes and also imbedded in the waxes. Appropriate extraction techniques depend on where the nutrients are localised as well as on the aim of the research. It has been clearly demonstrated that extraction with water or with chloroform and water for different durations may result in very different quantities and proportions of carbohydrates. The method used to characterise and localise carbohydrates on the leaf surface makes it possible to affirm (i) that they are really on the leaf surface, (ii) that their presence is not due to induced leaching during the experiment and (iii) that they are not necessarily due to the presence of epiphytic micro-organisms. They can have a photosynthetic origin and their appearance on the leaf surface may be very rapid after their biosynthesis. Their distribution varies greatly among plant species and may be heterogeneous on a single leaf.

Numerous properties of the cuticle play a role in the nutrient composition of the leaf surface environment. These properties are the renewal and change in wax composition after full maturity; its permeability, absorption, structure; plant species-specificity traits and its adaptation to environmental conditions. In this way lipids can be mechanically extracted from the leaf surface by spraying water. Nutrient ratios at the leaf surface can be different from the leaf tissues. Carbohydrates cross through cherry laurel leaf cuticle at different speeds. This selectivity was maintained after wax extraction with chloroform (30s and 100s). Carbohydrates in the waxes were accessible with chloroform according to the distribution and quantities of alkanoic acids.

The information provided by the nutrients on the leaf surface is complex, and integrates a great variety of factors: plant physiology and metabolism, plant species, leaf cuticle properties and adaptation of both plant and cuticle to environmental conditions. They may be related to photosynthesis, to leaf tissue chemical contents, to the chemical composition of waxes and consequently may be variable with the time of day. But the information can be also very stable and plant species-specific.

To understand the role of nutrients for different organisms (insects, epiphytic micro-organisms) it would probably also be necessary to pay attention to the notion of time: to the delay needed for a nutrient to reach the leaf surface after its biosynthesis, then to the stability of this information. Furthermore, for insects, and particularly for very small neonate larvae emerging on the leaf surface, the nutrient information must be very quickly detectable to avoid a rapid death of the organism. The fact that carbohydrates and inorganic nutrients (ions) were found together could accelerate the perception of useful information about the plant. Both chemical groups are very often detectable by several nervous cells localised on the same sensilla. Moreover, a single group of substances (for example, the eight amino acids which discriminated plant species) could give information both about the plant species and its nutritional status. Do nutrients on leaf surfaces play a similar role for epiphytic micro-organisms? Do they communicate information about plant species or variety or about the status of the defence mechanisms of the plant? Could these chemicals be involved in triggering mechanisms of attack by plant pathogenic micro-organisms? These pertinent questions await to be resolved.

REFERENCES

Alenas, I. and Skärby, L. 1988, Throughfall of plant nutrients in relation to crown thinning in a Swedish coniferous forest. *Water, Air, Soil Poll.* 38: 223-237.

Baker, E.A. and Hunt, G.M. 1981, Developmental changes in leaf epicuticular waxes in relation to foliar penetration. *New Phytol.* 88: 731-747.

Baker, E.A. and Hunt, G.M. 1986, Erosion of waxes from leaf surfaces by simulated rain. *New Phytol.* 102: 161-173.

Bernays, E.A. and Chapman, R.F. 1994, Chemicals in plants pp. 15-59 In: Bernays E.A. and Chapman, R.F., *Host Plant Selection by Phytophagous Insects.* Chapman and Hall, New York.

Blaker, T.W. and Greyson, R.I. 1988, Developmental variation of leaf surface wax of maize, *Zea mays. Can. J. Bot.* 66: 839-847.

Bucovac, M.J, Flore, J.A. and Baker, E.A. 1979, Peach leaf surfaces : changes in wettability, retention, cuticular permeability, and epicuticular wax chemistry during expansion with special reference to spray application. *J. Am. Soc. Hort. Sci.* 104 (5): 611-617.

Chamel, A, Gaillardon, P. and Gauvrit, C. 1991, La pénétration foliaire des herbicides. pp. 7-50 In: Scala, R. (ed.) *Les Herbicides.* I.N.R.A. éditions Versailles.

Chapman, R.F., and Bernays, E.A. 1989, Insect behaviour at the leaf surface and learning as aspect of host plant selection. *Experientia* 45: 215-222.

Derridj, S., Gregoire, V., Boutin, J.P., and Fiala, V. 1989, Plant growth stages in the interspecific oviposition preference of European corn borer and relation with chemicals present on leaf surface. *Entomol. Exp. Appli.* 53: 267-276.

Derridj, S., Anglade, P., Fiala, V., and Panouillé, A. 1990, Perspectives d'utilisation de critères biochimiques de la feuille dans la sélection de maïs défavorable à la ponte de pyrale (*Ostrinia nubilalis* Hbn.), pp. 1209-1216 In: *A.N.P.P., 2nd International Conference on Pests in Agriculture*, Versailles 1990.

Derridj, S., Boutin, J.P., Fiala, V., and Soldaat, L.L. 1996, Composition en métabolites primaires de la surface foliaire du poireau : étude comparative, incidence sur la sélection de la plante-hôte pour pondre par un insecte. *Acta Bot. Gallica* (in press).

Derridj, S., Wu, B.R., Stammitti, L., Garrec, J.P. and Derrien, A. 1996, Chemicals on the leaf surface, information about the plant available to the insect. *Entomol. Exp. Appl.* 80: 197-201.

Eigenbrode, S.D. and Espelie, K.E. 1995, Effects of plant epicuticular lipids on insect herbivores. *Ann. Rev. Entomol.* 40: 171-194.

Ferrandon, M., and Chamel, A. 1989, Foliar uptake and translocation of iron, zinc, manganese. Influence of chelating agents. *Plant Physiol Biochem.* 27(5): 713-722.

Fiala, V., Boutin, J.P., Barry, P. and Derridj, S. 1993, Les métabolites de la surface foliaire (phylloplan): présence et rôle dans les relations plante-insecte. *Acta Bot. Gallica* 140: 207-216.

Fiala, V., Glad, C., Martin, M., Jolivet, E. and Derridj, S. 1990, Occurence of soluble carbohydrates on the phylloplane of maize (*Zea mays* L.): variations in relation to leaf heterogeneity and position on the plant. *New Phytol.* 115: 609-615.

Goatley, J.L. and Lewis, R.W. 1966, Composition of guttation fluid from rye, wheat and barley seedlings. *Plant Physiol.* 41: 373-375.

Harr, J., Guggenheim, R., Boller, TH. and Oertlie, J.J. 1980, High pH- values on the leaf surfaces of commercial cotton varieties. *Coton Fibres Trop.* XXXV, 4: 379-384.

Jeffree, C.E., Grace, J. and Hoad, S.P. 1994, Spatial distribution of sulfate uptake by wind-damaged beech leaves. *NATO ASI G* 36 : 183-193.

Jermy, T. 1993, Evolution of insect plant relationships: a devil's advocate approach. *Entomol. Exp. Appl.* 66 : 3-12.

Leece, D.R. 1976, Composition and ultrasctuce of leaf cuticles from fruit trees relative to different foliar absorption. *Aust. J. Plant Physiol.* 3: 833-847.

Leonardi, S. and Flückiger, W. 1987, Short term canopy interactions of beech trees : mineral ion leaching and absorption during rainfall. *Tree Physiol.* 3: 137-145.

Long, W.G., Sweet, D.V. and Tukey, H.B. 1956, Loss of nutrients from plant foliage by leaching as indicated by radioisotopes. *Science.* 123: 1039-1040.

Lorenzoni, C. and Salamini, F. 1975, Glossy mutants of maize. V. Morphology of the epicuticular waxes, *Maydica XX*: 5-19.

Mengel, K., Breininger, M.TH., and Lutz, H.J. 1990, Effect of simulated acidic fog on carbohydrates leaching, CO_2 assimilation and development of damage symptoms in young spruce trees (*Picea abies* L. Karst). *Environ. Exp. Bot.* 30 (2): 165-173.

Morgan, J.V. and Tukey, H.B. 1964, Characterization of leachate from plant foliage. *Plant Physiol.* 590-593.

Price, C.E. 1982, A review of the factors influencing the penetration of pesticides through plant leaves. pp. 237-252. In: Cutler, D.F, Alvin, K.L, and Price, C.E. (eds.) *The Plant Cuticle,* Academic Press, London.

Renwik, J.A.A. and Chew, F.S. 1994, Oviposition behaviour in *Lepidoptere. Annu. Rev. Entomol.* 39: 377-400.

Schönherr, J. 1976, Water permeability of isolated cuticular membranes: the effect of cuticular waxes on diffusion of water. *Planta* 131: 159-164.

Schönherr, J. and Riederer, M. 1989, Foliar penetration and accumulation of organic chemicals in plant cuticles. *Rev. Environ. Contam. Toxicol.* 108: 1-64.

Schreiber, L. and Schönherr, J. 1993, Determination of foliar uptake of chemicals: influence of leaf surface microflora, *Plant Cell Environ.* 16: 743-748.

Soldaat, L.L., Boutin, J.P. and Derridj, S. 1996, Species specific composition of free amino acids on the leaf surface of four *Senecio* species . *J. Chemical. Ecol.* 22 (2): 1-2.

Städler, E. 1986, Oviposition and feeding stimuli in leaf surface waxes, pp. 105-121, In: Juniper, B.E. and Southwood, T.R.E. (eds.) *Insects and Plant Surface,* Edward Arnold, London.

Städler, E. 1992, Behavioral responses of insects to plant secondary compounds. pp. 45-88. In: Rosenthal, G. A., Berebaum, M.R.(eds.) *Herbivores: Their Interaction with Secondary Plant Metabolites,* San Diego Academic.

Städler, E. and Roessingh, P. 1990, Perception of surface chemicals by feeding and ovipositing insects. *Symp. of Biol. of Hungary* 39 : 71-86.

Stammitti, L., Garrec, J.P. and Derridj, S. 1995, Permeability of isolated cuticles of *Prunus laurocerasus* to soluble carbohydrates. *Plant Physiol. Biochem.* 33 (3): 319-326.

Stephanou, M. and Manetas, Y. 1995, Allelopathic and water concerning functions of leaf epicuticular exudates in the Mediterranean shrub *Dittrichia viscosa. Aust. J. Plant Physiol.* 22: 755-9.

Taylor, F.E., Davies, L.G. and Cobb, A.H. 1980, An analysis of the epicuticular wax of *Chenopodium album* leaves in relation to environmental change, leaf wettability and the penetration of the herbicide bentazon. *Ann. Rev. Appl. Biol.* 98: 471-478.

Tukey, H.B., Jr. 1970 The leaching of substances from plants. *Ann. Rev. Plant Physiol.* 21: 305-324.

Tukey, H.B., Jr. and Morgan, J.V. 1963, Injury to foliage and its effects upon the leaching of nutrients from above-ground plant parts. *Physiol. Plant.* 16: 557-564.

Tukey, H.B., Jr., Tukey, H.B. and Wittwer, S.H. 1958, Loss of nutrients by foliar leaching as determined by radioisotopes. *Proc. Am. Soc. Hortic . Sci.* 71: 496-506.

Turner, D.P., Broekhuizen, H.J. Van 1992, Nutrient leaching from conifer needles in relation to foliar apoplast cation exchange capacity. *Environ. Pollut.* 75: 259-263.

Van der Meijden, E., Van Zoelen, A.M. and Soldaat, L.L. 1989, Oviposition by the cinnabar moth, *Tyria jacobaeae,* in relation to nitrogen, sugars and alkaloids of ragwort, *Senecio jacobaea. Oikos* 54: 337-344.

Von Kunert, R., and Libbert, E. 1972, Beziehungen zwischen Planzen und epiphytischen Bacterien hinsichtlich ihres Auxinstoffwechsels. X. Die Exudation von Aminosäuren und Kohlenhydraten durch Maissprosse als Ernärungdlage für epiphytische Bakterien. *Biochem. Physiol. Pflanz.* 163: 524-535.

Von Scheffer, F., Stricker, G. and Kickuth, R. 1965, Organische Verbindungen in der Guttationsflüssigkeit einiger Wild und Kultur. *Pflanzenernaehr. Bodenkd.* 240-248.

Wollenweber, E. and Dietz, V.H. 1981, Occurence and distribution of free flavonoid aglycones in plants. *Phytochemistry* 20: 869-932.

Woodhead, S. and Chapman, R.F. 1986. Insect behaviour and the chemistry of plant surface waxes pp. 123-135. IN: Juniper, B. and Southwood, T.R.E. (eds.) *Insects and the Plant Surface,* Edward Arnold, London.

Wu, B.R., Derrien, A. and Derridj, S. 1995, Possible role of fatty acids in the permeability of the leaf cuticle to water soluble carbohydrates, p. 100 In: *Proc. 6th Inter. Symp. Microbiol. Aerial Plant Surface* September 1995, Bandol, France.

3

MICROBIAL ATTACHMENT TO PLANT AERIAL SURFACES

Martin Romantschuk, Elina Roine, Katarina Björklöf, Tuula Ojanen, Eeva-Liisa Nurmiaho-Lassila, and Kielo Haahtela

Department of Biosciences
Division of General Microbiology
P.O. Box 56, FIN-00014
University of Helsinki
Finland

INTRODUCTION

Plant aerial surfaces are colonised by a variety of microbes, including bacteria, fungi and yeasts. Some of these microbes are opportunistic pathogens, whereas others are specialised epiphytes. The outcome of the interaction between the microbe and the plant host is dependent on the plant-microbe combination. Often the host range for disease is more narrow than that for epiphytic colonisation.

Adhesion of appressoria and hyphae to the plant host has long been recognized as being important in the pathogenesis of phytopathogenic fungi. Fungal spore attachment as a first step in the interaction appears to have received increasing attention in recent studies. Likewise, disease caused by *Agrobacterium* and the onset of symbiosis with *Rhizobium* spp. have been shown to be dependent on attachment. Recently it has become evident that the initial interaction between a fungal spore or a bacterial cell (suspended in water or spread by wind) and the plant surface involves attachment that increases the chances of successful colonisation by the microbe and possible disease development (Epstein *et al.*, 1994; Romantschuk, 1992; Suoniemi *et al.*, 1995). At this early point in the infection process there are similarities between fungal and bacterial interactions with the host. An understanding of the adhesive mechanisms is essential, both in order to clarify the events leading to penetration and overcoming of the defence mechanisms of the plant, and to be able to assess the criteria for successful colonisation by a nonpathogen to be used as a bio-control agent. In both cases the microbe tries to avoid being dislodged by wind or running water. Despite the difference in size between bacterial cells and fungal spores, they share certain properties, such as sensitivity to physical shearing forces. Furthermore, the interaction probably needs to be dynamic, so that the microbial propagule is not immobilised and inactivated on the plant surface, but retains its capacity to divide and, in some cases, to move and spread along the solid support. As conditions change, differential induction of genes and of cell division may

Aerial Plant Surface Microbiology, edited by Morris et al.
Plenum Press, New York, 1996

also influence the attachment status of the microbial cells: compounds and structures active in attachment may be induced or repressed. Bacterial daughter cells in emerging microcolonies during a rapid growth phase may be more likely to dislodge and spread to neighbouring leaves and plants, or to the inside of the leaf, potentially resulting in disease. For both types of organisms the microbial mechanisms underlying attachment are only now starting to be unravelled.

Both microbial and plant surfaces have negative surface potential, which results in electrostatic repulsion keeping the surfaces apart. Spikes or pili are often found extending from the surface of the microbial particle that aims to bind to the plant or other surfaces. These structures may be present already on microbes prior to any induction by the presence of plant surfaces or otherwise favourable conditions, and help bridge the gap between the interacting bodies that are kept apart by electrostatic repulsion. These structures depend on attracting forces, such as van der Waals attraction, hydrophobic interaction, and/or hydrogen bonding to achieve contact (Bell, 1978; Jones, 1994; Romantschuk, 1992).

The immediate early interactions appear generally to be dependent on properties of and structures on the surface of the microbe. A similar degree of hydrophobicity between the microbial and the plant surface appears to increase the chances for hydrophobic interaction. Both surfaces or appendages on them are generally hydrophobic, and decreases in hydrophobicity reduces adherence both in the case of fungi such as *Colletotrichum lindemuthianum* (Young and Kauss, 1984) and a wide variety of bacteria (Loosdrecht *et al.*, 1987; Rosenberg and Kjelleberg, 1986).

In some cases the release of a pre-existing mucilage is triggered by hydration. After contact has been made, or in some cases directly following hydration of the particles, additional compounds are often either released or produced by the activated microbial spores (Matthysse *et al.*, 1995, and references therein; Braun and Howard, 1994a; Romantschuk, 1992). In most cases only indirect evidence is available regarding the composition of mucilaginous compounds. They are generally believed to consist of polysaccharides and/or (glyco-) proteins.

Much of the work on fungal interactions with plant surfaces is descriptive and based on electron microscopy. However, the tools for performing mutational and biochemical analyses are now available, which will aid in confirming or disproving current hypotheses. For bacteria, most of the work on attachment has been done with animal pathogens, where the mechanisms and structures involved have, in many cases, been characterized in detail. Several interesting revelations have been made by comparison of microbes infecting different types of hosts.

SPREAD OF MICROBIAL PROPAGULES

Microbes spread to plant surfaces by wind as aerosols or dry spores of drought resistant cells, or within rain, rain splash or irrigation water. Plant-colonising epiphytic and pathogenic bacteria are dislodged in great numbers as aerosols from plants (Lindemann *et al.*, 1982) and may travel long distances airborne. The range of efficient aerial dispersal of viable *Pseudomonas syringae* cells from inoculated oats was, however, not more than several meters (Lindow *et al.*, 1988). Similar results were obtained when modelling spread of bacteria by rain splash. Bacteria are dislodged efficiently from the leaf surface by rain, but do not on average travel great distances (Butterworth and McCartney, 1991). However, given time and the right conditions, bacteria will apparently spread within boundaries determined by their environmental fitness rather than limited by distances.

Many fungal plant pathogens produce dry windborne spores that spread efficiently (McCartney, 1991) and stay dormant but viable in dry conditions. Even when dry, spores may be entrapped or adhere passively and reversibly to surfaces. At this stage the spore is not locked in a certain sequence of events, but spread can continue, until imbibition leads to committed development.

Upon arrival on a suitable target surface the initial interaction appears in most cases to be rather nonspecific and is often mediated by hydrophobic interactions. Adhered fungal spores germinate and the fungus colonises and spreads as growing hyphae. Immobilisation of the spore does not limit the spread of the fungus on and into the target plant. Neither are bacteria, adhered by their flagellum or pilus, irreversibly immobilised. Initial attachment is not necessarily very firm; cells attach and detach, and cells divide giving rise to progeny that may or may not be attached. Expression of the structures required for attachment may be regulated by environmental stimuli (*e.g.* fimbriation induced and flagellation repressed). Cells expressing retractable pili, such as the type IV pilus of *P. aeruginosa* and many other bacteria (Strom and Lory, 1993) have been reported to be capable of spreading along solid surfaces by so called twitching motility (Hendriksen, 1983). Among the plant-specific bacteria this type of pilus is expressed by at least *Xanthomonas campestris* (van Doorn *et al.*, 1994: T. Ojanen, unpublished) and *P. syringae* (Roine *et al.*, 1996; D. Nunn, unpublished).

Below, several relatively well characterized cases of bacterial and fungal plant surface interactions will be discussed.

EPIPHYTIC GROWTH AND SURVIVAL OF BACTERIA

From various studies dealing with epiphytic growth of plant pathogenic and non-pathogenic bacteria (Beattie and Lindow, 1994 a, b; Wilson and Lindow, 1994; for review see Hirano and Upper, 1990) it can be concluded that plant defence mechanisms do not reach the outside surface of an unwounded plant, and therefore immobilisation of bacterial cells on a plant leaf surface is not followed by elimination of the attached organism. In humid conditions nutrients exuded by the leaf are likely to diffuse in the film of moisture, enabling bacteria to grow and form microcolonies, from which cells can spread even if the original colony forming unit is attached.

Large populations of epiphytic bacteria, including potential pathogens can develop on the surfaces of symptomless plants. These bacteria include strains that function as ice nuclei inducing frost injuries (Lindow *et al.*, 1978), or as disease inocula when the right conditions arise (Hirano and Upper, 1990). Such conditions occur for example when plants are wounded by a rain storm or by frost injury (Sule and Seemuller, 1987). The risk for disease correlates with the population level, a fact that can be used in predicting disease outbreak (Rouse *et al.*, 1985).

The traits influencing epiphytic performance are likely to be different from those required to cause plant disease. Several properties have been shown or suggested to be epiphytic fitness factors (Lindow, 1991). Among these are UV- and desiccation tolerance, production of surface active agents, motility and production of extracellular polysaccharides (EPS). Specialised epiphytic bacteria can survive in decomposing plant material, but often are not efficient colonisers of the soil, and thus likely to spread mainly from plant to plant by wind via aerosols released from the plants by water, irrigation or rain splash.

In dry conditions, the population size of epiphytically growing *P. syringae* cells initially drops, after which the population may recover partially and stabilises at numbers which are lower than in humid conditions (Beattie and Lindow, 1994a, b; Björklöf *et al.*, 1995b; Wilson and Lindow, 1994). Whether or not there is cell division during the stabilised conditions (fast enough only to replace dying cells) or if the bacteria merely survive in a resting state on the plant surface is unclear, and probably depends on the bacterium in question.

P. syringae cells that were considered to occupy protected microsites on or in the leaf were shown to be able to survive during harsh environmental conditions, while more exposed cells were killed (Beattie and Lindow, 1994a, b). Ability to localise, multiply, and/or survive in protected sites on the leaf was reduced in some mutants impaired in epiphytic fitness. The capability to stay in and possibly to move towards such sites should hypothetically be an epiphytic fitness factor. The mutants investigated by Beattie and Lindow (1994a) were not impaired in flagellar motility, but flagella may not be the only means for motility. Twitching motility has been described for various bacteria harbouring retractive pili such as the type IV pilus of P. aeruginosa. This form of motility is seen as a spreading of cells on solid surfaces such as agar plates (Henrichsen, 1983) and might also function on leaf surfaces, although this has not been tested. In P. aeruginosa, type IV pilus-mediated twitching motility has chemotactic characteristics (Darzins, 1994) which fits well with a requirement for directed motility. Although the phage ϕ6 specific pili of P. syringae are of type IV (Roine et al., 1996; see below), twitching motility has not yet been observed. This capacity may be induced only in certain, as yet unknown, conditions.

A bacterium located on a leaf starts to multiply when the surface is wetted through rain or dew and conditions are otherwise favourable for bacterial growth. The same may be true when a bacterial cell arrives at a plant leaf surface in a droplet of water. In either case the droplets on the surface of a leaf contain bacterial cells in suspension. As the leaves dry, the drops on the hydrophobic leaf surface withdraw to small sections of the leaf leaving other parts dry. The portion of the bacterial cells that are still in suspension consequently concentrate in the regions of the leaf that dry last. This is what probably happens after rain, and when dew drops dry in the morning sun. Stained bacterial cells can be seen to collect at the leaf-veins and along the edges of a drying droplet. In SEM the bacteria are seen in the crevices between the leaf cells, and attached to stomata, but also heavily aggregated at the leaf veins, filling the more depressed locations (Fig. 1). In some cases the dried remnants of a matrix, possibly polysaccharide, can be observed around or covering the aggregates of bacterial cells (Fig 2). Such polymeric layers have been observed for a number of bacteria adhering to plant surfaces (Leben and Whitmoyer, 1979).

In the drying process the resulting local concentration of bacteria is far higher than the average cell concentration per square unit of the plant surface. Although this may seem impractical from the point of view of utilization of the limited nutrients (Tukey, 1970), it apparently facilitates production of sufficient amounts of polysaccharide, or other matrix material which helps to resist drying around the bacterial mass. The bacteria may also be better protected from radiation and other environmental influences. Bacterial aggregates perhaps also affect the efficiency for ice nucleation in the case of INA$^+$ bacteria, although this has to our knowledge not been investigated. Very high conjugation frequencies have been observed on the bean leaf surface, both in high humidity conditions (Björklöf et al., 1995a) as well as in low humidity (Björklöf et al., 1995b). An apparent contributing factor is the high bacterial concentration in selected locations on the leaf. An uneven distribution of the cells facilitates mating-pair formation.

BACTERIAL ATTACHMENT

Plant Associated Bacteria that Express Type IV Pili

Type IV pili are proteinaceous filamentous appendages present on a wide range of gram-negative bacteria isolated from different environments (Strom and Lory, 1993). The

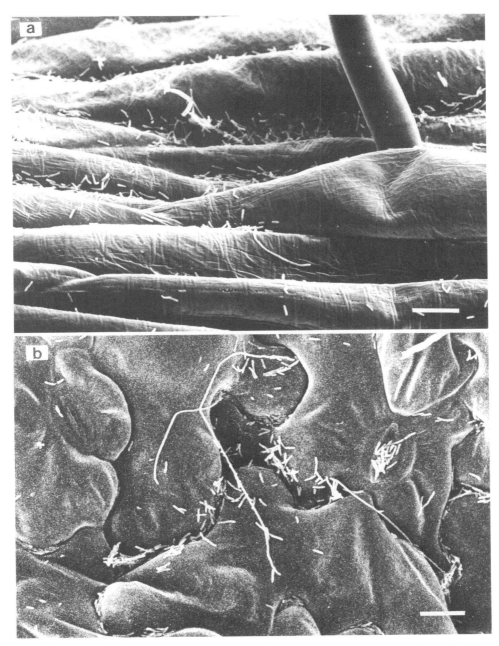

Figure 1. Cells of *Pseudomonas syringae* pathovar *phaseolicola* HB10Y on the lower surface of bush bean leaves. The bacteria were deposited from drying droplets mainly in the depressed regions of the leaf veins (a) and in crevices between cells and on stomata over the leaf surface. The bar denotes 10 μm.

biological functions of pili are all related to various forms of attachment: binding to host/substrata, pellicle formation, twitching motility, and functioning as a phage-receptor. The type IV pili bind to target cell surfaces in a reversible manner (Paranchych and Frost, 1988). The *P. aeruginosa* pilus binds to the glycolipid asialo-GM$_1$ (Gupta *et al.*, 1994), specifically to the carbohydrate sequence βGalNac(1-4)βGal (Sheth *et al.*, 1994), but also to a sialylated corneal cell surface glycoprotein (Hazlett *et al.*, 1995).

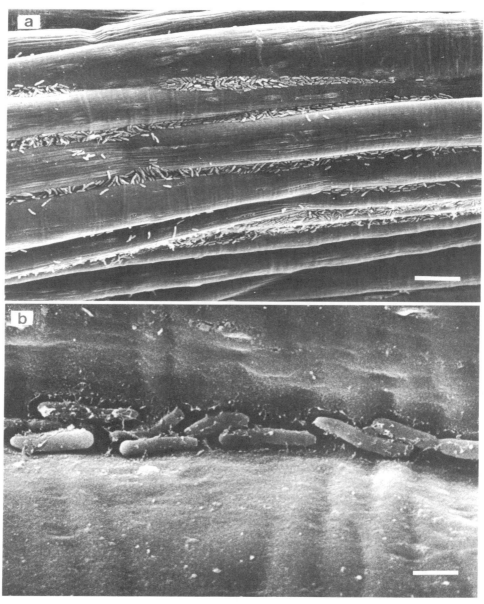

Figure 2. Cells of *Pseudomonas syringae* pathovar *syringae* R32 in depressions between cells of the bean leaf vein (a). With greater magnification (b) dried residues of what apparently is bacterial extracellular polysaccharide are seen on and surrounding the cells. The bar denotes 10μm in a, and 1μm in b.

The type IV pili, and their significance in the pathogenesis of the opportunistic human pathogen *P. aeruginosa,* have been extensively studied. The available information on the genes involved in formation and function of the pilus has been utilised in cases of isolation and characterization of the equivalent genes in plant pathogens. The plant-pathogenic bacteria that so far have been shown to express type IV pili are *Xanthomonas campestris* pv. *hyacinthi, X. campestris* pv. *vesicatoria*, and various pathovars of *P. syringae*. The pilin gene and the protein from *X. c. vesicatoria* and *X. c. hyacinthi* have been partially characterized (T. Ojanen unpublished; van Doorn *et al.*, 1994), and found to share homology with the type IV pilin and *pilA* gene of *P. aeruginosa*. The pilus of various pathovars of *P. syringae* functions as the receptor for the bacteriophage φ6 (Vidaver *et al.*, 1973; Romantschuk *et al.*, 1993). Several genes required for pilus biosynthesis in *P. syringae* pvs *phaseolicola* and *tomato* have been isolated and partially characterized (Roine *et al.*, 1996; D. Nunn, unpublished), and the results show that the φ6 receptor pilus is also of type IV. At least for some genes (*pilMNOP*) the degree of homology to the corresponding *P. aeruginosa* genes was very high (Martin *et al.*, 1993; 1995; Roine *et al.*, 1996). Many, but not all *P. syringae* pathovars are sensitive to the pilus-specific bacteriophage φ6 (Romantschuk *et al.*, 1993). It is quite possible that also those *P. syringae* strains that are φ6-resistant express the type IV pili, but with a pilin that does not function as receptor for the phage.

Phage-resistant bacterial mutants of φ6-sensitive *P. syringae* strains are generally non-piliated (Romantschuk and Bamford, 1985), and these mutants have a significantly reduced ability to attach to leaf surfaces (Romantschuk and Bamford, 1986) of both host and non-host plants. The relative decrease in attachment of the non-piliated mutants is roughly the same (3-5 fold) whether the strain is motile or not, but overall the attachment is more efficient for the motile strains (Romantschuk *et al.*, 1993). Whether the flagella also attach, or whether they just increase the frequency of impaction is not known.

Nonpiliated mutants of the bean pathogen *P. syringae* pv. *syringae* also displayed reduced efficiency in initiation of epiphytic colonisation when the inoculation was followed by flushing unattached bacteria from the plants. The sensitivity of the bacteria to rinsing decreased gradually, but continued to be higher for the non-piliated mutant for at least three days of incubation in high humidity conditions. After three days, the wild type bacterium resisted gentle rinsing, possibly by forming micro-colonies held together in pellicle-like structures (Suoniemi *et al.*, 1995). The ability to form a pellicle at the air-liquid interphase of stationary cultures has been associated with the presence of pili (Goochee *et al.*, 1987). The prediction held true for the motile bean pathogen *P. syringae* pv. *syringae* R32 (Suoniemi *et al.*, 1995). The wild type, and a super-piliated mutant readily formed a pellicle on the surface of stationary tube cultures in rich medium, whereas a non-piliated mutant did not, and subsequently, not being able to access the oxygen at the liquid surface, grew to a much lower density (Fig. 3). Neither the wild type, nor any of the piliation mutants of the non-flagellar strain *P. syringae* pv. *phaseolicola* HB10Y were able to form a pellicle (not shown), suggesting that for *P. syringae* flagellar motility is required, but not sufficient, for pellicle formation. Without flagella the cells cannot actively migrate toward the attractant oxygen, whereas without the capacity to attach to the walls of the test tube and to each other the cells cannot stay at the oxygenated interphase.

In attachment assays the cells of piliated strains of *P. syringae* pv. *phaseolicola* HB10Y were seen aggregated to the stomata. The cells can apparently also attach to each other, since the bacterial aggregates were more than one cell layer deep. No aggregates were seen when using non-piliated mutants (Romantschuk and Bamford, 1986). Cells of *P. syringae* pv. *syringae* pathogenic to bean and corn were seen evenly spread over the surface of the bean leaf (Korhonen *et al.*, 1986, Romantschuk *et al.*, 1993;). Also for these, and three additional pathovars (Romantschuk *et al.*, 1993) the attachment efficiency of non-piliated phage φ6-resistant mutants was substantially lower than for the wild type.

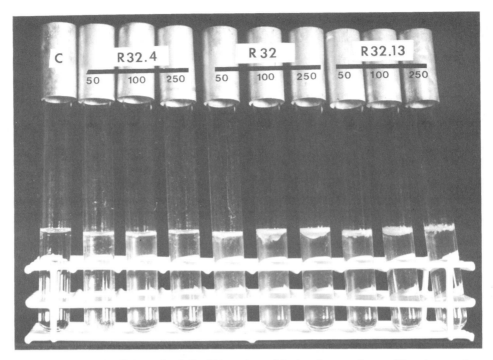

Figure 3. Formation of pellicles at the air-liquid interphase of 5 ml stationary cultures of *P. syringae* pathovar *syringae*. C, uninoculated control; R32.4, nonpiliated mutant strain; R32, wild type strain; R32.13, super-piliated mutant strain. The amount of overnight culture (μl) used as inoculum is indicated by the numbers below the bar.

Rhizobium and *Agrobacterium*

Attachment to plant cells is one of the early steps in the infection process of *Rhizobium* and *Agrobacterium*. The attachment is a two step process where the first and essential step is mediated by rhicadhesin, a Ca^{++} binding protein produced by both bacteria (Smit *et al.*, 1989; Swart *et al.*, 1993). The second step for both these related bacteria is dependent upon production of bacterial cellulose fibrils (Matthysse, 1983). The fibrils anchor the bacterial cells to each other and to the plant surface.

Among the phenotypes observed in mutants unable to produce cyclic β-1,2-glucan are non-motility, attachment deficiency, avirulence, production of a non-active rhicadhesin and reduced osmotolerance. Rhicadhesin activity, attachment and virulence on Kalanchoe leaves of *A. tumefaciens* could be restored by incubating the cells in high osmolarity and $CaCl_2$ (O'Connell and Handelsman, 1989; Swart *et al.*, 1994b). Alternatively, rhicadhesin produced in wild type cells restored the attachment when added separately (Swart *et al.*, 1993).

The receptors for bacterial attachment on the plant surface have not been firmly established, but a vitronectin-like glycoprotein present on the cell surface of a wide variety of plants was proposed to play a role (Wagner and Matthysse, 1992). Both vitronectin and antibodies raised against human vitronectin were able to reduce the attachment of *A. tumefaciens* to carrot cells in suspension. A putative 32 kDa receptor glycoprotein for rhicadhesin binding was isolated from pea roots (Swart *et al.*, 1994a). The authors suggest that the isolated protein might contain the common cell attachment determinant R-G-D that is present also in vitronectin. In fact, a hexapeptide containing this determinant did function as a receptor for rhicadhesin.

Translocation of Bacterial Macromolecules into Plant Cells

Recent findings regarding the nature of certain avirulence gene products indicate that their target of action is within the plant cell, while these proteins have not been seen secreted from the bacterial cell in *in vitro* studies. Products of *avrb6* and *pthA* show nuclear targeting signals in the protein sequence, and a fusion protein containing PthA was shown to migrate to the plant nucleus when expressed in the plant cells (Yang and Gabriel, 1995). The mode of action of the bacterial gene products is not known, but the notion that bacterial proteins are translocated into the plant cells presupposes that at some instance, there is a direct contact between the interacting cells, and thus attachment of bacterial cells to the target plant cells. The details of this interaction have not been worked out, but comparisons to other type III protein secretion systems (Salmond, 1994) can be made. Virulent bacteria of the genera *Yersinia*, *Shigella* and *Salmonella* secrete a number of virulence determinants (Yops, Ipas and Sips, respectively) by a type III secretion pathway (Rosqvist *et al.*, 1995) as described in the following chapter by Bonas and Van den Ackerveken (this volume). YopE and YopH of *Yersinia pseudotuberculosis* are translocated into target mammalian cells in a process that requires at least YopB and YopD (Håkansson *et al.*, 1993; Persson *et al.*, 1995; Rosqvist *et al.*, 1995), but the nature of the translocation apparatus is not clear. In analogy with this protein-translocation system, *Agrobacterium* translocates the T-DNA of the Ti-plasmid into host plant cells. Genes in the *virB* locus involved in the mating-pair formation have been characterized and found to share homology with conjugation genes of IncW and IncF -type plasmids. *virB2* is proposed to encode the structural pilin (Kado, 1994). Although there does not seem to be any protein sequence homology between the *Pseudomonas/Xanthomonas hrp* and the *Agrobacterium vir* genes, the translocation process and the structures involved may be principally similar. In both systems much work remains to be done in order to characterise the plant-bacterial interaction. For example, no knowledge regarding the plant components involved in the interaction during the macro-molecule translocation is available.

FUNGAL SPORE ATTACHMENT

Many fungal plant pathogens and endophytes produce dry windborne spores that attach nonspecifically or are entrapped on host surfaces. In suitable, high humidity conditions the spores imbibe -and in many cases release- water-soluble extracellular matrix materials before or during germination. The matrix increases the contact area between the fungal spore and the host surface and was suggested to contain surface active components with the same degree of hydrophobicity as the host leaf (Braun and Howard, 1994a; Jones, 1994). In several cases the composition of the matrix has been partially characterized and a role in adhesion to substrata has been demonstrated (Braun and Howard, 1994a, b; Clement *et al.* 1993a, b; Epstein *et al.*, 1994; Viret *et al.*, 1994).

Jones (1994) divides the early interactions of a fungal spore with the host plant into five stages: (1) initial passive attachment; (2) active, induced attachment; (3) spore differentiation and germination, with germ-tubes enveloped by a hyphal sheath; (4) development of appressoria; (5) penetration of the substratum/host. The first two stages, which are not always clearly distinguishable, will be addressed here.

In the case of fungal spores, the initial passive attachment (impaction, chemotaxis, entrapment, spore appendage mediated attachment) is often reversible; the spores are easily washed away. Attachment by a sticky cell wall or appendages is a form of passive attachment where the sticky structures may be exposed on the particle surface or on appendages extending from the surface. A mucilaginous sheath may be present on the spore surface or automatically released upon imbibition.

Active attachment mediated by extracellular adhesive material is thigmotropic and involves production and release of extracellular material. It presupposes an induced response by the attaching particle. Most common in the case of fungi is the induced production of an adhesive mucilage. This step may coincide with or precede germination of the spore.

Examples of Fungal Spore Attachment

Uredinospores of *Uromyces viciae-fabae* are hydrophobic and consequently aggregate in water suspensions. When a dry spore lands on a substratum the spore makes contact via spines. As the relative humidity increases sufficiently the spore imbibes, and an adhesive polymer is released which mediates surface attachment. This adhesive extracellular matrix (ECM) is produced during the initial phases of the infection process. The production is induced when the spores are exposed to a surface provided that the spore is imbibed. These early events are pre-programmed, and do not require gene expression in the spore (Clement *et al.*, 1993b). The initial adhesion appears not to be receptor-specific, but is more efficient on a hydrophobic surface than on a hydrophilic surface (Clement *et al.*, 1994). The composition of this ECM has recently been partially characterized (Clement *et al.*, 1993a). The ECM released before germination is composed of low molecular weight carbohydrates and some polypeptides. Active adhesion is induced in connection with germination when polymers are actively produced. This active phase binds the emerging germ tube to the host plant or substrate surface. During germination the matrix consists of increasing amounts of high molecular weight proteins some of which are glycosylated. The ability to spread on and attach to waxy surfaces correlates positively with the protein contents of the matrix. The specific role of certain degradative enzymes found in this matrix (phosphatase, esterase, protease, glycosidase) has not been clearly established (Clement *et al.*, 1993a).

For *U. appendiculatus* the events are probably similar. In SEM studies, pads of adhesive matrix have been observed to collect between the spore and the leaf surface. Formation of adhesion pads was independent of viability of the spores but firm adhesion appeared to be linked to production of esterases and cutinase (Deising *et al.*, 1992). These enzymes could cause erosion of the cuticle and waxes of the host plants. However, simply the increase in ECM production appears at least partly responsible for a stronger attachment in *U. viciae-fabae* (Clement *et al.*, 1993a). One or more proteins seem to be involved in the actual attachment of *U. appendiculatus* germlings, since treatment with pronase E prevented attachment and attached germlings could also be removed from surfaces with the same enzyme (Epstein *et al.*, 1987).

Cochliobolus heterostrophus (*Bipolaris maydis*) is a corn pathogen producing conidia that spread by wind and rain splash. After about 20 minutes in water the conidia begin to adhere non-specifically to a wide variety of artificial and host plant surfaces before germination. The attachment requires respiration and protein synthesis. Production of extracellular matrix and adherence is part of the active attachment phase. ECM produced just before germination forms pads that mediate the attachment which is not dependent of target surface hydrophobicity. This initial ECM was presumed to contain, but not consist exclusively of, proteins (Braun and Howard, 1994b).

Magnaporthe grisea is an ascomycete that causes rice blast. The conidia attach rapidly and immediately upon hydration to leaves and various other surfaces (Hamer *et al.*, 1988). The conidia attach more efficiently to hydrophobic surfaces and this attachment can be prevented by treatment with the lectin Concanavalin A (Con A). Con A apparently attaches to a polymeric material called spore tip mucilage (STM; Hamer *et al.*, 1988). When spores are hydrated the STM, which is stored in a periplasmic compartment, expands and ruptures the spore, releasing the mucilage which binds conidia to the leaf surface. The binding of Con A indicates the presence of α-mannoside or α-glucoside, but otherwise the composition of STM is unknown. The initial, essentially passive phase is followed by formation of germ tubes and appressoria, which are very strongly bound to the surface (Howard *et al.*, 1991).

The macroconidia of *Nectria haematococca* (*Fusarium solanii* f. sp. *cucurbitae*), a pathogen of cucurbits, attach to host and non-host plants as well as to both hydrophilic and hydrophobic surfaces. The macroconidia are dispersed by water and are nonadherent until exposure to an adhesion-inducing medium such as zucchini fruit extract. During the induction period the macroconidia produce a spore tip mucilage (STM) and a 90-kDa glycoprotein which both appear to be involved in adherence (Kwon and Epstein, 1993). However, mutants with reduced ability to adhere to zucchini fruits and to polystyrene still produced both the mucilage and the protein (Epstein *et al.*, 1994). Thus, the role of the mucilage and the protein as well as the nature of the mutation remains to be determined.

In endophytic interactions such as that of *Discula umbrinella* (*Apiognomonia errabunda*) with beech, the fungal conidia are covered by a sheath. This sheath apparently has exposed mannose and/or glucose residues on its surface since the lectin Con A strongly binds to the conidia, and inhibits binding of the conidia to the host surface. Based on results of enzymatic treatment of the conidia Viret *et al.* (1994) concluded that a glycoprotein is involved in adhesion.

CONCLUSION

The events of the micro-organism-plant interaction that precede the development of disease symptoms or the establishment of a symbiotic relationship are often prerequisite for disease or symbiosis. Although research efforts have focused primarily on the latter stages, the early stages of this interaction merit more detailed observation. This is particularly true when considering that the early and epiphytic stages may be more sensitive to biocontrol agents, for example. A variable degree of information is available regarding the role and mechanisms of attachment in the early interaction of microbes with the colonisable host plant. Utilising the knowledge of attachment in one system, may provide useful guidelines for the examination of new unknown combinations. This includes comparison between animal and plant pathogens.

Also, comparing different types of plant pathogens - foliar and root pathogens, bacteria and fungi - generate new ideas about key processes for investigation. In this review we have described several cases of plant-microbe interactions, specifically attachment at the early stages of colonisation. By comparing bacterial and fungal examples some similarities are apparent. Both bacteria and fungal spores produce extracellular fibrillar proteinaceous appendages. Whenever the function of such structures has been revealed, it has been attachment and/or motility. An extensive review of adhesion in fungi was recently published by Jones (1994).

The other common feature of plant associated bacteria and fungi is their capacity to produce extracellular polysaccharides. The bacterial extracellular polysaccharides are important virulence factors (Rudolph *et al.*, 1994), but are apparently also fitness factors for epiphytic pseudomonads. *Agrobacterium* and *Rhizobium* strains produce cellulose fibrils when introduced on the plant surface. This capacity is not unique for these bacteria, and might be more widespread among plant associated bacteria. The sticky polysaccharide-containing mucus formed by fungal spores is more complex than the bacterial EPS. Both the mucilage released by imbibed spores, and particularly the actively produced mucilage, contain glycoproteins apparently related to attachment and various enzymes that have been proposed to modify the plant surface to enhance attachment. Also plant specific bacteria secrete enzymes and other proteins, but only in the case of the filamentous structures (pili/fimbriae and flagella) has a role in epiphytic colonisation and attachment been shown.

For practical reasons different approaches have been used with different microbes. Fungal plant infections have been extensively studied by electron microscopy. In contrast to bacterial infections, the activity of an individual invading fungal propagule can be followed. Results obtained with fungi are on the other hand rather descriptive. With bacteria, further progress has been made with molecular genetics and biochemistry but the emphasis of these studies has been on genes with easily recognized phenotypes and plant responses. For fungi, mutational analysis has started (also in the case of attachment factors), and hopefully bacterial traits with less obvious, but still important functions will receive increasing attention in the future.

ACKNOWLEDGMENTS

This work was supported by the Academy of Finland and the Maj and Tor Nessling Foundation.

REFERENCES

Beattie, G.A. and Lindow, S.E. 1994a, Survival, growth, and localization of epiphytic fitness mutants of *Pseudomonas syringae* on leaves. *Appl. Environ. Microbiol.* 60:3790-3798.

Beattie, G.A. and Lindow, S.E. 1994b, Comparison of the behavior of epiphytic fitness mutants of *Pseudomonas syringae* under controlled and field conditions. *Appl. Environ. Microbiol.* 60:3799-3808.

Bell, G.I. 1978, Models for the specific adhesion of cells to cells. *Science* 200:618-627.

Björklöf, K. Suoniemi, A. Haahtela, K., Romantschuk, M. 1995a, High frequency of conjugation versus plasmid segregation of RP1 in epiphytic *Pseudomonas syringae* populations. *Microbiology* 141:2719-2727.

Björklöf, K., Nurmiaho-Lassila, E.L., Haahtela, K., Romantschuk, M. 1995b, Plasmid transfer of RP1 from *Pseudomonas syringae* CIT7 on the leaf surface of bean, p. 6 In: *Proc. 6th Inter. Symp. Microbiol. Aerial Plant Surfaces.* 11 - 15 September 1995, Bandol, France.

Braun, E.J. and Howard, R.J. 1994a, Adhesion of fungal spores and germlings to host plant surfaces. *Protoplasma* 181:202-212.

Braun, E.J. and Howard, R.J. 1994b, Adhesion of *Cochliobolus heterostrophus* conidia and germlings to leaves and artificial surfaces. *Exp. Mycol.* 18:211-220.

Butterworth, J. and McCartney, H.A. 1991, The dispersal of bacteria from leaf surfaces by water splash. *J. Appl. Bacteriol.* 71:484-496.

Clement, J.A., Butt, T.M. and Beckett, A. 1993a, Characterization of the extracellular matrix produced *in vitro* by urediniospores and sporelings of *Uromyces viciae-fabae. Mycol. Res.* 97:594-602.

Clement, J.A., Martin, S.G., Porter, R., Butt, T.M. and Beckett, A. 1993b, Germination and the role of extracellular matrix in adhesion of urediniospores of *Uromyces viciae-fabae* to synthetic surfaces. *Mycol. Res.* 97:585-593.

Clement, J.A., Porter, R., Butt, T.M. and Beckett A. 1994, The role of hydrophobicity in attachment of urediniospores and sporelings of *Uromyces viciae-fabae. Mycol. Res.* 98:1217-1228.

Darzins, A. 1994, Characterization of *a Pseudomonas aeruginosa* gene cluster involved in pilus biosynthesis and twitching motility: sequence similarity to the chemotaxis proteins of enterics an the gliding bacterium *Myxococcus xanthus. Mol. Microbiol.* 11:137-153.

Deising, H., Nicholson, R.L., Haug, M. Howard, R.J. Mendgen, K. 1992, Adhesion pad formation and the involvement of cutinase and esterase in the attachment of urediniospores to the host cuticle. *Plant Cell* 4:1011-1111.

van Doorn, J., Boonekamp, P. M. and Oudega, B. 1994, Partial characterization of fimbriae of *Xanthomonas campestris* pv. *hyacinthi. Mol. Plant Microbe Interact.* 7:334-344.

Epstein, L., Kwon, Y.H., Almond, D.E., Schached, L.M. and Jones, M.J. 1994, Genetic and biochemical characterization of *Nectria haematococca* strains with adhesive and adhesion-reduced macroconidia. *Appl. Environ. Microbiol.* 60:524-530.

Epstein, L., Laccetti, L. Staples, R.C. and Hoch, H.C. 1987, Cell-substratum adhesive proteins involved in surface contact responses of the bean rust fungus. *Physiol. Mol. Plant Pathol.* 30:373-388.

Goochee, C. F., Hatch, R. T. and Cadman, T. W., 1987, Some observations on the role of type 1 fimbriae in *Escherichia coli* autoflocculation. pp. 1024-1034. In: *Biotechnology and Bioengineering*, vol. XXIX,. New York: Wiley and Sons.

Gupta,, S.K., Berk, R.S., Masinick, S., Hazlett, L.D. 1994, Pili and LPS of Pseudomonas aeruginosa bind to the glycolipid asialo GM1. *Infec. Immun.* 62:4572-4579.

Håkansson, S., Bergman, T., Vanooteghem, J.C., Cornelis, G. and Wolf-Watz, H. 1993, YopB and YopD constitute a novel class of *Yersinia* Yop proteins. *Infec. Immun.* 61:71-80.

Hamer, J.E. Howard, R.J., Chumley, F. G. and Valent, G. 1988. A mechanism for surface attachment in spores of a plant pahtogenic fungus. *Science* 239:288-290.

Hazlett, L., Rudner, X., Masinick, S., Ireland, M. and Gupta, S. 1995, In the immature mouse, *Pseudomonas aeroginosa* pili bind a 57-kd (α2-6) sialylated corneal epithelial cell surface protein: a first step in infection. *Investigative Ophtalmology and Visual Science* 36, 634-643.

Henrichsen, J. 1983, Twitching motility. *Annu. Rev. Microbiol.* 37:81-93.

Hirano, S.S., and Upper, C.D. 1990. Population biology and epidemiology of *Pseudomonas syringae*. *Annu. Rev. Phytopathol.* 28:155-177.

Howard, R.J., Roach, D.H., Money, N.P. 1991, Penetration of hard substrates by a fungus employing enormous turgor pressures. *Proc. Natl. Acad. Sci USA* 88:11281-11284.

Jones, E.B.G. 1994, Fungal adhesion. *Mycol. Res.* 9:961-981.

Kado, C.I. 1994, Promiscuous DNA transfer system of *Agrobacterium tumefaciens*: Role of the virB operon in sex pilus assembly and synthesis. *Mol. Microbiol.* 12:17-22.

Korhonen, T.K., Haahtela, K., Romantschuk, M., and Bamford, D.H. 1986, The role of fimbriae and pili in the attachment of *Klebsiella*, *Enterobacter* and *Pseudomonas* to plant surfaces pp. 229-241. In: Lugtenberg, B. (ed.), NATO ASI Series, Vol. H4, *Recognition in Microbe-Plant Symbiotic and Pathogenic Interaction* Springer-Verlag Berlin, Heidelberg.

Kwon, Y.H. and Epstein, L. 1993, A 90 kDa glycoprotein associated with adhesion of *Nectria haematococca* macroconidia to substrata. *Mol. Plant-Microbe Interact.* 6:481-487.

Leben, C. and Whitmoyer, R.E. 1979, Adherence of bacteria to leaves. *Can. J. Microbiol.* 25:896-901.

Lindemann, J. Cosntantinidou, H.A., Barchett, W.R. and Upper, C.D. 1982, Plants as sources of airborne bacteria, including ice nucleation-active bacteria. *Appl. Environ. Microbiol.* 44:1059-1063.

Lindow, S.E. 1991. Determinants of epiphytic fittness in bacteria. pp. 295-314. In: Andrews, J.H. and Hirano, S.S. (eds.) *Microbial Ecology of Leaves*, NewYork: Springer-Verlag.

Lindow, S.E., Knudsen, G.R., Seidler, R.J., Walter, M.V., Lambou, V.W. Amy, P.S. Schmedding, D. Prince, V. and Hern, S. 1988, Aerial dispersal and epiphytic survival of *Pseudomonas syringae* during a pretest for the release of genetically engineered strains in the environment. *Appl. Environ. Microbiol.* 54:1557-1563.

Lindow, S.E., Arny D.C. and Upper C.D. 1978, Distribution of ice nucleation-active bacteria on plants in nature. *Appl. Environ. Microbiol.* 36:831-838.

Loosdrecht, M.C.M., Lycklema, J. Norde, W., Schraa, G. and Zehnder, J.B. 1987. The role of bacterial cell wall hydrophobicity in adhesion. *Appl. Environ. Microbiol.* 53:1893-1897.

Martin, P. R., M. Hobbs, P. D. Free, Y. Jeske, and J. S. Mattick. 1993. Characterization of *pilQ*, a new gene required for the biogenesis of type 4 fimbriae in *Pseudomonas aeruginosa*. *Mol. Microbiol.* 9:857-868.

Martin, P. R., Watson, A. A., McCaul, T. F. and Mattick, J. S. 1995, Characterization of a five-gene cluster required for the biogenesis of type 4 fimbriae in *Pseudomonas aeruginosa*. *Mol. Microbiol.* 16:497-508.

Matthysse, A.G. 1983, Role of bacterial cellulose fibrils in *Agrobacterium tumefaciens* infection. *J. Bacteriol.* 154:905-915.

Matthysse, A.G. 1995. Mechanism of cellulose synthesis in *Agrobacterium tumefaciens*. *J. Bacteriol.* 177:1076-1081.

McCartney, H.A. 1991, Airborne dissemination of plant fungal pathogens. *J. Appl. Bacteriol.* 70: 39S-48S.

O'Connell, K.P. and Handelsman, J. 1989, *chvA* locus may be involved in export of neutral cyclic β-1,2-linked D-glucan from *Agrobacterium tumefaciens*. *Mol. Plant-Microbe Interact.* 2:11-16.

Paranchych, W. and Frost, L. S. 1988, The physiology and biochemistry of pili. *Adv. Microb. Physiol.* 29:53-114.

Persson, C., Nordfelth, R., Holmström, A., Håkansson, S., Rosqvist, R. and Wolf-Watz, H. 1995, Cell-surface-bound *Yersinia* translocate the protein tyrosine phophatase YopH by a polarized mechanism into the target cell. *Molecular Microbiology.* 18:135-150.

Roine, E., Nunn, D., Paulin, L. and Romantschuk, M. 1996. Characterization of genes required for pilus expression in *Pseudomonas syringae* pathovar *phaseolicola. J. Bacteriol.* (in press).

Romantschuk, M. 1992, Attachment of plant-pathogenic becteria to the surface of plants. *Annu. Rev. Phytopatol.* 30:225-243.

Romantschuk, M., and D. H. Bamford. 1985. Function of pili in bacteriophage φ6 penetration. *J. Gen. Virol.* 66:2461-2469.

Romantschuk, M., and D. H. Bamford. 1986. The causal agent of halo blight in bean, *Pseudomonas syringae* pv. *phaseolicola*, attaches to stomata via its pili. *Microb. Pathogen.* 1:139-148.

Romantschuk, M., E.-L. Nurmiaho-Lassila, E. Roine, and A. Suoniemi. 1993. Pilus-mediated adsorption of *Pseudomonas syringae* to host and non-host plant leaves. *J. Gen. Microbiol.* 139:2251-2260.

Rosenberg, M. and Kjelleberg, S. 1986, Hydrophobic interactions: role in bacterial adhesion. *Adv. Microbiol. Ecol.* 9:353-393.

Rosqvist, R., Håkansson, S., Forsberg, A. and Wolf-Watz, H. 1995, Functional conservation of the secretion and translocation machinery for virulence proteins of Yersiniae, Salmonellae, and Shigellae. *EMBO J.* 14:4187-4195.

Rouse, D.I., Nordheim, E.V., Hirano, S.S. and Upper, C.D. 1985, A model relating the probability of foliar disease incidence to the population frequencies of bacterial plant pathogens. *Phytopathology* 75:505-509.

Rudolph, K.W.E, Gross, M., Ebrahim-Nesbat, F., Nöllenburg, M., Zomorodian, A., Wydra, K., Neugebauer, M., Hettwer, U., El-Shouny, W., Sonnenberg, B., Klement, Z. 1994, The role of extracellular polysaccharides as virulence factors for phytopathogenic pseudomonads and xanthomonads. pp 357-378. In: Kado, C.I. and Crosa, J.H. (eds.) *Molecular Mechanisms of Bacterial Virulence.* Kluwer Academic Publishers, Dordrecht.

Salmond, G.P.C. 1994, Secretion of extracellular virulence factors by plant pathogenic bacteria. *Annu. Rev. Phytopatol.* 32:139-148.

Sheth, H.B., Lee, K.K., Wong, W.Y., Srivastava, G., Hindsgaul, O., Hodges, R.S., Paranchych, W. and Irvin, R.T. 1994, The pili of *Pseudomonas aeruginosa* strains PAK and PAO bind specifically to the carbohydrate sequence βGalNac(1-4)βGal found in glycosphingolipids asialo GM_1 and asialo GM_2. *Mol. Microbiol.* 11:715-723.

Smit, G., Logman, T.J.J., Boerrigter, M.E.T.I., Kijne, J.W. and Lugtenberg, B.J.J. 1989, Purification and partial characterization of the Ca^{2+} dependent adhesin from *Rhizobium leguminosarum* biovar *viciae*, which mediates the first step in attachment of Rhizobiaceae cells to plant root hair tips. *J. Bacteriol.* 171:4054-4062.

Strom, M. S. and S. Lory. 1993. Structure-function and biogenesis of the type IV pili. *Annu. Rev. Microbiol.* 47:565-596.

Suoniemi, A., Björklöf, K., Haahtela, K. and Romantschuk, M. 1995, The pilus of *Pseudomonas syringae* pathovar *syringae* enhances initiation of bacterial epiphytic colonization of bean. *Microbiology* 141, 497-503.

Sule, S. and Seemuller, E. 1987, The role of ice formation on the infection of sour cherry leaves by *Pseudomonas syringae. Phytopathology*, 77:173-177.

Swart, S., Smit, G., Lugtenberg, B.J.J. and Kijne, J.W. 1993, Restoration of attachment, virulence and nodulation of *Agrobacterium tumefaciens chvB* mutants by rhicadhesin. *Mol. Microbiol.* 10:597-605.

Swart,S; Logman, T.J.J.; Smith, G., Lugtenberg, B.J.J., Kijne, J.W. 1994a, Purification and partial characterization of a glycoprotein from pea (*Pisum sativum*) with receptor activity for rhicadhesin, an attachment protein of Rhizobiaceae. *Plant Molecular Biology* 24:171-183.

Swart, S., Lugtenberg, B.J.J., Smit, G. and Kijne, J.W. 1994b, Rhicadhesin-mediated attachment and virulence of an *Agrobacterium tumefaciens chvB* mutant can be restored by growth in a highly osmotic medium. *J. Bacteriol.* 176:3816-3819.

Tukey Jr, H.B. 1970. The leaching of substances from plants. *Annu. Rev. Plant. Physiol.* 21, 305-24.

Vidaver, A. K., R. K. Koski, and J. L. Van Etten. 1973. Bacteriophage φ6: A lipid-containing virus of *Pseudomonas phaseolicola. J. Virol.* 11:799-805.

Viret, O., Toti, L., Chapela, I.H. and Petrini, O. 1994, The role of the extracellular sheath in recognition and attachment of conidia of *Discula umbrinella* (Berk and Br.) Morelet to the host surface. *New Phytol.* 127:123-131.

Wagner, V.T. and Matthysse, A.G. 1992, Involvement of vitronectin-like protein in attachment of *Agrobacterium tumefaciens* to carrot suspension culture cells. *J. Bacteriol.* 174:5999-6003.

Wilson, M. and Lindow, S.E. 1994, Inoculum density-dependent mortality and colonization of the phyllosphere by *Pseudomonas syringae*. *Appl. Environ. Microbiol.* 60:2232-2237.

Yang, Y.N. and Gabriel, D.W. 1995, *Xanthomonas* avirulence/pathogenicity gene family encoses functional plant nuclear targeting signals. *Molec. Plant-Microbe Interact.* 8:627-631.

Young, D. H. and Kauss, H. 1984, Adhesion of *Colletotrichum lindemutianum* spores to *Phaseolus vulgaris* hypocotyls and to polystyrene. *Appl. Environ. Microbiol.* 47:616-619.

4

BACTERIAL *hrp* AND AVIRULENCE GENES ARE KEY DETERMINANTS IN PLANT–PATHOGEN INTERACTIONS

Ulla Bonas and Guido Van den Ackerveken

Institut des Sciences Végétales
CNRS
Avenue de la Terrasse
91198 Gif-sur-Yvette
France

INTRODUCTION

Among the 1600 different species known in the bacterial kingdom only a small number are plant pathogenic. In fact, most pathogens can only infect a limited number of host plant species. On the other hand, many bacteria live in the plant's phyllosphere and rhizosphere without causing any harm. To be successful as a pathogen, *i.e.*, live on the expense of the host and cause damage that may even kill the host, the bacterium has to overcome the plant's physical barriers and defence responses. During evolution plant pathogenic bacteria have acquired multiple functions that enable them to colonise and multiply in living plant tissue. Due to this specialisation the host range is often limited to a few plant species. In the well studied plant pathogens *Agrobacterium tumefaciens, A. rhizogenes*, and *A. vitis* virulence genes are located on an endogenous plasmid and are responsible for the transfer of the so-called T-DNA region from the same plasmid to the plant nucleus. The integrated T-DNA contains genes for plant hormone metabolism and for the synthesis of opines which are expressed in the plant cell. Consequently, the transformed plant tissue develops into tumours or hairy roots; the opines are utilised by the surrounding bacteria. For excellent more comprehensive reviews on *Agrobacterium* see Zambryski (1992) and Winans (1992). This chapter will focus on subspecies of the gram-negative genera *Erwinia, Pseudomonas*, and *Xanthomonas* which cause numerous diseases in important crop plants. These bacteria have evolved tools different from *Agrobacterium* in order to infect the host plant.

Under natural conditions bacteria enter the plant through natural openings (stomata, hydathodes) or wounds. In general, two different types of interactions can be observed depending on the genotypes of both interacting organisms. In the compatible interaction, the bacterium is virulent and the plant susceptible, and the outcome is disease. In incompatible interactions the plant is resistant and the bacterium avirulent. In the incompatible interaction

Aerial Plant Surface Microbiology, edited by Morris et al.
Plenum Press, New York, 1996

bacteria often induce a hypersensitive reaction (HR) which is a rapid local necrosis of the infected plant tissue (Klement, 1982). It is important to note that saprophytic or non-phytopathogenic bacteria such as *Escherichia coli* and some strains of *Pseudomonas fluorescens* do not induce the HR and are unable to multiply in plant tissue. Under natural infection conditions the HR is microscopically small and can be induced by just one bacterial cell. Only when bacteria are introduced into plant tissue at high cell densities (about 10^7 colony forming units or more per ml) the HR is macroscopically visible as confluent necrosis which can be clearly distinguished from typical disease symptoms. The HR test on plant leaves has become an indispensable laboratory tool for the molecular genetic analysis of pathogens and their recognition by the plant.

Application of molecular genetics has led to considerable progress in our understanding of the molecular events taking place in plant-bacterial interactions. This chapter is dedicated to the analysis of basic pathogenicity and avirulence genes which will be described in some more detail for *Xanthomonas campestris* pv. *vesicatoria*, a pathogen of pepper and tomato which is studied in our laboratory.

hrp GENES

General Features of *hrp* Genes

It was a genetic approach that led to the identification of *hrp* (*h*ypersensitive *r*eaction and *p*athogenicity; phonetic "harp") genes. These genes, originally described for *Pseudomonas syringae* pv. *phaseolicola* by Lindgren *et al.* (1986), have been isolated from almost all major gram-negative plant pathogenic bacteria (except *Agrobacterium*). Briefly, mutants obtained by random chemical or transposon mutagenesis of a pathogenic wild-type strain were inoculated into susceptible and resistant host plants (or into non-host tobacco) and screened for both loss of ability to cause disease in susceptible plants and to induce the HR in resistant host or non-host plants. *hrp* mutants are not affected in genes for housekeeping functions as they still grow in minimal medium. A third characteristic of all *hrp* mutants is that they are unable to multiply *in planta* (reviewed by Willis *et al.*, 1991; Bonas 1994).

Complementation of a given *hrp* mutant with cosmid clones from a genomic library of the wild type strain and subsequent genetic analysis resulted in identification of large clusters of *hrp* genes from a number of different pathovars of *P. syringae,* and *Xanthomonas campestris*, and from *Pseudomonas solanacearum* and *Erwinia* spp. (reviewed in Bonas, 1994). In addition, homologous genes have been isolated from other species, *e.g.*, the so-called *wts* genes from *E. stewartii* (Coplin *et al.*, 1992; Laby and Beer, 1992), and a region containing pathogenicity genes from *X. c.* pv. *glycines* (Hwang *et al.*, 1992). The latter can complement some *hrp* mutants of *X. c.* pv. *vesicatoria* (Bonas, unpublished results). Interestingly, non-pathogenic xanthomonads that were originally isolated from diseased plants as opportunists together with pathogenic bacteria, do not contain *hrp* related DNA sequences (Stall and Minsavage, 1990; Bonas *et al.*, 1991).

In all cases, the *hrp* genes are organised in clusters of 22-35 kb. In addition, several smaller *hrp* loci have been described that are not linked to the large cluster present in the same bacterium, such as the *hrpX* locus that is conserved in *X. campestris* pathovars *campestris* (Kamoun *et al.*, 1992), *oryzae* (Kamdar *et al.*, 1993), and *vesicatoria* (K. Wengelnik and U. Bonas, unpublished data) and the *hrpM* locus in *P. s.* pv. *syringae* (Mukhopadhyay *et al.*, 1988). Besides being non-pathogenic and unable to induce the HR in tobacco, *P. syringae hrpM* mutants are also affected in mucous production.

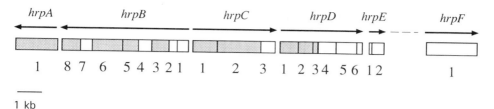

1 kb

Figure 1. Genetic and translational organisation of the *hrp* gene cluster of *X. c.* pv. *vesicatoria*. The arrows represent transcription units as determined by genetic analyses (Bonas *et al.,* 1991; Schulte and Bonas, 1992a). The boxes correspond to sequences of open reading frames (ORFs). Sequence similarities between putative Hrp proteins and proteins involved in type III secretion pathways are indicated by hatched boxes.

Transcriptional Organisation and Regulation of *hrp* Genes

Genetic studies using transposon-induced insertion mutants revealed that the *hrp* clusters contain at least 6 to 8 complementation groups. *hrp* genes of pathogens belonging to different genera share sequence homologies, however, the degree of conservation varies. The *hrp* regions of pathovars of *P. syringae* are highly conserved. Interestingly, a large part of the *hrp* regions in *X. c.* pv. *vesicatoria* and *P. solanacearum* are organised colinearly while these bacteria belong to different genera (see below). Whether certain *hrp* genes are pathovar- or host plant-specific has to await more sequence information.

We have sequenced the entire *hrp* cluster of *X. c.* pv. *vesicatoria*, the causal agent of bacterial spot disease on pepper and tomato. Based on genetic analyses and open reading frames (ORF) with a high coding probability we predict 21 genes in the 25 kb *hrp* cluster of *X. c.* pv. *vesicatoria*. Their transcriptional organisation is depicted in Figure 1. The loci *hrpA* and *hrpB* are transcribed from right to left; the other four loci are transcribed from left to right (Schulte and Bonas, 1992a). According to the locus (*hrpA-hrpF*), the ORFs were numbered consecutively. Nomenclature of *hrp* genes is not uniform for different bacterial species and will eventually be adapted for genes sharing homology in function.

Expression of *hrp* genes is controlled by environmental conditions. In general, expression of *hrp* loci is low when bacteria are grown in complex culture media. However, bacteria growing in the plant highly express *hrp* genes. Attempts to mimic the conditions present in the plant tissue resulted in the finding that certain minimal media without any plant-derived factor(s) were also able to induce *hrp* genes. This suggested that the bacteria have to experience some kind of starvation conditions for full expression of *hrp* genes. One of the first indications for *hrp* gene expression *in vitro* was the *hrp*-dependent expression of the avirulence gene *avrB* from the soybean pathogen *P. s.* pv. *glycinea* (Huynh *et al.,* 1989). The composition of *hrp*-inducing minimal media differs depending on the bacterium studied. Carbon source, concentration of organic nitrogen and phosphate, osmolarity, and pH have been found to influence gene expression. Detailed studies have been performed for *P. syringae* pathovars *syringae* (Huang *et al.,* 1991; Xiao *et al.,* 1992) and *phaseolicola* (Rahme *et al.,* 1992), *P. solanacearum* (Arlat *et al.,* 1992), *Erwinia amylovora* (Wei *et al.,* 1992b), *X. c.* pv. *campestris* (Arlat *et al.,* 1991), and *X. c.* pv. *vesicatoria* (Schulte and Bonas 1992a, 1992b).

hrp Regulatory Genes

Interestingly, *hrp* gene expression in different phytopathogenic bacteria is controlled by regulatory genes belonging to different classes. Understanding of regulation is most advanced for *P. syringae*. At least three loci are involved in complex positive regulation of the *P. syringae hrp* genes, *rpoN*, *hrpRS*, and *hrpL*. Fellay *et al.* (1991) showed that *hrp* gene expression in *P. s.* pv. *phaseolicola* is dependent on transcription factor sigma 54 (RpoN). The *hrpRS* locus encodes two proteins that are 60% identical to each other and which belong to the sigma 54 dependent activator family of two-component systems (Xiao *et al.*, 1994; Grimm *et al.*, 1995). *hrpL* encodes an alternative sigma factor which is regulated by *hrpRS*. Expression of *hrpL* is necessary and sufficient to activate the other *hrp* genes and a number of *avr* genes (Xiao *et al.*, 1994). A conserved sequence, the so-called "hrp" box (Fellay *et al.*, 1991), present in the promoters of *hrp* and *avr* genes (Salmeron and Staskawicz, 1993; Innes *et al.*, 1993; Shen and Keen, 1993) is sufficient to mediate control by *hrpL* (Xiao and Hutcheson, 1994). There is no "hrp" box sequence in *P. solanacearum* or *Xanthomonas hrp* gene promoters. However, another sequence motif, designated "PIP-box" (*plant-inducible promoter*) occurs in the promoter region of several *hrp* loci in *X. c.* pv. *vesicatoria* (Fenselau and Bonas, 1995).

The *hrp* loci of *P. solanacearum* and *X. c.* pv. *vesicatoria* are regulated by an AraC-type protein, i.e., different from the two-component regulators discussed above. The *hrpB* gene from *P. solanacearum*, which is part of the *hrp* cluster positively regulates four of the six *hrp* transcription units, as well as *popA* (Genin *et al.*, 1992). The *popA* gene is not a *hrp* gene, is located outside of the *hrp* cluster, and encodes a protein secreted in a Hrp-dependent way (see below; Arlat *et al.*, 1994). The *hrpX* gene of *X. c.* pv. *vesicatoria*, which was recently isolated and activates several *hrp* loci, is related to the *hrpB* gene from *P. solanacearum*, hence also belongs to the AraC/XylS family (Wengelnik and Bonas, 1996). In contrast to all other known *hrp* clusters, the *hrp* regulatory gene *hrpX* of *X. c.* pv. *vesicatoria* is not contained within the *hrp* gene cluster.

hrp Genes Encode a Type III Secretion System

For the *hrp* genes, other than regulatory genes, DNA sequence analysis has given some important clues as to what their biochemical function could be. The deduced amino acid sequences of *hrp* genes from *X. c.* pv. *vesicatoria* show significant similarity to Hrp proteins from other plant pathogenic bacteria (Table 1; see also for references). However, the degree of sequence similarity varies greatly. It appears that *hrp* genes in *X. c.* pv. *vesicatoria* are more closely related to *P. solanacearum* than to *P. syringae* and to *Erwinia*. Similarities were also found between HrpA1 and HrpB3 from *X. c.* pv. *vesicatoria* and NolW and NolT of the symbiotic bacterium, *Rhizobium fredii*. NolW and NolT mutants have a wider host range in nodulation of soybean indicating that they play a role in cultivar specificity (Meinhardt *et al.*, 1993).

An eye-opening finding was the relatedness of putative Hrp proteins to proteins in animal pathogens such as *Salmonella, Shigella,* and *Yersinia* spp. Since the first similarities found were to proteins from *Yersinia* spp., the yersiniae became a sort of "role model" for plant pathologists (Fenselau *et al.*, 1992; Gough *et al.*, 1992; Huang *et al.*, 1992). *Yersinia* is a mammalian pathogen which secretes proteinaceous virulence factors, called Yops (*Yersinia outer protein*; Michiels *et al.*, 1990; Michiels *et al.*, 1991). The Yops are hydrophilic proteins, essential for virulence, that are translocated without being processed. *Shigella flexneri* also secretes virulence factors without cleavage of a signal sequence, called Ipas (*invasion plasmid antigens*) (Hale, 1991). There is no homology between these virulence proteins and putative products of the *hrp* region in plant pathogenic bacteria. However, several components of the Yop and Ipa secretion systems (*sec*-independent; called "type III"), encoded in both *Yersinia* and *Shigella* by a large number of genes clustered on virulence plasmids, share significant similarity with Hrp proteins. Other Hrp related proteins, e. g., of *E. coli, Bacillus subtilis, Caulobacter* and *E. carotovora* (see Table 1) are important for the assembly of flagellae, and thus motility and chemotaxis. In summary, the common denominator in these different bacteria is the principle of a specialised protein secretion system. This finding led us and others to propose a Hrp protein secretion apparatus in plant pathogenic bacteria (Fenselau *et al.*, 1992; Gough *et al.*, 1992; Van Gijsegem *et al.*, 1993).

hrp-Mediated Protein Secretion

So far, only a few proteins have been identified as substrates of the Hrp secretion apparatus. Interestingly, these proteins function as inducers of the HR in non-host plants. The first bacterial HR-inducing protein identified was a cell-envelope associated protein, designated harpin, of *E. amylovora*, a pathogen of pear and apple (Wei *et al.*, 1992a; Wei and Beer, 1993). This harpin$_{Ea}$ is a glycine-rich and heat stable protein that induces the HR in the non-host plant tobacco. The corresponding gene, *hrpN*, is localized within the *E. amylovora hrp* cluster and thus might have a role in both HR induction on tobacco and pathogenicity on its host plants. Recently, a homologous gene was described for *Erwinia chrysanthemi* (Bauer *et al.*, 1995). There is no DNA homology between *hrpN* and sequences in other plant-pathogenic bacteria. Using an expression library of cloned *P. s.* pv. *syringae hrp* sequences in *E. coli,* He and co-workers identified harpin$_{Pss}$. This protein is encoded by the *hrpZ* gene in *P. s.* pv. *syringae* (He *et al.*, 1993), has HR-inducing activity on tobacco leaves, is also glycine-rich, and heat-stable. As with harpin$_{Ea}$ of *E. amylovora*, the biochemical function of harpin$_{Pss}$ in pathogenicity is unknown. Its product is secreted by *P. s.* pv. *syringae* in a HrpH dependent way. HrpH is highly related to proteins involved in secretion in other bacterial pathogens, the so-called PulD family of proteins (Huang *et al.*, 1992; see Table 1). A different HR-inducing protein has been isolated from culture supernatants of *P. solanacearum* grown in *hrp*-inducing medium. This protein, PopA1 (*P*seudomonas *out* protein; Arlat et al. 1994) induces the HR in tobacco and in certain *Petunia* lines, i. e., in non-host plants as with harpins. The Pop protein is heat-stable and glycine-rich, however, the sequence is entirely different from the harpins. The corresponding gene is located outside of the large *hrp* cluster of *P. solanacearum* and is not a *hrp* gene because mutations in *popA* do not affect pathogenicity on tomato or the HR on tobacco also indicating that more than one HR-inducing factor is produced. Arlat *et al.* (1994) demonstrated that secretion of PopA is dependent on *hrp* genes.

These findings show that Hrp proteins of *P. s.* pv. *syringae*, *Erwinia*, and *P. solanacearum* play a role in protein translocation. Since the secreted proteins have an effect on non-host plants the important question remains as to which proteins are secreted in the interaction with the host plant. The most likely candidates are virulence factors and race-specific elicitors. The next few years will certainly shed light on these major questions in plant pathology.

GENES INVOLVED IN HOST-RANGE

General Characteristics of Avirulence Genes

While *hrp* genes are basic pathogenicity genes, avirulence (*avr*) genes restrict the host range of pathogenic bacteria to certain plant lines/cultivars of the same species. Also the so-called non-host resistance can be due to the recognition of multiple *avr* genes resulting in avirulence of the bacterium on all lines of a given plant species. Avirulence genes have been genetically defined as genes in the pathogen that "correspond" to particular resistance genes in the plant. This so-called "gene-for-gene" concept was developed by Flor (1956) and has been confirmed by the physical isolation of avirulence genes (see review by Keen, 1992), and more recently, of some of the corresponding plant resistance genes (Staskawicz *et al.*, 1995). It is important to keep in mind that the outcome of the interaction is determined

Table 1. Protein sequence similarities between *X. c. pv. vesicatoria* Hrp proteins[1] and other bacterial proteins. The values give %similarity/identity calculated using GAP.

X. c. pv. vesicatoria	HrpA1	HrpB1	HrpB2	HrpB3	HrpB4	HrpB5	HrpB6	HrpB7	HrpB8	HrpC1	HrpC2	HrpC3
Pseudomonas solanacearum	HrpA[2] 66/48%	HrpK[3] 63/36%	HrpJ[3] 54/31%	HrpI[2] 70/59%	HrpH[3] 57/31%	HrpF[3] 52/33%	HrpE[3] 84/69%	HrpD[3] 42/24%	HrpC[3] 69/47%	HrpN[4] 74/50%	HrpO[2] 81/66%	HpaP[4] 54/31%
Pseudomonas syringae	HrpH[5] 52/29%			HrpC[6] 60/37%			HrpJ[47] 66/47%		HrpX[8] 53/24%	HrpY[8] 61/33%	HrpI[9] 62/38%	
Yersinia spp.	YscC[10] 55/34%			YscJ[10] 56/34%		LcrKc[11] 41/22%	YscN[12] 73/57%		YscT[13] 56/30%	YscU[13] 63/35%	LcrD[14] 79/46%	
Shigella flexneri	MxiD[15] 50/28%			MxiJ[16] 52/27%			Spa47[17] 65/45%		Spa29[18] 53/26%	Spa40[18] 55/27%	MxiA[19] 65/39%	
Salmonella typhimurium	InvG[20] 52/29%			PrgK[21] 56/28%			SpaL[22] 70/47% FliI[23] 65/48% InvC[24] 64/42%		SpaR[22] 54/26%	SpaS[22] 56/29% FlhB[26] 60/33%	InvA[25] 67/41% FlhA[26] 63/35%	
Bacillus subtilis							FlaA-ORF4[27] 68/50%		FliR[28] 52/24%	FlhB[28] 62/33%	FlhA[29] 63/37%	
Escherichia coli	SepD[30] 54/29%						β-F1[31] 53/28%		FliR[32] 50/23%		SepA[30] 65/39%	
Caulobacter crescentus									FliR[33] 50/26%			
Erwinia spp.	OutD[35] 47/25%								MopE[36] 50/23%		FlbF[34] 55/35%	
Rhizobium fredii	NolW[38] 51/29%			NolT[38] 61/41%							HrpI[37] 62/39%	

Table 1. (*Continued*).

References for the different proteins:

[1] Bonas et al. 1991; Fenselau and
    Bonas, 1995; and unpublished data
[2] Gough et al. 1992
[3] Van Gijsegem et al. 1995
[4] Gough et al. 1993
[5] Huang et al. 1992
[6] Preston et al. 1995
[7] Lidell and Hutcheson 1994
[8] Huang et al. 1995
[9] Huang et al. 1993
[10] Michiels et al. 1991
[11] Rimpiläinen et al. 1992
[12] Woestyn et al. 1994
[13] Bergman et al. 1994
[14] Plano et al. 1991
[15] Allaoui et al. 1993
[16] Allaoui et al. 1992
[17] Venkatesan et al. 1992
[18] Sasakawa et al. 1993
[19] Andrews and Maurelli 1992

[20] Kaniga et al. 1994
[21] Pegues et al. 1995
[22] Groisman and Ochman 1993
[23] Vogler et al. 1991
[24] Eichelberg et al. 1994
[25] Galan et al. 1992
[26] Minamino et al 1994
[27] Albertini et al. 1991
[28] Carpenter et al. 1993
[29] Carpenter and Ordal 1993
[30] Jarvis et al. 1995
[31] Saraste et al. 1981
[32] Malakooti et al. 1994
[33] Zhuang and Shapiro 1995
[34] Ramakrishnan et al. 1991
[35] Condemine et al. 1992
[36] Mulholland et al. 1993
[37] Wei and Beer 1993
[38] Meinhardt et al. 1993

by the genetic constitution of both interacting organisms. The same bacterium that is avirulent on a certain plant line (which is resistant to that strain) can be virulent on another plant line, *i.e.*, the bacterium is still a pathogen. In the incompatible interaction bacterial growth inside the plant is greatly reduced or null, and in many, but not all cases, an HR is induced in the resistant plant. Testing for the HR, which can be well distinguished from the typical disease symptoms appearing in the interaction with a susceptible plant, has been crucial for the isolation and study of avirulence genes.

The *avrA* gene from the soybean pathogen *P. s.* pv. *glycinea* was the first bacterial *avr* gene to be isolated (Staskawicz *et al.*, 1984). To date, more than 30 genes have been cloned, mainly from pathovars of *P. syringae* and from different *Xanthomonas* species. For an excellent and comprehensive overview see the recent review by Dangl (1994). Avirulence genes are located on the bacterial chromosome or on endogenous plasmids some of which have been shown to be conjugative (e.g., Bonas *et al.*, 1989). Many *avr* genes seem to be dispensable for the bacterium raising questions about their "value" for the pathogen, i.e., their "true" function in pathogenicity, features that are difficult to test for. Among the *avr* genes involved in bacterial fitness are the *avrBs2* gene from *X. c.* pv. *vesicatoria* which is highly conserved among strains belonging to different *X. campestris* pathovars (Kearney and Staskawicz, 1990), *pthA* from *X. citri* (Swarup *et al.,* 1991), *avrA* and *avrE* from *P. s.* pv. *tomato* (Lorang *et al.*, 1994), and *avrRpm1* from *P. s.* pv. *maculicola* which is recognized by certain *Arabidopsis* ecotypes (Ritter and Dangl, 1995).

Function of *avr* Genes

Sequence analysis of *avr* genes did not give any clue as to what their biochemical function might be. The predicted proteins are hydrophilic, and lack a typical N-terminal signal sequence and are, therefore, thought to be localized in the cytoplasm. Localisation of the Avr protein has been performed in a few cases and confirmed that the protein is in the cytoplasm of bacterial cells (Knoop *et al.*, 1991; Brown *et al.*, 1993). However, genetic data, for example, on the *avrBs3* family (discussed below) suggest that the avirulence gene-encoded protein is the signal molecule, although it is localized in the cytoplasm. The crucial question remains what governs the specific induction of the HR in the resistant host plant.

Only for one bacterial *avr* gene the "elicitor" of the plant's HR has been identified: syringolide is produced by bacteria expressing *avrD* and induces a specific HR in the interaction with the soybean cultivar carrying the corresponding resistance gene (Keen *et al.*, 1990). The protein product of the *avrD* gene seems to be an enzyme needed for the production of an elicitor, *i.e.*, in this case it is not the Avr protein itself that is being recognized.

The *avrBs3* Family

Among the many *avr* genes isolated, some are homologous and can be grouped into families. The *avrBs3* family of avirulence genes contains at least 11 members from different xanthomonads. The *avrBs3* gene has been isolated from a pepper pathogenic strain of *X. c.* pv. *vesicatoria* and mediates recognition by pepper plants carrying the *Bs3* resistance gene (Bonas *et al.*, 1989; Minsavage *et al.*, 1990). Homologous sequences occur in some but not all strains. One *avrBs3* homologous sequence, the *avrBs3-2* gene, was isolated and found to function as an avirulence gene on tomato (Canteros *et al.*, 1992; Bonas *et al.*, 1993). The most intriguing feature of *avrBs3* and *avrBs3-2* is the presence of 17.5 direct, nearly identical repeats of 102 bp in the internal region. The repeats and their order, however, are different in these genes. Analysis of derivatives carrying internal deletions demonstrated that the repeats determine the specificity of the *avrBs3* gene (Herbers *et al.*, 1992). Moreover, one *avrBs3* derivative, lacking four repeats, had lost the ability to induce the HR on *Bs3* plants, but instead induced an HR on *bs3* plants indicating that the susceptible allele of the *Bs3* gene (or a gene closely linked) can also function as a resistance gene (Herbers *et al.*, 1992).

How is specific recognition mediated? The biochemical function of members of the *avrBs3* family is still enigmatic. Interestingly, some genes have a dual function in that they also seem to play a role in virulence or bacterial spread, *e.g.*, *pthA* from *X. c.* pv. *citri* (Swarup *et al.*, 1991) and *avrb6* from *X. c.* pv. *malvacearum* (Yang *et al.*, 1994). Sequence comparisons among the five genes which have been sequenced to date (Bonas *et al.*, 1989; 1993; De Feyter *et al.*, 1993; Hopkins *et al.*, 1992; Yang *et al.*, 1995a) show that the N-terminal and C-terminal portions are highly conserved while the number of internal repeats and the sequence in the variable positions differ. Inspection of the sequence did not give any clue as to what their biochemical function might be. Interestingly, all sequenced *avrBs3* homologs contain three putative nuclear localisation signals (NLS) in the C-terminal region of the protein. The region containing these NLSs of the *pthA* gene has recently been shown to direct a reporter enzyme to the plant nucleus when expressed transiently in onion epidermal cells (Yang *et al.*, 1995a). The function of NLSs in HR induction remains to be clarified as it is not clear whether the avirulence protein is taken up by the plant cells.

avrBs3 Function Is *hrp*-Dependent

Interestingly, expression of *avrBs3* (and most likely that of the homologs) is constitutive, *i.e.*, *hrp*-independent. However, function is abolished in most *hrp* mutants (Knoop *et al.*, 1991). As the *hrp* genes encode a putative secretion apparatus, the signal molecule (elicitor) that is recognized by the resistant plant is thought to be transported by the Hrp secretion system. The signal molecule might even be the AvrBs3 protein itself as changes in specificity are easily obtained by changing the number of repeats. So far, it has not been possible to isolate an elicitor that is specifically recognized by the resistant plant. It might, therefore, very well be that the AvrBs3 protein is translocated - via the Hrp pathway - directly into the plant cell.

CONCLUSIONS

Considerable progress has been made over the last 10 years by isolating and studying bacterial genes involved in basic pathogenicity on susceptible host plants and specific recognition by resistant plants. We are, however, still far away from understanding how the bacterium can overcome the basic plant defence response and use the plant tissue for its own purposes, ultimately leading to disease. Are there suppressers of the plant's defence response produced by the bacterium as suggested by Brown *et al.* (1995)? Does the Hrp transport system secrete a whole set of different proteins, as is the case with the related secretion system in mammalian pathogens, and are these proteins also virulence factors? In the case of induction of the HR, *i.e.*, specific recognition mediated by matching resistance and avirulence genes, it will be interesting to find out which signals are produced and how they travel to the plant cell. The answers to these key questions are hopefully in reach. They might have a major impact on new strategies for plant disease control in agriculture.

ACKNOWLEDGMENTS

We thank our previous and our present co-workers - E. Huguet, M. Pierre, O. Rossier, J. Veuskens, and K. Wengelnik- for contributing to the data described here and for fruitful discussions. The research in our laboratory was supported in part by grants from the EEC (BIOT-CT90-0168) and an ATIPE from CNRS. G. V. d. A. is supported by an HFSP fellowship.

REFERENCES

Albertini A. M, Caramori, T., Crabb, W. D, Scoffone, F. and Galizzi, A. 1991, The *flaA* locus of *Bacillus subtilis* is part of a large operon coding for flagellar structures, motility functions and an ATPase-like polypeptide. *J. Bacteriol.* 173:3573-3579.

Allaoui, A., Sansonetti, P. J. and Parsot, C. 1992, MxiJ, a lipoprotein involved in secretion of *Shigella* Ipa invasins, is homologous to YscJ, a secretion factor of the *Yersinia* Yop proteins. *J. Bacteriol.* 174:7661-7669.

Allaoui, A., Sansonetti, P. J. and Parsot, C. 1993, MxiD, an outer membrane protein necessary for the secretion of the *Shigella flexneri* Ipa invasins. *Mol. Microbiol.* 7:59-68.

Andrews, G. P. and Maurelli, A. T. 1992, *mxiA* of *Shigella flexneri* 2a, which facilitates export of invasion plasmid antigens, encodes a homolog of the low-calcium response protein, LscD, of *Yersinia pestis*. *Infect. Immun* . 60:3287-3295.

Arlat, M., Gough, C. L., Barber, C. E., Boucher, C. and Daniels, M. J. 1991, *Xanthomonas campestris* contains a cluster of *hrp* genes related to the larger *hrp* cluster of *Pseudomonas solanacearum*. *Mol. Plant-Microbe Interact.* 4:593-601.

Arlat, M., Gough, C. L., Zischek, C., Barberis, P. A., Trigalet, A. and Boucher, C. A. 1992, Transcriptional organization and expression of the large *hrp* gene cluster of *Pseudomonas solanacearum*. *Mol. Plant-Microbe Interact.* 5:187-193.

Arlat, M., Van Gijsegem, F., Huet, J. C., Pernollet, J. C. and Boucher, C. A. 1994, PopA1, a protein which induces a hypersensitive-like response on specific *Petunia* genotypes, is secreted via the Hrp pathway of *Pseudomonas solanacearum*. *EMBO J.* 13:543-553.

Bauer, D. W., Wei, Z.-M., Beer, S. V. and Collmer, A. 1995, *Erwinia chrysanthemi* harpin$_{Ech}$: an elicitor of the hypersensitive response that contributes to soft-rot pathogenesis. *Mol. Plant-Microbe Interact.* 8: 484-491.

Bergman, T., Erickson, K., Galyov, E., Persson, C. and Wolf-Watz, H. 1994, The *lcrB* (*yscN/U*) gene cluster of *Yersinia pseudotuberculosis* is involved in Yop secretion and shows high homology to the *spa* gene clusters of *Shigella flexneri* and *Salmonella typhimurium*. *J. Bacteriol.* 176: 2619-2626.

Bonas, U. 1994, *hrp* genes of phytopathogenic bacteria. *Curr. Top. Microbiol. Immunol.* 192:79-98.

Bonas, U., Conrads-Strauch, J. and Balbo, I. 1993, Resistance in tomato to *Xanthomonas campestris* pv. *vesicatoria* is determined by alleles of the pepper-specific avirulence gene *avrBs3*. *Mol. Gen. Genet.* 238:261-269.

Bonas, U., Schulte, R., Fenselau, S., Minsavage, G.V., Staskawicz, B. J. and Stall, R. E. 1991, Isolation of a gene cluster from *Xanthomonas campestris* pv. *vesicatoria* that determines pathogenicity and the hypersensitive response on pepper and tomato. *Mol. Plant-Microbe Interact.* 4:81-88.

Bonas, U., Stall, R. E. and Staskawicz, B. J. 1989. Genetic and structural characterization of the avirulence gene *avrBs3* from *Xanthomonas campestris* pv. *vesicatoria*. Mol. Gen. Genetics 218:127-136.

Brown, I., Mansfield, J. and Bonas, U. 1995, *hrp* genes in *Xanthomonas campestris* pv. *vesicatoria* determine ability to suppress papillae deposition in pepper mesophyll cells. *Mol. Plant-Microbe Interact.* 8:825-836.

Brown, I., Mansfield, J., Irlam, I., Conrads-Strauch and Bonas, U. 1993, Ultrastructure of interactions between *Xanthomonas campestris* pv. *vesicatoria* and pepper, including immunocytochemical localization of extracellular polysaccharides and the AvrBs3 protein. *Mol. Plant-Microbe Interact.* 6:376-386.

Canteros, B., Minsavage, G., Bonas, U., Pring, D. and Stall, R. 1991, A gene from *Xanthomonas campestris* pv. *vesicatoria* that determines avirulence in tomato is related to *avrBs3* . *Mol. Plant-Microbe Interact.* 4:628-632.

Carpenter, P. B. and Ordal, G. W. 1993, *Bacillus subtilis* FlhA: a flagellar protein related to a new family of signal-transducing receptors. *Mol. Microbiol.* 7:735-743.

Carpenter, P. B., Zuberi, A. R. and Ordal, G. W. 1993, *Bacillus subtilis* flagellar proteins FliP, FliQ, FliR and FlhB are related to *Shigella flexneri* virulence factors. *Gene* 137:243-245.

Condemine, G., Dorel, C., Hugouvieux-Cotte-Pattat, N. and Robert-Beaudouy, J. 1992, Some of the *out* genes involved in the secretion of pectate lyases in *Erwinia chrysanthemi* are regulates by *kdgR*. *Mol. Microbiol.* 6:3199-3211.

Coplin, D. L., Frederick, R. D., Majerczak, D. R. and Tuttle, L. D. 1992, Characterization of a gene cluster that specifies pathogenicity in *Erwinia stewartii*. *Mol. Plant-Microbe Interact.* 5:81-88.

Dangl, J. L. 1994, The enigmatic avirulence genes of phytopathogenic bacteria. *Curr. Top. Microbiol. Immunol.* 192:99-118.

De Feyter, R., Yang, Y., abd Gabriel, D. W. 1993, Gene-for-genes interactions bewteen cotton R genes and *Xanthomonas campestris* pv. *malvacearum avr* genes. *Mol. Plant-Microbe Interact.* 6:225-237.

Eichelberg, K., Ginocchio, C. C. and Galan, J. E. 1994, Molecular and functional characterization of the *Salmonella typhimurium* invation genes *invB* and *invC*: homology of InvC to the F_0F_1 ATPase family of proteins. *J. Bacteriol.* 176: 4501-4510.

Fellay, R, Rahme, L. G., Mindrinos, M. N., Frederick, R. D., Pisi, A. and Panopoulos, N. J. 1991, Genes and signals controlling the *Pseudomonas syringae* pv. *phaseolicola*-plant interaction, pp. 45-52 In: Hennecke, H. and Verma, D.P.S. (eds.) *Molecular Genetics of Plant-Microbe Interactions*, Kluwer Academic Publishers, Dordrecht, Netherlands.

Fenselau, S., Balbo, I. and Bonas, U. 1992, Determinants of pathogenicity in *Xanthomonas campestris* pv. *vesicatoria* are related to proteins involved in secretion in bacterial pathogens of animals. *Mol. Plant-Microbe Interact.* 4:593-601.

Fenselau, S. and Bonas, U. 1995, Sequence and expression analysis of the *hrpB* pathogenicity operon of *Xanthomonas campestris* pv. *vesicatoria* which encodes eight proteins with similarity to components of the Hrp, Ysc, Spa and Fli secretion systems. *Mol. Plant-Microbe Interact.* 8:845-854.

Flor, H. 1956. The complementary genetic systems in flax and flax rust. *Adv. Genet.* 8:29-54.

Galán, J. E., Ginocchio, C. and Costeas, P. 1992, Molecular and functional characterization of the *Salmonella* invasion gene *invA*: homology of InvA to members of a new protein family. *J. Bacteriol.* 174:4338-4349.

Genin, S, Gough, C. L., Zischek, C. and Boucher, C. A. 1992, The *hrpB* gene encodes a positive regulator of pathogenicity genes from *Pseudomonas solanacearum*. *Mol. Microbiol.* 6:3065-3076.

Gough, C. L., Genin, S., Zischek, C. and Boucher, C. A. 1992, *hrp* genes of *Pseudomonas solanacearum* are homologous to pathogenicity determinants of animal pathogenic bacteria and are conserved among plant pathogenic bacteria. *Mol. Plant-Microbe Interact.* 5:384-389.

Gough, C. L., Genin, S., Lopes, V. and Boucher, C. A. 1993, Homology between the HrpO protein of *Pseudomonas solanacearum* and bacterial proteins implicated in a signal peptide independent secretion mechanism. *Mol. Gen. Genet.* 239:378-392.

Grimm, C., Aufsatz, W. and Panopoulos, N. J. 1995, The *hrpRS* locus of *Pseudomonas syringae* pv. *phaseolicola* constitutes a complex regulatory unit. *Mol. Microbiol.* 15:155-165.

Groisman, E. A. and Ochman, H. 1993, Cognate gene clusters govern invation of host epithelial cells by *Salmonella typhimurium* and *Shigella flexneri*. *EMBO J.* 12:3779-3787.

Hale, T. L. 1991, Genetic basis for virulence in *Shigella* species. *Microbiol. Rev.* 55:206-224.

He, S. Y., Huang, H. C. and Collmer, A. 1993, *Pseudomonas syringae* pv. *syringae* Harpin$_{Pss}$: a protein that is secreted via the Hrp pathway and elicits the hypersensitive response in plants. *Cell* 73:1255-1266.

Herbers, K., Conrads-Strauch, J. and Bonas, U. 1992, Race-specificity of plant resistance to bacterial spot disease determined by repetitive motifs in a bacterial avirulence protein. *Nature* 356:172-174.

Hopkins, C. M., White, F. F., Choi, S.-H., Guo, A. and Leach, J. 1993, Identification of a family of avirulence genes from *Xanthomonas oryzae* pv. *oryzae*. *Mol. Plant-Microbe Interact.* 5:451-459.

Huang, H. C., Hutcheson, S. W. and Collmer, A. 1991, Characterization of the *hrp* cluster from *Pseudomonas syringae* pv. *syringae* 61 and Tn*phoA* tagging of genes encoding exported or membrane-spanning Hrp proteins. *Mol. Plant-Microbe Interact.* 4:469-476.

Huang, H. C., He, S. Y., Bauer, D. W. and Collmer, A. 1992, The *Pseudomonas syringae* pv. *syringae* 61 *hrp*H product, an envelope protein required for elicitation of the hypersensitive response in plants. *J. Bacteriol.* 174: 6878-6885.

Huang, H. C., Xiao, Y., Lin, R. H., Lu, Y., Hutcheson S. W. and Collmer, A. 1993, Characterization of the *Pseudomonas syringae* pv. *syringae* *hrpJ* and *hrpI* genes: homology of HrpI to a superfamily of proteins associated with protein translocation. *Mol. Plant-Microbe Interact.* 6:515-520.

Huang, H.-C., Lin, R.-H., Chang, C.-J., Collmer, A. and Deng, W.-L. 1995, The complete *hrp* gene cluster of *Pseudomonas syringae* pv. *syringae* 61 includes two blocks of genes required for harpinPss secretion that are arranged colinearly with *Yersinia ysc* homologs. *Mol. Plant-Microbe Interact.* 8 (in press).

Huynh, T.V., Dahlbeck, D. and Staskawicz, B. J. 1989. Bacterial blight of soybean: regulation of a pathogen gene determining host cultivar specificity. *Science* 245:1374-1377.

Hwang, I., Lim, S. M. and Shaw, P. D. 1992, Cloning and characterization of pathogenicity genes from *Xanthomonas campestris* pv. *glycines*. *J. Bacteriol.* 174:1923-1931.

Innes, R. W., Bent, A. F., Kunkel, B. N., Bisgrove, S. R. and Staskawicz, B. J. 1993, Molecular analysis of avirulence gene avrRpt2 and identification of a putative regulatory sequence common to all known *Pseudomonas syringae* avirulence genes. *J. Bacteriol.* 175:4859-4869.

Jarvis, K. G., Giron, J. A., Jerse, A. E., McDaniel, T. K., Donnenberg, M. S. and Kaper, J. B. 1995, Enteropathogenic *E. coli* contains a putative type III scretion system necessary for the export of proteins involved in attaching and effacing lesion formation. *Proc. Natl. Acad. Sci. U.S.A.* (in press)

Kamdar, H. V., Kamoun, S. and Kado, C. I. 1993, Restoration of pathogenicity of avirulent *Xanthomonas oryzae* pv. *oryzae* and *X. campestris* pathovars by reciprocal complementation with the *hrypXo* and *hrpXc* genes and identification of HrpX function by sequence analyses. *J. Bacteriol.* 175:2017-2025.

Kamoun, S, Kamdar, H. V., Tola, E. and Kado, C. I. 1992, Incompatible interactions between crucifers and *Xanthomonas campestris* involve a vascular hypersensitive response: role of the *hrpX* locus. *Mol. Plant-Microbe Interact.* 5:22-33.

Kaniga, K., Bossio, J. C. and Galan, J. E. 1994, The *Salmonella typhimurium* invasion genes *invF* and *invG* encode homologues of the AraC and PulD family of proteins. *Mol. Microbiol.* 13: 555-568.

Kearney, B. and Staskawicz, B. J. 1990, Widespread distribution and fitness contribution of *Xanthomonas campestris* avirulence gene avrBs2. *Nature* 346:385-386.

Keen, N. T. 1992, The molecular biology of disease resistance. *Plant Mol. Biol.* 19:109-122.

Keen, N. T., Tamaki, S., Kobayashi, D., Gerhold, D., Stayton, M., Shen, H., Gold, S., Lorang, J., Thordal-Christenson, H., Dahlbeck, D. and Staskawicz, B. J. 1990, Bacteria expressing avirulence gene D produce a specific elicitor of the soybean hypersensitive reaction. *Mol. Plant-Microbe Interact.* 3:112-121.

Klement, Z. 1982. Hypersensitivity, pp. 149-177 In: *Phytopathogenic Prokaryotes.* vol. 2, Mount, M.S. and Lacy, G.H. (eds.) Academic Press, New York.

Knoop, V., Staskawicz, B. J. and Bonas, U. 1991, The expression of the avirulence gene *avrBs3* from *Xanthomonas campestris* pv. *vesicatoria* is not under the control of *hrp* genes and is independent of plant factors. *J. Bacteriol.* 173:7142-7150.

Laby, R. J. and Beer, S. V. 1992, Hybridization and functional complementation of the *hrp* gene cluster from *Erwinia amylovora* strain Ea321 with DNA of other bacteria. *Mol. Plant-Microbe Interact.* 5:412-419.

Lidell, M. C. and Hutcheson, S. W. 1994, Characterization of the *hrpJ* and *hrpU* operons of *Pseudomonas syringae* pv. *syringae* Pss61: Similarity with components of enteric bacteria involved in flagellar biogenesis and demonstration of their role in harpin$_{Pss}$ secretion. *Mol. Plant-Microbe Interact.* 7:488-497.

Lindgren, P. B., Peet, R. C. and Panopoulos, N. J. H. 1986. Gene cluster of *Pseudomonas syringae* pv. *phaseolicola* controls pathogenicity on bean plants and hypersensitivity on nonhost plants. *J. Bacteriol.* 168:512-522.

Lorang, J. M., Shen, H., Kobayashi, D., Cooksey, D. and Keen, N. T. 1994, *avrA* and *avrE* in *Pseudomonas syringae* pv. *tomato* PT23 play a role in virulence on tomato plants. *Mol. Plant-Microbe Interact.* 7:208-215.

Malakooti, J., Ely, B. and Matsumura, P. 1994, Molecular characterization, nucleotide sequence and expression of the *fliO, fliP, fliQ and fliR* genes of *Escherichia coli. J. Bacteriol.* 176:189-197.

Meinhardt, L. W., Krishnan, H. B., Balatti, P. A. and Pueppke, S. G. 1993, Molecular cloning and characterization of a sym plasmid locus that regulates cultivar-specific nodulation of soybean by *Rhizobium fredii* USDA257. *Mol. Microbiol.* 9:17-29.

Michiels, T., Wattiau, P., Brasseur, R., Ruysschaert, J. M. and Cornelis, G. 1990, Secretion of Yop proteins by Yersiniae. *Infect. Immun.* 58:2840-2849.

Michiels, T., Vanooteghem, J.-C., Lambert de Rouvroit, C, China, B., Gustin, A., Boudry, P. and Cornelis, G. R. 1991, Analysis of *virC*, an operon involved in the secretion of Yop proteins by *Yersinia enterocolitica. J. Bacteriol.* 173:4994-5009.

Minamino, T., Iino, T. and Kutskake, K. 1994, Molecular characterization of the *Salmonella typhimurium flhB* operon and its protein. *J. Bacteriol.* 176:7630-7637.

Minsavage, G. V., Dahlbeck, D., Whalen, M. C., Kearney, B., Bonas, U., Staskawicz, B. J. and Stall, R. E. 1990, Gene-for-gene relationships specifying disease resistance in *Xanthomonas campestris* pv. *vesicatoria*-pepper interactions. *Mol. Plant-Microbe Interact.* 3:41-47.

Mukhopadhyay, P, Williams, J., and Mills, D. 1988. Molecular analysis of a pathogenicity locus in *Pseudomonas syringae* pv. *syringae. J. Bacteriol.* 170:5479-5488.

Mulholland, V. , Hinton, J. C. D., Sidebotham, J., Toth, I. K., Hyman, L. J., Pérombelon, M. C. M., Reeves, P. J. and Salmond, G. P. C. 1993, A pleiotropic reduced virulence (Rvi⁻) mutant of *Erwinia carotovora* subspecies *atroseptica* is defective in flagella assembly proteins that are conserved in plant and animal bacterial pathogens. *Mol. Microbiol.* 9:343-356.

Pegues, D. A., Hantman, M. J., Behlau, I. and Miller, S. I. 1995, PhoP/PhoQ transcriptional repression of *Salmonella typhimurium* invasion genes: evidence for a role in protein secretion. *Mol. Microbiol.* 17: 169-181.

Plano, G. V., Barve, S. S. and Straley, S. C. 1991, LcrD, a membrane-bound regulator of the *Yersinia pestis* low-calcium response. *J. Bacteriol.* 173:7293-7303.

Preston, G. M., Huang, H.-C., He, S. Y. and Collmer, A. 1995, The HrpZ proteins of *Pseudomonas syringae* pvs. *syringae, glycinea* and tomato are encoded by an operon containing *Yersinia ysc* homologs and elicit the hypersensitive response in tomato but not soybean. Mol. Plant Microbe Interact. (in press).

Rahme, L. G., Mindrinos, M. N. and Panopoulos, N. J. 1992, Plant and environmental sensory signals control the expression of *hrp* genes in *Pseudomonas syringae* pv. *phaseolicola. J. Bacteriol.* 174:3499-3507.

Ramakrishnan, G., Zhao, J-L. and Newton, A. 1991, The cell cycle-regulated flagellar gene *flbF* of *Caulobacter crescentus* is homologous to a virulence locus (*lcrD*) of *Yersinia pestis. J. Bacteriol.* 173:7283-7392.

Rimpiläinen, M., Forsberg, A. and Wolf-Watz, H. 1992, A novel protein, LcrQ, involved in the low-calcium response of *Yersinia pseudotuberculosis* shows extensive homology to YopH. *J. Bacteriol.* 174:3355-3363.

Ritter, C. and Dangl, J. L. 1995, The *avrRpm1* gene of *Pseudomonas syringae* pv. *maculicola* is required for virulence on *Arabidopsis. Mol. Plant-Microbe Interact.* 8:444-453.

Salmeron, J. M. and Staskawicz, B. J. 1993, Molecular characterization and *hrp* dependence of the avirulence gene *avrPto* from *Pseudomonas syringae* pv. tomato. *Mol. Gen. Genet.* 239:6-16.

Saraste, M., Gay , N. J., Eberle, A. , Runswick, M. J. and Walker, J. E. 1981. The *atp* operon: nucleotide sequences of the genes for the a, b and g subunits of *Escherichia coli* ATP synthase. *Nucl. Acids. Res.* 9:5287-5296.

Sasakawa, C., Komatsu, K., Tobe, T., Suzuki, T. and Yoshikawa, M. 1993, Eight genes in region 5 that form an operon are essential for invasion of epithelial cells by *Shigella flexneri. J. Bacteriol.* 175:2334-2346.

Schulte, R. and Bonas, U. 1992a. Expression of the *Xanthomonas campestris* pv. *vesicatoria hrp* gene cluster, which determines pathogenicity and hypersensitivity on pepper and tomato, is plant inducible. *J. Bacteriol.* 174:815-823.

Schulte, R. and Bonas, U. 1992b. A *Xanthomonas* pathogenicity locus is induced by sucrose and sulfur-containing amino acids. *Plant Cell* 4:79-86.

Shen, H. and Keen, N. T. 1993, Characterization of the promoter of avirulence gene D from *Pseudomonas syrinage* pv. tomato. *J. Bacteriol.* 175:5916-5924.

Stall, R. E. and Minsavage, G. V. 1990, The use of *hrp* genes to identify opportunistic xanthomonads, pp. 396-374 In: Klement, Z. (ed.) *Plant Pathogenic Bacteria*, Proceedings of the Seventh International Conference on Plant Pathogenic Bacteria, Budapest, Hungary, 1989, Akadémiai Kiadó, Budapest.

Staskawicz, B. J., Dahlbeck, D. and Keen, N. 1984. Cloned avirulence gene of *Pseudomonas syrinage* pv. *glycinea* determines race-specific incompatibility on *Glycine max* (L.) Merr. *Proc. Natl. Acad. Sci. USA* 81:6024-6028.

Staskawicz, B. J., Ausubel, F. M., Baker, B. J., Ellis, J. G., Jones, J. D. G. 1995, Molecular genetics of plant disease resistance. *Science* 268:661-667.

Swarup, S., De Feyter, R., Brlanski, R. H. and Gabriel, D. W. 1991, A pathogenicity locus from *Xanthomonas citri* enables strains from several pathovars of *X. campestris* to elicit canker-like lesions on citrus. *Phytopathology* 81:802-809.

Van Gijsegem, F., Genin, S. and Boucher, C. 1993, Conservation of secretion pathways for pathogenicity determinants of plant and animal bacteria. *Trends Microbiol.* 1:175-180.

Van Gijsegem, F., Gough, C. , Zischek, C. , Niqueux, E., Arlat, M., Genin, S., Barberis, P., German, S., Castello, P. and Boucher, C. A.. 1995, The *hrp* gene locus of *Pseudomonas solanacearum* that controls the production of a type III secretion system, encodes eight proteins related to components of the bacterial flagellar biogenesis complex. *Mol. Microbiol.* 15:1095-1114.

Venkatesan, M. M., Buysse, J. M. and Oaks, E. V. 1992, Surface presentation of *Shigella flexneri* invasion plasmid antigens requires the products of the *spa* locus. *J. Bacteriol.* 174:1990-2001.

Vogler, A. P., Homma, M., Irikura, V. M. and Macnab, R. M. 1991, *Salmonella typhimurium* mutants defective in flagellar filament regrowth and sequence similarity of FliI to F0F1, vacuolar and archaebacterial ATPase subunits. *J. Bacteriol.* 173:3564-3572.

Wei, Z. M., Laby, R. J., Zumoff, C. H., Bauer, D. W., He, S. Y., Collmer, A., Beer, S. V. 1992a. Harpin, elicitor of the hypersensitive response produced by the plant pathogen *Erwinia amylovora. Science* 257:85-88.

Wei, Z. M., Sneath, B. J. and Beer, S. V. 1992b. Expression of *Erwinia amylovora hrp* genes in response to environmental stimuli. *J. Bacteriol.* 174:1875-1882.

Wei, Z. M. and Beer, S. V. 1993, HrpI of *Erwinia amylovora* functions in secretion of harpin and is a member of a new protein family. *J. Bacteriol.* 175: 7958-7967.

Wengelnik, K., and Bonas, U. 1996, HrpXv, an AraC-type regulator, activates expression of five of the six loci in the *hrp* cluster of *Xanthomnas campestris* pv. *vesicatoria. J. Bacteriol.* 178: 3462-3469.

Willis, D. K., Rich, J. J. and Hrabak, E. M. 1991, *hrp* genes of phytopathogenic bacteria. *Mol. Plant-Microbe Interact.* 4:132-138.

Winans, S. C. 1992, Two-way chemical signaling in *Agobacterium*-plant interactions. *Micobiol. Rev.* 56:12-31.

Woestyn, S., Allaoui A., Wattiau, P. and Cornelis, G. R. 1994, YscN, the putative energizer of the *Yersinia yop* secretion machinery. *J. Bacteriol.* 176: 1561-1569.

Xiao, Y. X., Lu, Y., Heu, S. G. and Hutcheson, S. W. 1992, Organization and environmental regulation of the *Pseudomonas syrinage* pv. *syrinage* 61 *hrp* cluster. *J. Bacteriol.* 174:1734-1741.

Xiao, Y. and Hutcheson, S. W. 1994, A single promoter sequence recognized by a newly identified alternate sigma factor directs expression of pathogenicity and host range determinants in *Pseudomonas syrinage. J. Bacteriol.* 176:3089-3091.

Xiao, Y., Heu, S., Yi, J., Lu, Y. and Hutcheson, S. W. 1994, Identification of a putative alternate sigma factor and characterization of a multicomponent regulatory cascade controlling the expression of *Pseudomonas syrinage* pv. *syrinage* Pss61 *hrp* and *hrmA* genes. *J. Bacteriol.* 176:1025-1036.

Yang, Y., De Feyter, R. mand Gabriel, D.W. 1994, Host-specific symptoms and increased release of *Xanthomonas citri* and *X. campestris* pv. *malvacearum* from leaves are determined by the 102-bp tandem repeats of *pthA* and *avrb6*, respectively. *Mol. Plant-Microbe Interact.* 3:345-355.

Yang, Y. and Gabriel, D. W. 1995a. *Xanthomonas* avirulence/pathogenicity gene family encodes functional plant nuclear targeting signals. *Mol. Plant-Microbe Interact.* 8:627-631.

Yang, Y. and Gabriel, D. W. 1995b. Intragenic recombination of a single plant pathogen gene provides a mechanism for the evolution of new host specificities. *J. Bacteriol.* 177:4963-4968.

Zambryski, P. C. 1992, Chronicles from the *Agobacterium*-plant cell DNA transfer story. *Annu. Rev. Plant Physiol. Plant Mol. Biol.* 43:465-490.

Zhuang, W. Y. and Shapiro, L. 1995, Caulobacter FlQ and FliR membrane proteins, required for flagellar biogenesis and cell division, belong to a family of virulence factor export proteins. *J. Bacteriol.* 177:343-356.

5

HOST RESPONSE TO INTRODUCTION OF ANTAGONISTIC YEASTS USED FOR CONTROL OF POSTHARVEST DECAY

Samir Droby,[1] Edo Chalutz,[1] Michael E. Wisniewski,[2] and Charles L. Wilson[2]

[1] Department Postharvest Science
Institute for Technology and Storage
ARO, The Volcani Center
P.O. Box 6, Bet Dagan
Israel
[2] USDA
ARS, Appalachian Fruit Research Station
45 Wiltshire Road
Kearneysville, West Virginia 25430

INTRODUCTION

Increasing public concern about the extensive use of synthetic chemicals for the control of plant diseases has led to intensified research efforts world-wide to develop alternative control methods that are safe for humans and the environment. Following the realisation of the risks involved in long term use of pesticides in agriculture, many countries have adapted a policy aimed at minimising the use of synthetic chemical pesticides by establishing vigorous research programs aimed at the development of environmentally sound control measures.

Losses from postharvest spoilage of fruits and vegetables have been managed primarily by applying fungicides. This practice has proved to be the most effective in maintaining a low rate of decay and extending the postharvest life of produce. Consumer concerns about fungicide residues in their food, however, have resulted in pressure to restrict or ban the use of some fungicides on freshly harvested fruits and vegetables. Some countries that import fruits and vegetables do not allow residues of certain fungicides, such as N-[(trichloromethyl)thio]-4-cyclohexene-1, 2-dicarboximide (captan). Other countries, such as Japan, allow no residues of any postharvest fungicide on fruits. As a result, most major fungicides used for postharvest diseases control have been forced off the market, withdrawn voluntarily or identified to undergo critical (and costly) recertification.

Even greater impetus has been given to the removal of pesticides (particularly fungicides) from our food chain by a recent American National Academy of Science report

Aerial Plant Surface Microbiology, edited by Morris et al.
Plenum Press, New York, 1996

entitled, "Pesticides in the Diets of Infants and Children". This report emphasised that children are at particularly greater risk from synthetic pesticides on our fruits and vegetables because: (a) they consume greater quantities of those commodities containing the greatest amounts of pesticides, and (b) their cellular activity and metabolism makes them more susceptible than adults to synthetic pesticide carcinogenicity (National Research Council, 1993).

In addition to risks associated to human health and the environment, many of the fungicides widely used for the control of postharvest decay have limited effectiveness because of the development of resistance in the pathogens, particularly to benzimidazoles (Eckert and Ogawa, 1985; Eckert and Ogawa, 1988). Thus, their is clearly an urgent need to develop new control methods which are effective, safer than fungicides to human health, and perceived to be so by the public. Physical means have a potential to answer part of this need. The application of various sanitation techniques to reduce inoculum pressure, the use of heat treatments, cold storage, irradiation and controlled and modified atmospheres are some examples. In addition, natural plant products and nonselective fungicides (sodium carbonate, sodium bicarbonate, active chlorine and sorbic acid) are among the approaches currently being evaluated for the control of postharvest diseases (Eckert, 1991). Harvesting and handling techniques that minimise injury to the commodity, along with storage conditions that are optimum for maintaining host resistance (Sommer, 1982; Sommer, 1985), can also aid in suppressing disease development after harvest.

Biological control of postharvest diseases is a relatively new research area. The potential of this approach as a measure of postharvest decay control has been discussed in several reviews (Wilson and Pusey, 1985; Wilson and Wisniewski, 1989; Jeffries and Jeger, 1990; Droby et al., 1991b; Wilson, et al., 1991; Wisniewski and Wilson, 1992; Wilson and Wisniewski, 1994). In the past five years, substantial progress has been made in identifying and developing potential biological alternatives to synthetic fungicides for the control of postharvest diseases of fruits and vegetables (Wilson and Wisniewski, 1994). A number of antagonistic micro-organisms have been discovered which have the potential to effectively control postharvest diseases (Droby and Chalutz, 1994a). Some of this technology has been patented and commercial products such as Aspire[TM] (Ecogen Corporation, Langhorne, PA, USA), Biosave 10 and Biosave 11 (Ecoscience Inc., Worcester, MA, USA) have been registered for commercial use.

In this review, recent findings regarding the use of epiphytic, naturally occurring yeasts for control of postharvest diseases of fruits and vegetables will be discussed. Special emphasis is given to the interactions taking place at surface wounds - the site of infection and interaction between the antagonist, the host tissue and the pathogen. The importance of these interactions in the mechanism of action of yeast antagonists is also discussed.

THE ROLE OF EPIPHYTIC MICROFLORA IN FRUIT RESISTANCE TO PATHOGENS

It is assumed that biocontrol of plant diseases occurs naturally on aerial plant surfaces and is the main reason why crops are not entirely destroyed during their cultivation. This assumption is supported by work with epiphytic populations on leaf surfaces (Blakeman and Fokkema, 1982; Andrews, 1990; Andrews, 1992; Fokkema, 1992).

We are accustomed to thinking of fruit resistance to pathogen infection and development in terms of the presence of various resistance mechanisms such as pre-formed antifungal compounds, phytoalexins, lignin formation and deposition of cell wall materials (Brown and Barmore, 1983; Brown, 1989; Hahlbrock and Scheel, 1989; Sequeira, 1990).

Although these are integral components of resistance, little attention has been given to the possible role of naturally occurring micro-organisms in resistance of fruits to the development of postharvest diseases. Can we consider epiphytic microflora on the surface of harvested commodities as part of the resistance mechanism in freshly harvested fruits and vegetables? If we assume that resistance of the fruit is multidimensional then the epiphytic microflora should be considered as the external layer of resistance which may act to suppress disease development.

The treatment that fruits and vegetables receive after harvest profoundly affects their epiphytic microbial populations. The dumping of a commodity into biocidal solutions and spraying them with waxes and fungicides likely eliminates a number of micro-organisms from fruit and vegetable surfaces as well as introducing new ones. In support of this assumption, it has been commonly observed that washed commodities develop more rot than unwashed commodities (Chalutz *et al.*, 1988; Chalutz and Wilson, 1990). Washing harvested fruits and vegetables is a common postharvest practice aimed at obtaining clean and "marketable" produce with high value. Washing may affects the resistance of a commodity due to the removal of protective natural waxes and the introduction of surface wounds which are the sites of infection of many postharvest pathogens. In addition, the possible removal of naturally occurring microbial populations which may play a role in suppressing the establishment and proliferation of a pathogen should not be disregarded. Chalutz and Wilson (1990) found that when citrus fruit were washed, dried and stored, the fruit developed a higher level of decay compared with unwashed fruit. They also found that bacteria and yeasts were predominant when undiluted washings from citrus fruit were plated out. It was only after the washings were diluted that fungal rot pathogens began to appear on the Petri plates. This suggested that there may be an epiphytic microbial population on citrus that is naturally suppressive to rot pathogens.

Storage conditions may affect postharvest pathogens as well as antagonists and other epiphytic microflora on fruit surface in a differential manner. Tronsmo and Hofun (1984) found that the way carrots were stored affected the type of diseases that developed. In an ice bank cooler Mycocentrospora and Rhizoctonia rots were markedly reduced as compared with cold room storage. *Botrytis cinerea,* however, caused severe damage in the former type of storage. An indication of natural biocontrol on the surface of harvested commodities has been provided by fungicides that affect micro-organisms other than the pathogen against which they are directed. In several studies, diseases other than those being targeted appear to have increased because of a negative effect by the fungicide on resident antagonists that had suppressed that pathogen. For example: *Rhizopus* spp. on strawberry after the use of benomyl to control *Botrytis cinerea* (Dennis, 1975; Jordan, 1973) and *Colletotrichum gloeosporioides* (Laville, 1971) and *Phytophthora syringae* (Upstone, 1977) and *A. tenuis* (Spalding, 1970) on pome fruit after the use of benzimidazoles to control *Gloeosporium* spp. and *P. expansum.*

It has been suggested that resistance of fruits and vegetables to postharvest diseases may be achieved using biological control through two basic approaches: 1) use and management of the natural, beneficial microflora that already exist on fruit and vegetable surfaces, and 2) artificial introduction of micro-organisms antagonistic to postharvest pathogens. Our knowledge of methods to manipulate naturally occurring populations of micro-organisms in a beneficial manner is meagre (Wilson, 1989). To effectively manipulate epiphytic microbial populations of fruits and vegetables we need more information on the microbial ecology of fruits and vegetables including species composition and population dynamics. Virtually all the studies on biocontrol published to date have involved the artificial introduction of large numbers of a known antagonistic micro-organism. Fundamental questions on the ecology and population dynamics of particular components of the epiphytic microflora of agricultural commodities have not been addressed in the past due to lack of appropriate research tools to track and identify specific species and strains in natural microbial populations. In recent years, powerful and diverse molecular techniques have been developed which could resolve this problem, and provide information that has been difficult to obtain (Williams *et al.*, 1990). Recently, the random amplified polymorphic DNA (RAPD) technique was used successfully to determine composition of epiphytic yeast populations of citrus fruit (Droby *et al.*, 1995b).

Since the early 1970's, research on the phyllosphere of crop plants has developed rapidly. The microflora of leaves and ripening parts of arable crops has received much attention but little information is available on the composition and development of epiphytic populations of ripening perishable fruits. Preharvest factors that affect fruit quality, such as mineral nutrition and maturity at harvest, as well as pesticide use, may profoundly affect the composition of epiphytic microbial populations on the surfaces of fruits and vegetables (Andrews, 1982; Turner *et al.*, 1985; Andrews and Kenerley, 1987). Also, the nutritional composition of fruits and vegetable surfaces may selectively promote certain micro-organisms during storage (Morris and Rouse, 1985).

Preharvest application of biocontrol agents might prove useful against postharvest diseases of fruits and vegetables which develop from infections occurring at various stages of fruits development (*e.g.* latent infections), as well as through wounds. This approach could be used as a tool to manipulate epiphytic populations and change patterns of surface wound colonisation. Recent reports indicate the possibility of reduction of postharvest decay caused by *C. gleoesporioides* on avocado and mangoes by a preharvest spray with *Bacillus subtilis*, an antibiotic producing bacterium (Korsten *et al.*, 1991). Koomen and Jeffries (1993) have also demonstrated the feasibility of control of anthracnose on mango fruit with *B. cereus* and *Pseudomonas fluorescens*. Droby *et al.* (1993c) suggested the possible use of preharvest application of a single antagonist (*Pichia guilliermondii,* isolate US-7) to reduce the development of postharvest decay of citrus fruit, and a reduction of postharvest *Rhizopus* rot of table grapes was also achieved by a preharvest spray of the yeast *P. guilliermondii* (Ben-Arie *et al.*, 1991). To fully explore the potential of this approach, however, obtaining data on the composition of epiphytic populations before and after the introduction of a single antagonist is crucial.

THE USE OF INTRODUCED ANTAGONISTS FOR THE CONTROL OF POSTHARVEST DISEASES

In order to select a successful antagonist a thorough understanding of pathogen etiology should be available. Detailed knowledge on the ecology of the pathogen, the infection process and the epidemiology of the disease is critical for determining the type of antagonist needed to interrupt the establishment of the pathogen in the tissue.

Postharvest decay of fruits and vegetables may be the result of: 1) infections taking place at early stages of fruit development such as infections through flower parts or infections of unripe fruit which remain dormant until after harvest (latent infections); and 2) infections through surface wounds inflicted during harvest, packing and transport. Latent infections are difficult to control with the use antagonists since they may occur at different stages of fruit development. Thus, antagonists which interfere with the infection process should presumably be present on the fruit surface before infection occurs. Examples of latent infections on fruits include *Colletotrichum* infections of avocado, mangoes, bananas and papayas (Eckert and Ogawa 1985); stem end rots of mangoes, avocado and citrus; *Botrytis cinerea* infections of grapes, and strawberries (Snowdon, 1990); and *Monilinia fructicola* on stone fruits (Kable, 1971). On the other hand, many pathogens responsible for postharvest decay are able to infect host tissue only through surface wounds resulting from either mechanical or physiological injuries. Postharvest losses resulting from wound-pathogens are very significant in most commodities (Snowdon, 1990).

Research conducted in the past decade on biocontrol of postharvest diseases has concentrated mainly on searching for micro-organisms that are antagonistic to wound pathogens. Typically, infection of wounds by spores of the pathogen is very rapid (often within 24 hr). Thus, rapid colonisation and growth of the antagonist at the wound site is a key characteristic of a successful antagonist.

Fokkema (1992) suggested that biological control of infections in fresh wounds has the advantage of the absence of resident microflora competing with the antagonists. In addition, favourable nutritional conditions as well as a suitable micro-environment also favour the antagonist. This may be true, however, only when the antagonist is introduced immediately after wounding. Mercier and Wilson (1994) reported that fresh wounds of apple fruits were colonised within 2 to 4 days by a characteristic microbial flora, reaching levels of about 10^5 to 10^6 cfu/wound for fungi and 10^3 to 10^6 cfu/wound for bacteria. The predominant micro-organisms were *Aureobasidium pullulans*, *Sporobolomyces roseus*, *Erwinia* spp., *Pseudomonas* spp., and *Gluconobacter* spp. Micro-organisms, such as *A. pullulans* and *S. roseus*, are most likely to be a major part of the resident microflora on intact fruits (Davenport, 1976). Introduction of the yeast antagonist *Candida oleophila* to freshly made wounds resulted in a film of yeast cells on the wound surface. Under commercial conditions the treatment of fruits with biocontrol agents sometimes may be delayed for a few days after harvest and then the antagonist may encounter natural populations higher than those on fresh wounds, as well as pathogens that have already started the infection process.

The first step before initiation of a screening program for potential antagonists should be to determine the features needed from the biocontrol agent. In general, features required are: (1) good coloniser of fruit surfaces and wounds; (2) rapid growth in surface wounds; (3) effective utiliser of nutrients at low concentrations present in the wound and (4) able to survive and develop better than the pathogen on the surface of the commodity and at infection sites under a wide range of temperature, pH and osmotic conditions.

The use of micro-organisms naturally occurring on the surface of fruit or vegetables has been usually preferred for the control of postharvest diseases by most investigators (Janisiewicz, 1988; Janisiewicz and Roitman, 1988; Stretch, 1989; Chalutz and Wilson, 1990; Roberts, 1990a; Roberts, 1990b; Droby *et al.*, 1991b; Gullino *et al.*, 1991; Lurie *et al.*, 1995). In particular, yeasts naturally occurring on fruits and vegetables have been targeted by many workers as potential antagonists of postharvest diseases. As reviewed by Janisiewicz (1988), yeasts can colonise a surface for long periods under dry conditions, produce extracellular polysaccharides that enhance their survival and restrict both colonisation sites and the flow of germination cues to fungal propagules, use available nutrients rapidly to proliferate, and are affected minimally by pesticides. While the mechanism of action of yeast antagonists in inhibiting postharvest pathogens has not been fully understood, there are several lines of evidence to indicate that they do not rely on the production of antibiotic substances (Droby and Chalutz, 1994b). This feature would be an advantage since the use of antibiotic producing antagonists on food could be problematic with regard to the development of antibiotic-resistant strains of animal and plant pathogens.

A method for the isolation and screening of yeast antagonists has been developed by Wilson *et al.* (1993). The method is based on the application of fruit surface washings to fruit wounds, which are subsequently challenged with spores of the pathogen. After a few days of incubation, micro-organisms are isolated from the non-infected wound sites. Interestingly, it was found that yeast colonies were usually predominant in the growth medium. Following random selection according to colour and morphological features, pure cultures were tested for antagonism in an *in vivo* test as described by Droby *et al.* (1989). Such isolation procedure favours the selection of yeasts which are rapid colonisers of wounds.

CHARACTERISTICS OF AN "IDEAL ANTAGONIST" OF POSTHARVEST DISEASES

Antagonists will have to meet certain criteria to be successfully developed for commercial use on harvested crops. Characteristics of an "ideal antagonist" (Wilson and Wisniewski, 1989) are: (1) genetically stable; (2) effective at low concentrations; (3) not fastidious in its nutrient requirements; (4) able to survive adverse environmental conditions (including low temperature and controlled atmosphere storage); (5) effective against a wide range of pathogens on a variety of fruits and vegetables; (6) amenable to production on an inexpensive growth medium; (7) amenable to a formulation with a long shelf life; (8) easy to dispense; (9) does not produce metabolites that are deleterious to human health; (10) resistant to pesticides; (11) compatible with commercial processing procedures and (12) non-pathogenic to the host commodity.

ANTAGONIST-HOST-PATHOGEN INTERACTIONS AT THE WOUND SITE: A CASE STUDY

The opportunities for fruit infection may be limited since the presence of the pathogen, an entry site, physiological susceptibility of the host, and favourable nutrient and environmental conditions are required simultaneously. Thus, infection can be markedly reduced by either reducing the inoculum of the pathogen through sanitation and fungicides or through the manipulation of one or more of the above factors to reduce host susceptibility. In most cases, the key factor in fruit susceptibility to postharvest pathogens is the presence of mechanical and/or physiological injuries in the peel. Any interference with the course of infection through wounds would reduce disease development. In this regard, the use of antagonistic micro-organisms to target the pathogen at the court of infection has proved to be of great potential (Droby *et al.*, 1993b).

A full understanding of the interactions taking place at the wound site between the antagonist cells, the host tissue and the pathogen is a prerequisite for the development of successful biocontrol strategies. A complex interaction involving host resistance and wound responses, as well as interaction with other micro-organisms, has to be taken into consideration. This conceptualisation raises some critical questions: (1) What are the effects of antagonists on "wound healing" and host resistance? (2) How important and widespread are the direct effects of antagonists on pathogens? (3) How do incidental micro-organisms or mixtures of antagonists affect pathogen/antagonist interactions? and (4) How does nutrient/chemical composition at the wound site affect the antagonist, other microflora, the infection process, and the wound response?

ANTAGONIST-PATHOGEN INTERACTIONS

Indirect Interactions

To inhibit infection, the antagonist must be present at the wound site prior to arrival of the pathogen or within a short period thereafter. Recent studies on biological control of postharvest diseases of fruits and vegetables have reported the use of antagonistic micro-organisms that multiply rapidly, colonise the wound and out-compete the pathogen for nutrients and space. The yeast *Pichia guilliermondii* (formerly identified as *Debaryomyces hansenii*), *Cryptococcus laurentii, Aureobasidium pullulans, Candida* spp. *Sporobolomyces roseus* and the bacteria *Enterobacter cloacae, Pseudomonas cepacia* and *P. syringae,* have all been reported to rapidly and extensively colonise the wound site (Droby *et al.,* 1989; Wilson and Wisniewski, 1989; Wisniewski *et al.,* 1989; Roberts, 1990a; Roberts, 1990b; Shefelbine and Roberts, 1990; Gullino *et al.,* 1991; Smilanick *et al.,* 1992; Smilanick *et al.,* 1993; Janisiewicz *et al.,* 1994; Lurie *et al.,* 1995).

Several reports on the interaction between epiphytic micro-organisms have shown that bacteria and yeasts are able to take up nutrients from dilute solutions more rapidly, in greater quantity, and with greater efficiency than can the germ tubes of fungal pathogens. This should result in a marked reduction in the amount of exogenous nutrients available for the pathogen if the antagonists colonise prior to the pathogen (Brodie and Blakeman, 1976; Fokkema, 1981; Fokkema, 1986).

In our studies (Droby *et al.,* 1989), we observed that isolate US-7 of *P. guilliermondii* multiplied very rapidly at the wound site and increased in numbers by 1 to 2 orders of magnitude within 24 h while the pathogen spores had just started to germinate and grow. The growth of the yeast antagonist at the wound site is shown in Figure 1.

Several lines of evidence support the assumption that inhibition of pathogen development by the antagonist involves competition for nutrients (Droby *et al.,* 1989). Such competition was demonstrated by *P. guilliermondii* in culture, when both the antagonist and the pathogen were co-cultured in minimal synthetic medium or in wound leachate solutions. The efficacy of the yeast could be markedly reduced by the addition of nutrients to the spore suspension used for inoculation. Similarly, *E. cloacae,* an antagonistic bacterium, inhibited germination of *Rhizopus stolonifer* spores through nutrient competition (Wisniewski *et al.,* 1989). In both studies, indirect evidence was provided to demonstrate the role of competition for nutrients as part of the mode of action of these two antagonists: (1) inhibition of spore germination or growth of the pathogen during co-culturing with the antagonist was demonstrated; (2) inhibition of the pathogen was dependent on the concentration of the antagonist propagules, and (3) partial or complete reversal of inhibition could be achieved by the addition of exogenous nutrients.

In most reports on biological control of postharvest diseases of fruits and vegetables, a quantitative relationship has been demonstrated between the antagonist concentration and the efficacy of the biocontrol agent. Thus, a delicate balance apparently exists at the wound

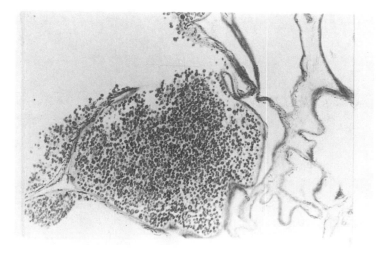

Figure 1. Growth of the yeast antagonist *Pichia guilliermondii* in a surface wound of apple.

site between the number of antagonist cells and the pathogen propagules which affects the outcome of the interaction and determines whether or not the wound becomes a site of infection. Manipulation of the initial concentration of the antagonist cells and/or the fungal spores clearly affects infection. On the other hand, we have shown that the number of antagonist cells at the wound site will not always determine its efficacy. Our data suggested that active multiplication and growth of US-7 cells was required for this yeast to exhibit its biocontrol activity. This was demonstrated by using a mutant of *P. guilliermondii* which lost its biocontrol activity against *P. digitatum* on grapefruit and against *B. cinerea* on apples, even when applied to the wound at concentrations as high as 10^{10} cells/ml (Droby *et al.*, 1991a). This mutant was produced by chemical mutagenesis and selected for its inability to grow in minimal synthetic medium, but grew normally in rich undefined medium. The cell population of this mutant remained constant at the wound sites during the incubation period, while that of the wild type increased 60- to 100-fold, within 24 hours. Failure of the mutant to inhibit spore germination of the pathogen in culture on a minimal salt medium suggested that this mutant lost its ability to utilise some nutrients and grow in culture as well. This could be the reason for its lack of efficacy.

While competition for nutrients is likely to be the general phenomenon in the interaction between micro-organisms on the phylloplane, the demonstration of nutrient competition as a mechanism of antagonism assumes that the pathogenic fungus depends on an external source of nutrients for germination and penetration into the host tissue. This assumption is very difficult to demonstrate. In this regard it would be worthwhile to mention that while in macroecology it has long been axiomatic that competition for nutrients plays a central role, Andrews (1992) more recently stated that except under certain conditions, competition is probably not a major force on the phylloplane, or at least that evidence for its role under natural conditions is either lacking or largely indirect.

In summary, rapid growth and extensive colonisation of the wound site by the antagonist are important features of many reported postharvest biocontrol agents. Active growth of antagonist cells, presumably depletes nutrients and/or space available to the pathogen and thus reduces pathogen growth rate and the incidence of infection. For effective levels of biocontrol activity, the antagonist cells must probably reach a critical number to inhibit or prevent the establishment of the pathogen in the fruit or vegetable tissue. In most cases, therefore, application of biocontrol agents, after the establishment of the pathogen in the tissue, results in much reduced efficacy, the degree of which depends on the length of time that elapses between the initiation of the infection by the pathogen and application of the antagonist.

Direct Interactions

(1) Attachment. Direct parasitism by the antagonist of the pathogens' propagules has been reported to play a role in biological control against soil-borne and foliar diseases. In this regard, studies has been conducted with *Trichoderma* (Elad *et al.*, 1983). In the postharvest arena, very little information is available on biological control agents that directly parasitise pathogens. Wisniewski *et al.* (1991), however, have shown that the cells of the yeast *P. guilliermondii* which were in direct contact with *B. cinerea* hyphae appeared to be lying within a depression of the hyphal cell wall. *P. guilliermondii* similarly attached to hyphae of *Penicillium digitatum* (Figure 2). In addition, the yeast appeared firmly imbedded within these depressions and surrounded by an extracellular matrix. This attachment was blocked when the yeast cells or the pathogen hyphae were exposed to compounds that affect protein integrity, or when respiration was inhibited. This suggested that the protein(s) involved in attachment are most likely located on the yeast cell surface and/or

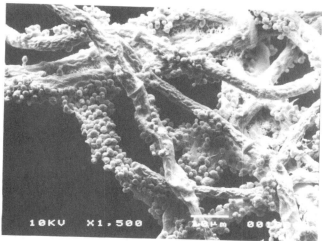

Figure 2. Attachment of cells of *Pichia guilliermondii* to the mycelium of *Penicillium digitatum*.

within the surrounding extracellular matrix. The close contact of the yeast cells with the fungal cell wall would also facilitate the efficient uptake and depletion of nutrients from immediate vicinity surrounding the fungus.

(2) Production of Cell-Wall Hydrolases. P. guilliermondii was also found to exhibit high levels of β-(1-3)-glucanase activity when cultured on various carbon sources or on cell walls of several fungal pathogens (Wisniewski *et al.*, 1991).The fastidious attachment of this yeast antagonist to fungal cell walls would enhance the effectiveness of any cell wall hydrolases secreted by the yeast to the extracellular matrix. When yeast cells were dislodged from the hyphae, a concave appearance of the hyphal surface and partial degradation of the cell wall of *B. cinerea* was also observed at the attachment sites. Thus, the firm attachment of the yeast cell along with the production of hydrolases may be responsible for the observed degradation of the fungal cell wall. The ability to produce high levels of β-glucanase by the yeast has been suggested to be associated with the firm attachment of the yeast cells to fungal hyphae and pitting observed in some areas on fungal mycelium (Wisniewski *et al.* 1991). In addition, the kinetics of β-(1-3)-glucanase production as related to growth may suggest that firm attachment of the yeast cells to fungal hyphae and to plant tissue might be facilitated by extensive production of β-(1-3)-glucanase. (See Romantschuk *et al*, this volume, for further details about microbial attachment.)

Isolate US-7 of *P. guilliermondii* investigated by Wisniewski *et al.* (1991) appears to produce and secrete high levels of exo-β-(1-3)-glucanase into the growth medium. Monitoring the kinetics of an exo-β-(1-3)-glucanase production during growth of the yeast revealed that the activity detected in the yeast culture filtrate reached the highest activity after 36-48 h growth at 25 C, and correlated with the growth rate of yeast in NYDB medium. When tested in culture filtrate of a non-antagonistic yeast isolate, β-(1-3)-glucanase activity was much lower as compared with that of US-7.

In further studies of the glucanase activity of US-7 an exo-β-(1-3)-glucanase with molecular weight of 45 KDa was identified (Avraham, 1994). The exoglucanase activity was confirmed by its ability to hydrolyse various substrates. A similar secretable exo-β-(1-3)-

glucanases, with the same molecular weight, has been found in *Candida albicans* (Chambers *et al.*, 1993).

Characterization of the optimal conditions for enzyme activity has shown that the exo-β-(1-3)-glucanase produced by the yeast *P. guilliermondii* exhibited its maximal activity at pH 4.0 - 5.0. Interestingly, however, the exoglucanase retained about 80% of its activity over a wide range of pH (3.0 - 6.0). Low activity was evident at extreme acidic and basic conditions. Optimal temperature for its activity was 50 C. In addition, enzymatic activity was not affected by glucose up to a concentration of 50 mM at pH 5. At glucose concentrations of 500 or 1400 mM (pH=5) the activity was reduced by 35% and 64%, respectively.

Optimal activity of extracellular β-glucanases of fungal origin has been reported to be at pH values of 4 -6 (Pistons *et al.*, 1993) and readily inhibited (50% inhibition) at 8 mM of glucose (Heupel *et al.*, 1993). Shoseyov *et al.* (1988) have reported that β-glucosidase isolated from *Aspergillus niger*, which is considered relatively tolerant to high glucose levels, was totally inhibited at a concentration of 100 mM. The ability of exo-β-(1-3)-glucanases isolated from the yeast *P. guilliermondii* to tolerate a wide range of pH and relatively high concentration of glucose may indicate that osmotolerance has an important role in survival and reproduction of the yeast under diverse environmental conditions. This may contribute to its ability to antagonise other organisms on the fruit surface and in wounds.

(3) Production of Extracellular Materials. In our early studies, conducted to evaluate antagonist-pathogen interactions at the wound site, extensive production of extracellular materials was found to be present around yeast cells in the wound (Wisniewski *et al.*, 1988). To study the possible role of this extracellular material in the mode of action by which *P. guilliermondii* inhibits postharvest pathogens, we have recently observed that an extracellular material extracted from the surface of the yeast cells exhibited antifungal activity against several postharvest pathogens when tested *in vitro* (Droby *et al.*, 1995a). Both spore germination and germ tube elongation were inhibited. These results suggested the possible involvement of this extracellular material in the interaction of the yeast with the pathogen. In addition, the extracellular material also inhibited infection and development of green mold decay, caused by *P. digitatum*, on wounded grapefruits (Droby *et al.*, 1993d; Droby *et al*, 1995a).

Thus, biocontrol activity of *P. guilliermondii*, and possibly other yeast biocontrol agents, may be dependent not only on its ability to rapidly colonise the wound site and compete for nutrients, but may also depend on its ability to attach firmly to hyphae of the pathogen and to produce extracellular materials as well as cell-wall degrading enzymes.

ANTAGONIST-HOST INTERACTIONS

Induction of Resistance Mechanisms

Induced resistance has been recognized as an important and manageable form of resistance in vegetative plant tissues (Dixon and Harrison, 1990; Kuc, 1990; Kuc and Strobel, 1992). Induction of resistance by micro-organisms was shown to be an effective way to protect vegetative plant tissue from fungal pathogens (Kuc and Strobel, 1992). Resistance can be induced in plants by a variety of plant pathogens and nonpathogenic micro-organisms (Kuc and Strobel, 1992). Similar mechanisms of resistance may operate in harvested fruit and vegetable tissues. Unlike vegetative plant tissue, harvested commodities are senescing rather than developing. The senescence process generally reduces resistance responses. These factors may have contributed to a lack of emphasis on disease resistance in harvested fruits and vegetables. In our experience, induced resistance holds promise as a new technology for the control of postharvest diseases. Both physical and biological agents elicit resistance responses in harvested fruits and vegetables (Droby *et al.*, 1993a; Wilson *et al.*, 1993; Wilson *et al.*, 1994). Heat treatment, wounding, gamma radiation, UV-C light, antagonists, attenuated strains, and natural compounds have all been suggested as elicitors of resistance in harvested crops. In this review, we will discuss induction of resistance by antagonists of postharvest diseases. Indeed, some reports have indicated that certain postharvest biocontrol agents may interact with the host tissue, in particular with wounded surfaces, leading to enhanced wound healing processes. Direct evidence to support this possible mode of action, however, is lacking.

As indicated above, several non-antibiotic-producing yeast antagonists of wound pathogens are most effective when their application occurs prior to the inoculation of the wound with the pathogen. Application of antagonist cells after inoculation with the pathogen results in decreased efficacy. Chalutz and Wilson (1990) reported that the longer the time elapsing between infection of *P. digitatum* and application of the antagonist (*P. guilliermondii*), the more that biocontrol efficacy was reduced. At 25 C, only 30% reduction of the incidence of disease was observed when the antagonist was applied 7 hr after inoculation and no reduction was evident when antagonist application occurred at 24 hr after inoculation. This is in comparison to greater than 90% reduction in disease incidence when the antagonist and pathogen were applied simultaneously. While this trend has been demonstrated in laboratory studies, in larger scale, semi-commercial tests reduction of efficacy due to application of the yeast after inoculation or natural infection was lower than exhibited in laboratory tests. These observations suggested the possibility that application of the yeast cells may induce resistance processes in the peel tissue. To test this hypothesis, we examined the production of ethylene in yeast-treated tissues. When cell suspensions of the US-7 yeast antagonist were placed on surface wounds of grapefruit, pomelo, table grapes or carrot root tissue enhanced ethylene production was evident in all tissues (Droby and Chalutz., 1994b). In carrot root discs, which were used as a model system, the application of the yeast antagonist resulted not only in enhancement of ethylene production but also the accumulation of phenols or phenol-like materials - associated with increased resistance - which exhibited absorption at 280 nm (Figure 3).

When cultured *in vitro*, the yeast cells themselves did not produce ethylene. In addition, induction of ethylene was demonstrated in grapefruit peel discs treated with isolated extracellular material of the yeast *P. guilliermondii* (Figure 4), indicating its possible role as an elicitor of ethylene.

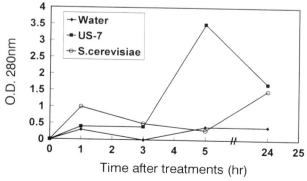

Figure 3. Induction of phenols or phenol-like materials in carrot root discs in response to treatment with the yeast antagonist US-7 (*Pichia guilliermondii*) and baker's yeast (*Saccharomyces cerevisiae*).

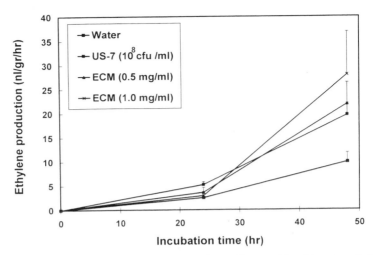

Figure 4. Induction of ethylene production by cells of *Pichia guilliermondii* and its extracellular material in disks of grapefruit peel.

Involvement of ethylene in the induction of resistance processes in grapefruit and carrot roots was demonstrated in the past (Chalutz *et al.*, 1969; Mercier *et al.*, 1993), possibly through the induction of the activity of phenyalanine ammonia-lyase (PAL), an enzyme which catalyses the branch point step reactions of the shikimic acid pathway, leading to the synthesis of phenols, phytoalexins and lignins. All these compounds have been associated with induced resistance processes. In citrus fruit, ethylene production and PAL activity was induced following application of an effective yeast antagonist to disks of the peel. Exogenously applied ethylene to the discs or to whole grapefruits also induced resistance to *P. digitatum* infections (Droby *et al.*, 1994). In addition, application of US-7 to lemon wounds enhanced production of the phytoalexin scoparone (Rodov *et al.*, 1994). Thus, the induction of both ethylene, and scoparone, by fruit tissues in response to the application of a yeast antagonist, suggest the involvement of host resistance mechanisms in the biocontrol activity of the yeast. The nature and mechanism of this induction is yet to be elucidated. The yeast antagonist *Candida saitoana,* was also found to induce chitinase and cause deposition of papillae along host cell walls in apple tissue (A. El-Ghaouth and C.L. Wilson, unpublished), providing another example of induction of host resistance.

Induced resistance in plants by non-pathogenic micro-organisms has been studied in many plant systems (Kuc and Strobel, 1992). Potato tuber tissue was shown by Hammerschmidt (1984) to deposit lignin-like materials more rapidly in response to incompatible races of *Phytophthora infestans* than to compatible ones. Hammerschmidt *et al.* (1984) has also demonstrated the use of nonpathogenic *Cladosporium cucumerinum* for induction of lignin and hydroxyproline in cucumber tissue. The use of non-pathogenic micro-organisms to induce resistance in the harvested commodity has not yet been fully explored. The possibility that non-pathogenic micro-organisms induce defence mechanisms leading to degradation of plant tissues after harvest is discussed in the chapter by Morris and Nguyen-the (this volume). However, it should be possible to screen for epiphytic micro-organisms that induce beneficial resistance processes in the wound as a part of a strategy to prevent the entry of postharvest pathogens.

CONCLUSIONS

The search for nonchemical, environmentally friendly and safe control methods for postharvest diseases of fruits and vegetables will continue. Although significant progress has already been made towards the development of alternative control methods, the application of the technologies for commercial use has not been fully realised yet. Successful commercial implementation of biocontrol procedures will greatly depend on overcoming problems related to inadequate levels of control and consistent performance. For postharvest biological control to be a success we should increase our efforts to: (1) enhance the activity of antagonists through additives, formulations and genetic manipulation; (2) develop an integrated disease management approach in which more than one strategy is included (3) understand how antagonists interact with the plant host, the environment, and other organisms and (4) better understand the ecology of specific antagonists and the role of natural epiphytic microbial populations in suppressing postharvest disease.

Concern was raised recently regarding health and safety aspects related to the mass introduction of antagonists on our food. Some of the antagonists reported to effectively control postharvest diseases have also been reported to be an opportunistic pathogens on immuno-compromised humans. This might pose an obstacle in the public perception of this technology. However, we must remember that these antagonists are indigenous to agricultural commodities, and humans are continuously exposed to them. Although these antagonists are introduced in large numbers to the surface of the commodity, they will survive and grow only in very restricted sites on the fruit surface (*e.g.* surface wounds). On the intact fruit surface, antagonist population would reach the level of natural epiphytic microflora in a very short period of time after their introduction. Thus, in spite of the rigorous tests needed to verify their safety to humans and the environment, introduction of antagonists to fruits and vegetables to control postharvest diseases may find its commercial use as an alternative to the use of synthetic fungicides.

REFERENCES

Andrews, J.H. 1982, Effects of pesticides on non-target microorganisms on leaves, pp. 283-304, In: Blakeman, J.P. (ed.) *Microbial Ecology of the Phylloplane*, Academic Press, New York.

Andrews, J.H. 1990, Biological control in the phyllosphere-realistic goal or false hope? *Can. J. Plant Pathol.* 12:300-307.

Andrews, J.H. 1992, Biological control in the phyllosphere. *Annu. Rev. Phytopathol.* 30:603-635.

Andrews, J.H. and Kenerley, C.M. 1987, The effect of a pesticide program on non-target epiphytic microbial population of apple leaves. *Can. J. Microbiol.* 24:1058-1072.

Avraham, A. 1994, Isolation and Biochemical Characterization of Exo-β-(1-3)-glucanase from *Pichia guilliermondii*. *M.Sc. Thesis*, Faculty of Agriculture, the Hebrew University of Jerusalem.

Ben-Arie, R., Droby, S., Zutkhi, J., Cohen, L., Weiss, B., Sarig, P., Zeidman, M., Daus, A., and Chalutz, E. 1991, Preharvest and postharvest biocontrol of Rhizopus rot of table grapes with yeasts. *Proc. Int. Workshop Biological Control of Postharvest Diseases of Fruits and Vegetables.*, Shepardstown, WV, USA. USDA-ARS Pub. 92:100-113.

Blakeman, J.P. and Fokkema, N.J. 1982, Potential for biological control of plant diseases on the phylloplane. *Annu. Rev. Phytopathol.* 20: 176-192.

Brodie, I.D.S. and Blakeman, J.P. 1976, Competition for exogenous substrates *in vitro* by leaf surface microorganisms and germinating conidia of *Botrytis cinerea. Physiol. Plant Pathol.* 9:227-239.

Brown, G.E,. 1989, Host defenses at the wound site on harvested crops. *Phytopathology* 79:1381-1384.

Brown, G.E., Barmore, C.R. 1983, Resistance of healed citrus exocarp to penetration by *Penicillium digitatum. Phytopathology* 73:691-694.

Chalutz, E., Droby, S. and Wilson, C.L. 1988, Microbial protection against postharvest diseases of citrus fruit. *Phytoparasitica* 16: 195-196.

Chalutz, E. and Wilson, C.L. 1990, Postharvest biocontrol of green and blue mold and sour rot of citrus by *Debaryomyces hansenii*. *Plant Dis*. 74:134-137.

Chalutz, E., DeVay, J.E. and Maxie, E.C. 1969, Ethylene induced isocoumarin formation in carrot root tissue. *Plant Physiol*. 44:235-241.

Chambers, R.S., Brougton, M.J., Cannon, R.D., Carne, A., Emerson, G.W. and Sullivan, P.A. 1993, An exo-β-(1-3)-glucanase of *Candida albicans*: purification of the enzyme and molecular cloning of the gene. *J. Gen. Microbiol*. 139:325-334.

Davenport, R.R. 1976, Distribution of yeasts and yeast-like organisms from aerial surfaces of developing apples and grapes. pp. 325-359 In: Dickinson, C.H. and Preece, T.F. (eds.) *Microbiology of Aerial Plant Surfaces*. Academic Press, London.

Dennis, C. 1975, Effect of pre-harvest fungicides on spoilage of soft fruit after harvest. *Ann. Applied Biol*. 81:227-234.

Dixon, R.A. and Harrison, M.J. 1990, Activation , structure and organization of genes involved in microbial defense in plants. *Adv. Genetics* 28:165-234.

Droby, S., Chalupovicz, L., Chalutz, E., Wisniewski, M.E. and Wilson, C.L. 1995a, Inhibitory activity of yeast cell wall materials against postharvest fungal pathogens. *Phytopathology* (abstract, in press)

Droby, S. and Chalutz, E. 1994a, Successful biocontrol of postharvest pathogens of fruits and vegetables. *Proc. Brighton Crop Protection Conference - Pests and Diseases* 1994:1265-1272.

Droby, S., Chalutz, E., Cohen, L., Weiss, B. and Wilson C.L. 1991a, Nutrition competition as a mechanism of action of biological control agents of postharevst diseases. *Proc. Int. Workshop on Biological Control of Postharvest Diseases of Fruits and Vegetables*. Shepherdstown, WV, Sept. 1990. USDA-ARS Publication 92: 142-160

Droby, S. and Chalutz, E. 1994b, Mode of action of biocontrol agents for postharvest diseases. pp. 63-75 In: Wilson, C. L. and Wisniewski, M.E. (eds.) *Biological Control of Postharvest Diseases of Fruits and Vegetables-Theory and Practice*. CRC Press, Boca Raton Fla.

Droby, S., Chalutz, E., Horev, B., Cohen L., Gaba, V., Wilson, C.L. and Wisniewski, M.E. 1993a, Factors affecting UV-induced resistance in grapefruit against the green mould decay caused by *Penicillium digitatum*. *Plant Pathol*. 42:418-424.

Droby, S., Chalutz, E. and Wilson, C.L. 1991b, Antagonistic microorganisms as biological control agents of postharvest diseases of fruits and vegetables. *Postharvest News and Information* 2: 169-173.

Droby, S., Chalutz, E., Wilson, C.L. and Wisniewski, M.E. 1989, Characterization of the biocontrol activity of *Debaryomyces hansenii* in the control of *Penicillium digitatum* on grapefruit. *Can. J. Microbiol*. 35: 794-800.

Droby, S., Chalutz, E., Wilson, C.L. and Wisniewski, M.E. 1993b, Biological control of postharvest diseases: A promising alternative to the use of synthetic fungicides. *Phytoparasitica* 20: 149-153.

Droby, S., Hofstein, R., Wilson, C.L., Wisniewski, M., Fridlender, B., Cohen, L., Weiss, B., Daus, A., Timar, D. and Chalutz, E. 1993c, Pilot testing of *Pichia guilliermondii*: A biocontrol agent of postharvest diseases of citrus fruit. *Biological Cont*. 3:47-52.

Droby, S., Lischinsky, S., Cohen, L., Manulis, S., Mehra, R.K. and Eckert, J.W. 1995b, Epiphytic yeasts of citrus fruit tolerant to extreme conditions are effective antagonists of green mild decay. *Phytopathology* (abstract, in press).

Droby, S., Robin, D., Chalutz, E. and Chet, I. 1993d, Possible role of glucanase and extracellular polymers in the mode of action of yeast antagonists of postharvest diseases. *Phytoparasitica* 2:167.

Eckert, J.W. 1991, Role of chemical fungicides and biological agents in postharvest disease control. *Proc. Int. Workshop Biological Control of Postharvest Diseases of Fruits and Vegetables*., Shepardstown, WV, USA. USDA-ARS Publication. 92:14-30

Eckert, J.W. and Ogawa, J.M. 1985, The chemical control of postharvest diseases: subtropical and tropical fruits. *Annu. Rev. Phytopathol*. 23: 421-454.

Eckert, J.W. and Ogawa, J.M. 1988, The chemical control of postharvest diseases: Deciduous fruits, berries, vegetables and root/tuber crops. *Annu. Rev. Phytopathol*. 26:433-469.

Elad, Y., Chet, I., Boyle, P. and Henis, Y. 1983, Parasitism of *Trichoderma* spp. on *Rhizoctonia solani* and *Sclerotium rolfsii* - scanning electron microscopy and fluorescence microscopy. *Phytopathology* 73:85-88.

Fokkema, N.J. 1992, The phyllosphere as an ecological neglected milieu: A plant pathologist's point of view. pp. 3-18 In: Andrews, J.H. and Hirano, S.S.(eds.) *Microbial Ecology of Leaves*. Springer- Verlag, New York.

Fokkema, N.J. and Van den Heuvel, J. 1986, *Microbiology of the Phyllosphere*. Cambridge University Press, Cambridge, UK.

Fokkema, N.J. 1981, Fungal leaf saprophytes, beneficial or detrimental? pp. 433-454 In: Blakeman, J.P. (ed.), *Microbial Ecology of the Phylloplane*. Academic Press, London.

Gullino, M.L., Aloi, C., Palitto, M., Benzi, D. and Garibaldi, A. 1991, Attempts at biocontrol of postharvest diseases of apple. *Med. Fac. Landbouw. Rijksuiv. Gent.* 56: 195.

Hahlbrock, K. and Scheel, D. 1989, Physiology and molecular biology of phenylpropanoid metabolism. *Annu. Rev. Plant Physiol. Plant Mol. Biol.* 40: 347-469.

Hammerschmidt, R. 1984, Rapid deposition of lignin in potato tuber tissue as a response to fungi non-pathogenic on potato. *Physiol. Plant Pathol.* 24:33-42.

Hammerschmidt, R. and Lamport, D.T.A., Muldoon, E.P. 1984, Cell wall hydroxyproline enhancement and lignin deposition as an early event in the resistance of cucumber to *Cladosporium cucumerinum*. *Physiol. Plant Pathol.* 24:43-47.

Heupel, C., Schlochtermeier, A and Schrempf, H. 1993, Characterization of an intercellular β-glucosidase from *Streptomyces reticuli*. *Enzyme Microbiol. Technol.* 15: 127-132.

Janisiewicz, W.J. 1988, Biological control of diseases of fruit, pp. 153-165, In: Mukergi, K.G. and Grag, K.L.(eds.) *Biocontrol of Plant Diseases*. vol II. CRC Press, Boca Raton, FL.

Janisiewicz, W. and., Roitman, J. 1988, Biological control of blue mold and gray mold on apple and pear with *Pseudomonas cepacia*. *Phytopathology* 78:1697-1700.

Janisiewicz, W.J., Peterson, D.L. and Bors, R. 1994, Control of storage decay of apples with *Sporobolomyces roseus*. *Plant Dis.* 78:466-470.

Jeffries, P. and Jeger, M.J. 1990, The biological control of postharvest diseases of fruit. *Postharvest News and Information* 1:365-368.

Jordan, V.W.L. 1973, The effects of prophylactic spray programs on the control of pre- and postharvest diseases of strawberry. *Plant Pathol.* 22:67-68.

Kable, P.F. 1971, Significance of short-term latent infections in the control of brown rot in peach fruits. *Phytopathol. Z.* 70:173.

Koomen, I. and Jeffries, P. 1993, Effect of antagonistic microorganisms on the postharvest development of *Colletotrichum gloeosporioides* on mango. *Plant Pathol.* 42: 230-237.

Korsten, L., Villiers, E.E., Jager, E.S., Cook, N. and Kotze, J.M. 1991, Biological control of avocado postharvest diseases. *South African Avocado Growers' Association Yearbook* 14: 57-59.

Kuc, J. and Strobel, N.E. 1992, Induced resistance using pathogens and nonpathogenic. pp. 295-303 In: Tjamos, E.C, Papavizas, G.C. and Cook, R.J.(eds.), *Biological Control of Plant Diseases - Progress and Challenges for the Future*. Plenum Press, New York.

Kuc, J. 1990. Immunization for the control of plant diseases. pp. 355-373, In: Hornby, D.(ed.), *Biological Control of Soil-borne Plant Pathogens*. C.A.B. International, Wallingford, UK.

Laville, E. 1971, Evolution des pourritures d'entreposage des argrumes avec l' utilisation de nouveaux fongicides de traitement apres recolte. *Fruits* 26:301-304.

Lurie, S., Droby, S., Chalupowicz, L. and Chalutz, E. 1995, Efficacy of *Candida oleophila* strain 182 in preventing *Penicillium expansum* infection of nectarine fruits. *Phytoparasitica* 23:231-234.

Mercier, J., Arul, J., Ponnampalam, R. and Boulet, M. 1993, Induction of 6-methoxymellein and resistance to storage pathogens in carrot slices by UV-C. *J. Phytopathol.* 137:44-55.

Mercier, J. and Wilson, C.L. 1994, Colonization of apple wounds by naturally occurring microflora and introduced *Candida oleophila* and their effect on infection by *Botrytis cinerea* during storage. *Biol. Con.* 4:138-144.

Morris, C.E. and Rouse, D.I. 1985, Role of nutrients in regulating epiphytic bacterial populations. pp. 63-83, In: Windels, E.E. and Lindow, S.E. (eds.), *Biological Control on the Phylloplane*, American Phytopathological Society, St. Paul, MN.

National Research Council, Board of Agriculture 1993, Pesticide in the diets of infants and children. National Academy Press, Washington DC.

Piston, S.M., Seviour, R.J. and McDougall, B.M. 1993, Noncellolytic fungal β-glucanases: Their physiology and regulation. *Enzyme Microbiol. Technol.* 15:178-192.

Roberts, R.G. 1990a, Biological control of gray mold of apple by *Cryptococcus laurentii*. *Phytopathology* 80:526-530.

Roberts, R.G. 1990b, Biological control of mucor rot of pear by *Cryptococcus laurentii*, *C. flavus* and *C. albidus*. *Phytopathology* 80: 1051.

Rodov, V., Ben-Yehoshua, S., D'hallewin, G., Castia, T. and Farg, D. 1994, Accumulation of phytoalexins scoparone and scopoletin in citrus fruit subjected to various postharvest treatments. *Acta Hort.* 381:517-523.

Sequeira, L. 1990, Induced resistance: Physiology and Biochemistry. pp. 663-678 In: Baker, R.R. and Dunn, P.E. (eds.) *New Directions in Biological Control: Alternatives for Suppressing Agricultural Pests and Diseases.* Wiley-Liss Inc., New York, USA.

Shefelbine, P.A. and Roberts, R.G. 1990, Population dynamics of *Cryptococcus laurentii* in wounds in apple and pear fruit stored under ambient or controlled atmospheric conditions. *Phytopathology* 80:1020.

Shoseyov, O., Bravdo, B., Ikan R. and Chet, I. 1988, Endo β-glucosidase from *Aspergillus niger* grown on glycoside-containing medium. *Phytochem.istry* 27:1973-1976.

Smilanick, J.L. and Dennis-Arrue, R. 1992, Control of green mold of lemons with *Pseudomonas* species. *Plant Dis.* 76:481-485.

Smilanick, J.L., Denis-Arrue, R., Bosch, J.R., Gonzales, A.R., Henson, D.J. and Janisiewicz, W.J. 1993, Biocontrol of postharvest brown rot of nectarines and peaches by *Pseudomonas* species. *Crop Protec.* 12:513-520.

Snowdon, A.L. 1990, *A Color Atlas of Post-Harvest Diseases and Disorders of Fruits and Vegetables.* vol. 1 and 22, CRC Press, Boca Raton, FL.

Sommer, N.E. 1985, Strategies for control of postharvest diseases of selected commodities, pp. 83-99 IN: Kader, A.A., Kasmire, R.F., Mitchel, G.F., Reid, M.S., Sommer, N.F. and Thompson, J.F. (eds), *Postharvest Technology of Horticultural Crops.* Cooperative Extension, University of California, Davis, USA.

Sommer, N.F. 1982, Postharvest handling practices and postharvest disease of fruit. *Plant Dis.* 66:357-364.

Spalding, D.H. 1970, Postharvest use of benomyl and thiabendazole to control blue-mold rot development in pears. *Plant Dis. Rep.* 54:655-656.

Stretch, A.W. 1989, Biological control of blueberry and cranberry fruit rots (*Vaccinium corymbosu* L. and *Vaccinium macrocarpon* Ait.). *Acta Hort.* 241:301-306.

Tronsmo, A. and Hoftun, H. 1984, Storage and distribution of carrots. Effect on quality of long term storage in ice bank cooler and cold room and of different packing materials during distribution. *Acta Hort.* 163: 143-150.

Turner, S.M., Newmank, E.I. and Campbell, R.,1985, Microbial population of ryegrass root surfaces: Influence of nitrogen and phosphorus supply. *Soil Biol. Biochem.* 17: 711-715.

Upstone, M. 1977, Evaluation of chemicals for control of *Phytophthora* fruit rot in stored apples. *Proc. Br. Crop Prot. Conf. Pest Dis.* 1: 203.

Williams, J.G.K., Kubelik, A.R., Livak, K.J., Rafalski, J.A. and Tingey, S.V. 1990, DNA polymorphism amplified by arbitrary primers are useful as genetic markers. *Nucl. Acids Res.* 18: 6531-6535.

Wilson, C.L. 1989, Managing the microflora of harvested fruits and vegetables to enhance resistance. *Phytopathology* 79: 1387-1390.

Wilson, C.L. 1989, Managing the microflora of harvested fruits and vegetables to enhance resistance. *Phytopathology* 79: 1387-1390.

Wilson, C.L. and Chalutz, E. 1989, Postharvest biological control of Penicillium rots of citrus with antagonistic yeasts and bacteria. *Sci. Hort.* 40:105-112.

Wilson, C.L. and El-Ghaouth, A. 1993, Multifaceted biological control of postharvest diseases of fruits and vegetables. pp. 181-185, In: Lumsden, R.D. and Vaughn, J.L. (eds.), *Pest Management: Biologically Based Technologies.* American Chemical Society Press, Washington, D.C.

Wilson, C.L., El-Ghaouth, A., Chalutz, E., Droby, S., Stevens, C., Lu, J.Y., Khan, V. and Arul, J. 1994, Potential of induced resistance to control postharvest diseases of fruits and vegetables. *Plant Dis.* 78:837-844.

Wilson, C.L. and Pusey, P.L. 1985, Potential for biological control of postharvest plant diseases. *Plant Dis.* 69:375-378.

Wilson, C.L. and Wisniewski, M.E. 1989, Biological control of postharvest diseases of fruits and vegetables: An emerging technology. *Annu. Rev. Phytopathol.* 27:425-441.

Wilson, C.L. and Wisniewski, M.E. 1994, *Biological Control of Postharvest Diseases of Fruits and Vegetables - Theory and Practice.* CRC Press, Boca Raton, FL.

Wilson, C.L., Wisniewski, M.E., Biles, C.L., McLaughlin, R., Chalutz, E. and Droby, S. 1991, Biological control of postharvest diseases of fruits and vegetables: alternatives to synthetic fungicides. *Crop Protec.* 10: 172-177.

Wilson, C.L., Wisniewski, M.E., Droby, S. and Chalutz, E. 1993, A selection strategy for microbial antagonists to control postharvest diseases of fruits and vegetables. *Sci. Hort.* 40:105-112.

Wisniewski, M.E., Biles, C., Droby, S., McLaughlin, R., Wilson, C. and Chalutz, E. 1991, Mode of action of the postharvest biocontrol yeast, *Pichia guilliermondii.* I. Characterization of the attachment to *Botrytis cinerea. Physiol. and Mol. Plant Pathol.* 39: 245-258.

Wisniewski, M.E. and Wilson, C.L. 1992, Biological control of postharvest diseases of fruits and vegetables: Recent advances. *HortSci.* 27:94-98.

Wisniewski, M.E. and Wilson, C.L. and Chalutz, E. 1988, Biological control of postharvest diseases of fruit: Inhibition of *Botrytis* rot on apple by an antagonistic yeast. *Proc. Ann. Mtg. Electron Microsc. Soc. Am.* 46:290-291.

Wisniewski, M.E., Wilson, C. and Hershberger, W. 1989, Characterization of inhibition of *Rhizopus stolonifer* germination and growth by *Enterobacter cloacae. Can. J. Bot.* 67:2317-2323.

6

COEVOLUTION BETWEEN PLANTS AND PATHOGENS OF THEIR AERIAL TISSUES

Donald D. Clarke

Division of Environmental and Evolutionary Biology
Graham Kerr Building
Glasgow University
Glasgow, G12 8QQ
United Kingdom

INTRODUCTION

The term coevolution was coined by Ehrlich and Raven (1964) to describe the influences that plants and herbivorous insects have on each other's evolution, but it is now accepted that most, if not all, ecologically intimate associations between organisms are coevolving associations (Pirozynski and Hawksworth, 1988). One of the largest group of coevolving associations includes the associations between plants and phytopathogens and this review examines a subset of this group, associations between plants and phytopathogens of their aerial tissues. It is concerned with phytopathogens of native plants, where natural selection will be the driving force for coevolution, and considers both the extent to which associations may coevolve to establish long term and stable relationships and the fitness traits of both pathogen and host which are required to maintain such relationships. The account is based largely on our studies of the associations between native *Senecio* spp. and their powdery mildew and rust parasites.

THE LONGEVITY OF HOST/PATHOGEN ASSOCIATIONS

There is very good evidence that some associations between phytopathogens and the aerial surfaces of plants, such as powdery mildew and rust fungi, are very ancient, *e.g.*, there are records of powdery mildews and rusts on cereal crops dating back to the first written records, almost 5000 years ago, including references in the Old Testament of the Bible, and while the genotypes of both the host and parasite may have changed over the years, the same species are clearly still associated. Although records are only available on native plants and their parasites from around the early to the middle years of the last century, some of these associations, *e.g.*, those between *Erysiphe fischeri* or *Coleosporium tussilaginis* and certain *Senecio* spp. must be just as old, if not much older than those between pathogens and

Aerial Plant Surface Microbiology, edited by Morris et al.
Plenum Press, New York, 1996

cultivated cereals. Clearly, coevolution of a host and a parasite can establish an association which is capable of long term survival.

However, not all associations are old and some may be extremely young. For example, although the rust, *Puccinia lagenophorae*, is now common on *S. vulgaris* and *S. squalidus* throughout Britain, it was not recorded in Europe until the early 1960's (Wilson and Henderson, 1966) and the now common and widespread blister rust (*Albugo* sp.) on *S. vulgaris* is even more recent, being first recorded on this host in Britain in 1978 (Preece and Francis, 1987) but has become relatively common throughout Britain over the last 10 years. *P. lagenophorae* originates from Australia, where it parasitises several genera of the Compositae including species of *Lagenophora*, as well as a native species of *Senecio, S. lautus* and although the origins of the *Albugo* sp. are uncertain, it is probably derived from (and has been considered to be) *A. tragopogonis*, the blister rust fungus which attacks species of the composite genus *Tragopogon*.

A number of other cases of the establishment of new associations between plants and fungal parasites are known and if new associations establish relatively commonly then it is likely that, as predicted by Ehrlich and Raven (1964), extinctions also occur. Unfortunately, extinctions are less easily noted than new associations and it is not possible to cite specific examples. Instances where the frequency of occurrence of a parasite has changed however are known and where it is falling it is possible that the association is heading towards extinction. For example, *Coleosporium tussilaginis* is no longer recorded on *S. vulgaris* as frequently as it was 30 years ago. Since *Coleosporium tussilaginis* (although almost certainly a different pathotype) is still very common on *Tussilago farfara*, the decline is unlikely to be related to environmental factors but probably relates to changes in the host population. It is possible of course that the decline may be normal and temporary with the association having a fluctuating, rather than stable equilibrium. It is also possible that the decline is the direct result of competition with the new rust, *P. lagenophorae*, but if so, it is not obvious how the two rusts are competing.

Although some associations between native plants and their parasites have established quite recently, others, and perhaps the majority, are likely to be of considerable longevity. For an association to survive in the long term, coevolution must optimise the fitness of each organism, since the maximisation of fitness for either may threaten the survival of the other, and thus the survival of the association. Fitness is generally defined as the relative ability of an organism to persist over time and is measured as reproductive output, since genotypes which have the greatest reproductive output are likely to contribute the greatest number of progeny to subsequent generations. The following sections will consider the fitness traits of both host and parasite which may have evolved during the establishment of a long term relationship.

THE EVOLUTION OF HOST FITNESS

Infection by a pathogenic fungus should impose a selection pressure on the host population for those genotypes which are best able to cope with the infection, *i.e.*, those with some form of resistance and/or tolerance, since such genotypes are likely to have greater reproductive output than less resistant or less tolerant genotypes.

In relation to resistance, the evolution of three general types will be considered, two forms of complete or near complete resistance which will be referred to as non-host resistance and race-specific resistance, respectively; and one form of resistance which is less than total which will be referred to as incomplete, or quantitative resistance. Non-host resistance is that resistance possessed by all genotypes of a plant species which prevents them from being parasitised by any isolate of most of the thousands of species of potential parasites they come into contact with, *e.g.*, the resistance of *S. vulgaris* to all the *formae speciales* of the cereal mildew *E. graminis* or of the resistance of cereals to *E. fischeri*. Race-specific resistance refers to that resistance possessed by certain genotypes of a plant species which prevents them from becoming infected by avirulent but not virulent races of a parasite. Finally, incomplete, or quantitative resistance is that resistance which some genotypes of plant species possess which restricts the development within them of those virulent races of parasites which can infect. Incomplete resistance, by definition, is never complete.

Classification of resistance into these three types is to an extent artificial since some levels of race-specific resistance against avirulent races which is less than complete may be phenotypically indistinguishable from high levels of incomplete resistance against virulent races. In addition, although incomplete resistance to virulent races is generally non-race specific, cases of race-specific action have been recorded (Bevan, Crute and Clarke, 1993c). Finally, clear distinctions between the three forms of resistance are not easily made in relation to either their genetic control or the physiological or biochemical basis of their mechanism. Nevertheless, for the purposes of the following discussion, the differences are considered distinct enough to make the classification useful.

In addition to resistance, an ability to tolerate infection could also be an important attribute of host fitness. Tolerance of infection refers to the extent to which a host's growth and reproductive output are reduced by infection; those host genotypes whose growth and reproductive output are reduced the least by a given level of infection being the most tolerant.

Non-Host Resistance

Most plant species, when introduced into a new region, have been found to have total, non-host, resistance to most parasites within that region even though it is unlikely that either have ever come into contact with each other before. For example, wherever crops have been introduced throughout the world their common parasites are not species picked up from the local flora but are species introduced from the centre of origin of the crop. The fact that plants introduced into a new region possess resistance to most of the fungal pathogens of that region, suggests either that non-host resistance to some potential pathogens is fortuitous, being due to fitness traits not directly involved in resistance, or that plants possess a general system of resistance that operates against and excludes all but a few parasites. The latter is the most likely explanation since resistance to parasites must have been an important component of plant fitness throughout the whole of plant evolution. Even the first primitive eukaryotic autotrophs evolving in aquatic environments, long before the establishment of any land flora, will have been vulnerable to attack by heterotrophic organisms and some of the cellular reactions that evolved in response to such attacks, honed by selection over the millennia, may persist to this day in all species of plants as part of a general defence system. If there is a general form of non-host resistance, then it is likely that the underlying mechanisms and genetic control systems will be rather similar in all groups of plants. However, it is also likely that non-host resistance has continued to evolve to the present day leading to the evolution of additional and unique systems in different groups of plants. For example, environmental changes could lead to changes in the frequency and intensity with which infection by a parasite occurs with the result that the selection pressures on the host become so intense that either the host becomes extinct or only host genotypes which have total resistance to the parasite can survive. In the later case the resistance would be effectively non-host resistance and the parasite would become extinct unless it could survive on some other host species.

A genetic analysis of non-host resistance would help to differentiate between systems which may be common to all groups of plants and unique systems which have evolved relatively recently. However, virtually nothing is known of the genetic basis of most non-host resistance because a genetic analysis requires the production of hybrids between different plant species and this is generally not possible. The only non-host resistance currently amenable to analysis is that of a plant species to a parasite where that plant species is closely enough related to the host species for hybrids to be produced. For example, the non-host resistance of *S. squalidus* to *E. fischeri* has been analysed (Campbell, 1990) since, although classified in a different tribe of the genus to the host species *S. vulgaris*, the two plants will form hybrids. By backcrossing the hybrid onto *S. vulgaris* for several generations it has been possible to transfer some of the non-host resistance of *S. squalidus* to *E. fischeri*, into *S. vulgaris*. Resistance in the progeny obtained by selfing the final backcross generation to *E. fischeri* showed continuous segregation, indicative of polygenic, rather than oligogenic control (Campbell, 1990). Some forms of non-host resistance could have evolved from a race specific, gene-for-gene, type of interaction by host individuals accumulating more R-genes than any isolate of the parasite has loci at which matching virulence can accumulate. However, the evolution of complete non-host resistance to a parasite will, unless the parasite can survive on another host species, condemn it to immediate extinction. Thus non-host resistance cannot be a component of a host fitness trait of any coevolved association. On the other hand, for a parasite to establish a new association, that parasite must either be unaffected by, or overcome, or avoid non-host resistance and establish some form of basic compatibility. How such an event may occur or what it may involve is completely unknown but it must be a very early, if not the first, step in the evolution of a new association.

Race-Specific Resistance

Basic Features of Race-Specific Resistance. Race-specific resistance is resistance which operates against some races of a parasite but not others. It has been shown to be controlled by specific interactions between two sets of genes, one set in the host determining specific resistance and one set in the parasite determining specific avirulence. The interaction is commonly referred to as a gene-for-gene interaction (Flor, 1956). Most race-specific resistance genes in crops have been derived from wild species related to that crop, particularly wild species found near the centres of diversity of the crop and parasite. It has been suggested that the race-specific activity of resistance genes in crop plants might be artefactual and result from the transfer of the genes into the crop without any of the associated polygenic or other controlling elements likely to be present in the wild species. If this were the case the resistance of wild plants might not be race-specific (Day, 1974). However, race-specific resistance to powdery mildews and rust fungi has been found in several wild plants, e.g., to *Puccinia coronata* in *Avena sterilis* and *A. barbata* (Dinoor, 1977), to *Melampsora lini* in *Linum marginale* (Burdon and Jarosz 1991) and to *E. fischeri* in *Senecio vulgaris* (Harry and Clarke, 1986; Bevan and Clarke 1993b).

Evolution of Race-Specific Resistance. A state of basic compatibility must exist between all genotypes of a host and all genotypes of its parasites, otherwise it would be difficult to see how, within associations with an established gene-for-gene interaction, mutation from avirulence to virulence at a single gene locus in a parasite isolate could convert that isolate from a non-pathogen to a completely adapted pathogen. Thus, race-specific resistance must have evolved after the establishment of basic compatibility between a host and parasite. There is certainly both observational and experimental evidence from studies on the recently established associations between *S. vulgaris* and both *P. lagenophorae* and the *Albugo* sp. to indicate that these associations involve basic compatibility only. Firstly, a small sample of 50 lines of groundsel that had been found to include a wide range of specific-resistance phenotypes to *E. fischeri* were all found to be susceptible to a single pustule isolate of the rust *P. lagenophorae*. Furthermore, when exposed to natural infection in the field, these same lines all became relatively heavily infected by both *P. lagenophorae* and the *Albugo* sp., yet showed clear evidence of specific resistance to *E. fischeri*. It seems unlikely that this 50 plant sample of *S. vulgaris*, which possessed so many specific resistance factors to *E. fischeri*, would not also possess some specific resistance to *P. lagenophorae* and/or the *Albugo* sp. if such resistance were present. On the other hand some progenies of up to 30 plants from single plants of *S. squalidus*, when grown in small adjacent plots in the field, exhibited a complete range of reaction type to natural infection by *P. lagenophorae*. All the plants in some progenies developed heavy infections quite early in the season. Other progenies were more variable with some very susceptible plants, and a few highly resistant plants which developed one or two small pustules only late in the season. These latter plants possessed very high levels of resistance to the rust but we have yet to determine if this resistance is race-specific to an avirulent race or simply a very high level of incomplete resistance to an otherwise virulent isolate. Clearly, studies on this association and the associations between *S. vulgaris* and *P. lagenophorae* and the *Albugo* sp. could provide valuable clues regarding the early stages of evolution of host-parasite interactions and possibly of the development of gene-for-gene interactions.

Once a new association between a host and parasite has established, any mutation which confers complete resistance on a host line, would give it a selective advantage over all susceptible lines so that the resistant line would increase in frequency in the population. As the frequency of the resistant line increased in the population it would impose an increasing selection pressure on the parasite population for any virulent mutants able to overcome that resistance. These new virulent mutants, because of their selective advantage over the original race, now avirulent on the resistant line, would themselves increase in frequency, putting an increasing selection pressure on the host population for new mutations for resistance. And so the opposed changes would continue, each new resistance gene in the host being followed by a matching new virulence gene in the parasite.

Although this evolutionary pathway is conjecture there is powerful evidence to support some aspects of it, particularly in relation to the responses of the parasite population to the introduction of novel resistance genes in its host population. Thus it has been repeatedly shown that the introduction of new resistant cultivars of a crop can lead to the appearance of new virulence phenotypes leading to rapid changes in the virulence structure of the associated parasite population (Wolf and McDermott, 1994). Unfortunately, there is little or no evidence to indicate how frequently mutations for new resistance arise in host populations and how rapidly such resistance may spread.

The Role of Race-Specific Resistance in Host Fitness. Studies on native plant species with a well established system of race-specific resistance to a pathogen may not provide much information about how that resistance evolved, but could indicate the likely end result of the evolutionary pathways involved and the effectiveness of the resistance. A series of studies of the association between *S. vulgaris* and *E. fischeri* has provided quite a lot of information on the distribution of race-specific resistance in the host population and of specific virulence in the parasite (Harry and Clarke, 1986; Bevan *et al.*, 1993a, 1993b). These studies were facilitated by the ease with which different races, each differentiating different resistance phenotypes of the host, could be collected from *E. fischeri* populations; among a sample of only 24 mildew isolates collected at random from naturally infected plants, 19 different races were found (Bevan *et al.*, 1993a). Since the work started, over 50 distinct specific resistance factors have been identified in a relatively small sample of *S. vulgaris* and the fact that such a large number were found in this host sample suggests that there are many more, probably hundreds, and possibly thousands still to be identified in the species. Each specific resistance factor appears to be controlled by a single major gene, with each gene showing simple Mendelian inheritance (Harry and Clarke, 1987; Campbell, 1990). The resistance genes may be present singly or in any combination in different plant lines.

A study of two populations of *S. vulgaris*, one from Glasgow, Scotland and one about 480 km south at Wellesbourne, in the midlands of England, revealed that groundsel populations are extremely heterogeneous for race-specific resistance (Bevan *et al.*, 1993b). In this study, the reactions of plants sampled at each site were determined to up to ten different races of the mildew, five collected from Glasgow and five from Wellesbourne. At both Glasgow and Wellesbourne, between 80 and 90% of the plants were susceptible to all ten races of the mildew but a proportion were resistant to one or other of the races. The frequency of resistance to each race in each population ranged from 1% to 10% with the exception of resistance to one of the Glasgow mildew races, which was present in 37% of the plants sampled at Wellesbourne. Although both the Glasgow and Wellesbourne populations tended to be dominated by one or two of the resistance phenotypes they were highly heterogeneous for resistance when the less frequent resistance phenotypes were considered. This was particularly evident at Wellesbourne where ten different resistance phenotypes, including the phenotype which was susceptible to all races, were recorded amongst 75 plants growing within an area of one metre square. Such a high level of heterogeneity in specific-resistance is probably common to rusts and mildews in wild plants, *e.g.* race-specific resistance to oat crown rust (*Puccinia coronata*) was found to be widespread in the wild oats, *Avena sterilis* and *A. barbata*, in Israel where both hosts are common components of the native flora in some regions (Dinoor, 1977). Similar heterogeneity in race specific resistance was also found in the same two wild oats and in the wild barley, *Hordeum spontaneum* to *E. graminis* in Israel (Wahl *et al.*, 1978).

Despite the extreme heterogeneity of race-specific resistance in *S. vulgaris* populations, the resistance seems to provide limited protection, since each isolate of *E. fischeri* has such complex virulence that it is able to attack 90% or more of the host population; even groundsel plants with large numbers of specific resistance factors are generally vulnerable to some isolates of the mildew and such plants are not common in *S. vulgaris* populations (Harry and Clarke, 1986; Bevan *et al.*, 1993b).

Race-specific resistance must have evolved in *S. vulgaris* in response to selection pressures imposed by the mildew, but this resistance, in its turn, has imposed strong selection pressures on the mildew population for isolates with matching virulence. The capacity of individual isolates of the mildew to accumulate matching virulence effectively ensures that race-specific resistance can only be a minor component of the survival strategy of *S. vulgaris*. Clearly other systems must play more important roles in enabling the host to cope with mildew attack.

Incomplete or Quantitative Resistance and Tolerance

Infection by a virulent isolate of a pathogen would be expected to impose a selection pressure on the host population not just for race-specific resistance but for any attribute that will enable the host to cope better with the infection, *e.g.* incomplete resistance and tolerance. Incomplete resistance includes any property of the host which reduces infection by a virulent isolate, although never to the extent that infection is inhibited completely. Tolerance, in contrast to resistance, does not involve factors which affect parasite development but encompasses any property of the host which enables it to cope better with an infection (Shafer, 1971; Clarke, 1986). To date, tolerance has received relatively little attention and very little is known of its underlying physiology. Both incomplete resistance and tolerance are probably controlled by polygenic systems but different systems almost certainly control each.

The selection pressures imposed by virulent isolates, would be expected to favour both high levels of incomplete resistance and high levels of tolerance. However, studies of the reactions of *S. vulgaris* to *E. fischeri* have shown that while tolerance may be an important component of the survival strategy of the host (Ben-Kalio and Clarke, 1979; Harry and Clarke, 1992) incomplete resistance is less important, since plants with high levels of such resistance are not as common in populations of *S. vulgaris* as plants with low levels (Bevan *et al.*, 1993b). This is not surprising since levels of incomplete resistance which restrict spore output to below the threshold value required for effective transmission would put isolates at risk of extinction just as much as new R-genes can. Clearly the relative lack of high levels of incomplete resistance could be due to the strong selection pressures such resistance imposes on the parasite population for isolates able to overcome it. On the other hand, incomplete resistance in crop plants has generally proved to be durable since parasite isolates which are able to overcome it are rarely recorded, and there is no reason to believe that it should not be equally durable in natural pathosystems. It is possible that the infrequency of high levels of incomplete resistance is due to factors other than the ability of pathogen isolates to adapt to that resistance; host lines with high levels of incomplete resistance could simply be less fit than lines with low levels.

High levels of incomplete resistance would certainly confer reduced fitness on a host if the energy costs of that resistance were greater than the energy costs involved in supporting certain levels of parasite development. Unfortunately, very little work has been done on the energy costs to the host of any form of resistance. Smedegaard-Petersen and Stølen (1981) measured yield losses of up to 7% in cultivated barleys in response to inoculation with avirulent races of *E. graminis* f. sp. *hordei* even though the barley showed no visible signs of infection. On the other hand, considerable tissue damage (necrosis and chlorosis) can occur with high levels of incomplete resistance, *e.g.*, in cereals to virulent races of rusts (Roelfs, 1988) and powdery mildews (Moseman, 1956) and in *S. vulgaris* to *E. fischeri* (Harry and Clarke, 1986). Furthermore, this damage occurs at a much earlier stage of infection than with low levels of incomplete resistance. Thus the costs of high levels of such resistance are likely to be much greater than those involved in hypersensitive resistance and are probably greater than those involved in supporting the development of significant levels of parasite development. An indication that the costs of intermediate levels of incomplete resistance with low levels of parasite development may be greater than those due to low levels of such resistance with the associated higher levels of parasite development has been obtained from a study of the effects of mildew infection on wild (*Avena fatua*) and cultivated (*A. sativa*) oats (Sabri and Clarke, 1996). In this study, one cultivar of *A. sativa* was found to support significantly less mildew growth than another cultivar, or a line of wild oat, but to suffer more rapid tissue damage and a greater reduction in growth and yield than either.

For the reasons discussed in the next section, quite high parasite development is likely to be essential to ensure effective transmission and thus survival of powdery mildew and rust fungal parasites of plants in natural vegetation. Surprisingly there have been few attempts to determine the actual level of reproductive output of a rust or mildew on any host plant that may be required to ensure effective transmission. The only study involving a wild plant is that of Sabri and Clarke (1996) which showed that spore production by *E. graminis*, on wild oat as well as on a cultivar of cultivated oat, amounted to more than 10% of the dry weight of the infected plant. Such high levels of parasite development may be essential to ensure sufficient parasite transmission to maintain the parasite population and while the energy costs to the host resulting from this development will be much higher than those involved in hypersensitive resistance the difference between these costs and those of high levels of partial resistance may not be so great.

The costs of parasite development to the host will certainly impose a selection pressure on the host population for those host genotypes which suffer the least from infection.

The fittest host genotypes will clearly be those whose responses are only those which are an inevitable consequence of the development of the parasite and any genotype which suffers additional losses, due to any form of intolerance reaction, will be less fit and should be selected against. There is much observational evidence (Tarr, 1972) and some experimental evidence to indicate that wild plants are more tolerant of infection than crop plants which suggests that tolerance of infection could be a component of the survival strategy which enables wild plants to cope with parasitic infections. For example, studies of the reactions of *S. vulgaris* to infection by *E. fischeri* (Ben-Kalio and Clarke, 1979; Harry and Clarke, 1992) and of wild oats to *E. graminis* (Sabri and Clarke, 1996) have shown that heavy infections do not reduce plant growth and yield as much as lower levels of infection by similar parasites reduce the growth of crop plants.

The Effects of Resistance on the Survival of the Parasite

High levels of resistance may maximise host fitness but they will threaten the survival of the parasite and thus the survival of the association. As indicated above, a significant level of infection may be essential for the survival of the parasite. On the other hand too high a level of infection will threaten the survival of the host and so some level of partial resistance, together with tolerance, must be a major component of the host's survival strategy. The level of partial resistance which is likely to evolve will be that which maximises the fitness of the infected host. However the level of infection which develops is not determined solely by the partial resistance of the host but is partly a function of the level of aggressiveness of the parasite. The evolution of parasite aggressiveness, as a component of parasite fitness, is considered in the following section.

PARASITIC FITNESS AND ITS EVOLUTION

For long term survival, powdery mildew and rust fungal parasites of the aerial parts of plants, must, during their host's growing season, produce sufficient spores on infected plants to be able to transfer and establish infections on uninfected plants. If there are periods when susceptible host tissue (including alternate hosts) is not available for infection, or environmental conditions are unsuitable, parasites must also be capable of producing resting structures. During the summer and autumn months when *E. fischeri* infections are common-est in Britain, *Senecio* spp. other than *S. vulgaris*, including *S. viscosus* and *S. sylvaticus*, can carry infections and on parts of the European continent *S. vernalis* is also a host. Thus the survival of *E. fischeri* is not wholly dependent upon the survival of *S. vulgaris* since it could continue to survive on the other hosts. Resting structures are probably not required for the long term survival of *E. fischeri* in Britain because growing plants of *S. vulgaris* occur throughout the year in most areas and indeed the sexual stage has never been reported in this country. However, the sexual stage does occur in Sweden (Junell, 1967) and also in Switzerland (Blumer, 1967) and it may be that in these regions *E. fischeri* does not survive throughout the year as growing mycelium, particularly during the cold winter months.

Transfer of powdery mildew and rust fungi from infected to uninfected individuals is mainly effected by asexually produced spores which are wind borne and capable of long distance dispersal. Wind borne dispersal is random, and because susceptible plants in natural communities are dispersed among resistant genotypes of the same species as well as non-host species, spore wastage can be high. Thus to ensure that sufficient transmission of spores necessary for the maintenance of their populations occurs, powdery mildew and rust fungi must invest heavily in spore production. In theory, the extent of this investment could vary according to the frequency of susceptible host plants in the community, since a lower spore production per plant would be required to ensure new contacts for a host that is common than for one that is less common. However, there is no obvious evolutionary process which could adjust spore production specifically to host frequency, since natural selection should always favour those parasite isolates which produce the maximum number of viable and infective spores per plant. Clearly, very aggressive isolates which debilitate the host to the extent that the host's survival is threatened cannot be the fittest.

Plants have indeterminate growth and although the different growth stages may vary in resistance, all are generally susceptible to infection to some degree by virulent isolates. All infections above a certain level will inhibit plant growth to some extent, with the degree of inhibition varying with the level of infection and the rate at which it develops. Very aggressive isolates, which develop very heavy infections very rapidly and which inhibit all host growth subsequent to infection would be limited to those host energy resources available at the time of infection, while less aggressive isolates which do not inhibit continued growth would have the additional resources of any tissues produced subsequent to infection. Very aggressive isolates may have a higher rate of spore production in the short term than less aggressive isolates, but less aggressive isolates are likely to have additional host tissue to infect and therefore could produce the highest number of spores in total. Even if the less aggressive isolates have a lower reproductive output than the more aggressive isolates, they could still be more fit because by spreading spore production over a longer period their effective transmission could be the greater.

A study of spore production by *E. graminis* on cultivated oat *(A. sativa)* and wild oat *(A. fatua)* has shown that high initial rates of spore production are not necessarily associated with high total spore production, either per leaf, or per plant (Sabri and Clarke, 1996). In this study, spore production on the third leaf of one cultivated oat was found to occur at a slightly higher rate than on another cultivar, or on a line of wild oat, but the leaf senesced much more rapidly so that spore production ceased earlier and total production was little more than half that on the other two oats. Furthermore the heaviest infections of *E. fischeri* on *S. vulgaris* do not prevent the continued expansion of new leaves and dry matter accumulation by *S. vulgaris* (Ben-Kalio and Clarke, 1979; Harry and Clarke, 1992) indicating that the common isolates at least have limited aggressiveness.

Natural selection would only be expected to favour the less aggressive isolates in situations where host plants are generally infected by a single isolate so that direct competition between isolates does not occur. When plants are infected by more than one isolate, the isolates would be in competition and there is experimental evidence to indicate that natural selection then favours the most aggressive isolate (Kinkel, this volume). However, because, as indicated above, very aggressive isolates threaten their hosts' survival, multiple infections would be expected to be rare in natural pathosystems. Unfortunately, no information is available on the relative frequencies of single versus multiple infections of host plants in any natural pathosystem.

It is possible that mechanisms operate to reduce the incidence of multiple infections. For example, aggressive isolates, by reducing their host's reproductive output, will reduce the frequency of the host in its community and thus the probability of multiple infections. As the number of multiple infections tends towards zero so competition between isolates will become less common and highly aggressive isolates will become less fit.

If aggressive isolates become common the infection pressures on the host population will increase, threatening the survival of the host and thus its own survival and that of the association. The host could only survive in such circumstances by evolving high levels of partial resistance and ultimately total, non-host, resistance. However, because of seasonal changes in climate it is often the case that parasites are not active throughout the whole of their host's growing season. For example, *E. fischeri* infections are most common in the late summer and early autumn, whereas its main host, *S. vulgaris*, is common throughout the year. Periods of high infection pressure alternate with long periods of low infection pressure. The frequencies of single versus multiple infections are likely to vary throughout the year, with singly infected plants being more common during periods of low infection pressure and less common during periods of high infection pressure. The relative fitness of highly aggressive isolates would also be likely to vary, being high during periods of high infection pressure, when multiple infections may be common, but low during periods of low infection pressure, when plants are mainly singly infected. Furthermore, host genotypes which are particularly affected during periods of high infection pressures may recover during periods when infection pressures are low. *S. vulgaris* has substantial representation in the seed bank in the soil (Roberts, 1964) and so genotypes which, as growing plants, are decimated by infection during periods of high infection pressure will survive in the seed bank and their presence in the seed bank could be maintained by seed produced on plants growing during periods of low infection pressure. Finally, the relative fitness of different host resistance phenotypes could vary with environmental factors. For example, susceptible genotypes, of *S. vulgaris*, although much less fit than resistant genotypes during periods of high infection pressure, could be more fit than resistant genotypes during the winter when infection pressures are low.

CONCLUSIONS

Clearly, there are many gaps in our knowledge of the fitness attributes which ensure the long term survival of associations between plants and phytopathogens of their aerial tissues and how these attributes may have evolved. This final section attempts to pinpoint those aspects for further research where advances may help our understanding significantly.

Long term survival of an association between a phytopathogen and its host plant, is dependent upon each partner producing sufficient reproductive propagules to ensure their own survival. Any evolutionary change in either partner which threatens the survival of the other also threatens the association. Clearly, the fittest isolates of the parasite will be those which are aggressive enough to produce sufficient spores to maintain the population, but not so aggressive that they reduce the hosts reproductive output below that required for the host to maintain its population. There would appear to be constraints on the evolution of parasite aggressiveness in natural pathosystems but nothing is known about what these may or how they evolve. We need to know the relative frequencies of singly versus multiply infected plants and how these frequencies change throughout the year. We also need to know if the fitness of aggressive isolates changes with infection pressure.

Survival of the infected host clearly depends upon the evolution of levels of partial resistance to control that infection to permit adequate host reproduction. However, no work has been done on the extent to which host fitness in natural pathosystems varies with levels of partial resistance. It is possible that the costs to the host of partial resistance relative to parasite development are important factors but until we know more about the values of such costs it is not possible to draw firm conclusions.

REFERENCES

Ben-Kalio, V.D. and Clarke, D.D. 1979, Studies on tolerance in wild plants: effects of *Erysiphe fischeri* on the growth and development of *Senecio vulgaris*. *Physiological Pl. Path.* 14: 203-211.

Bevan, J.R., Crute, I.R. and Clarke, D.D. 1993a, Variation for virulence in *Erysiphe fischeri* from *Senecio vulgaris*. *Pl. Path.* 42: 622-635.

Bevan, J.R., Clarke, D.D. and Crute, I.R. 1993b, Resistance to *Erysiphe fischeri* in two populations of *Senecio vulgaris*. *Pl. Path.* 42: 636-646.

Bevan, J.R., Clarke, D.D. and Crute, I.R. 1993c, Diversity and variation in expression of resistance to *Erysiphe fischeri* in *Senecio vulgaris*. *Pl. Path.* 42: 647-653.

Blumer, S. 1967, *Echte mehltaupilze (Erysiphaceae)*. Gustav Fischer Verlag, Jena. 436pp.

Burdon, J.J. and Jarosz A.M. 1991, Host-pathogen interactions in natural populations of *Linum marginale* and *Melampsora lini*. 1, Patterns of resistance and racial variation in a large host population. *Evolution*. 45: 205-217

Clarke, D.D. 1986, Tolerance of parasites and disease in plants and its significance in host-parasite interactions, pp. 161-198 In: Ingram, D.S. and Williams, P.H. (eds.) *Adv. Pl. Pathol., Vol 5.*, Academic Press, London,

Campbell, F. 1990, Genetic interactions between *Erysiphe fischeri* (Blumer) and members of the genus *Senecio*. *Ph.D. thesis*, University of Glasgow.

Day, P.R. 1974, *Genetics of Host-Parasite Interaction*. W.H. Freeman, San Francisco.

Dinoor, A. 1977, Oat crown rust resistance in Israel. *Ann. N. Y. Acad. Sci.* 287: 357-366.

Ehrlich, P.R. and Raven, P.H. 1964, Butterflies and plants: a study in coevolution. *Evolution*. 18: 586-608.

Flor, H.H. 1956, The complementary genic systems in flax and flax rust. *Adv. Genetics*. 8: 29-54.

Harry, I.B. and Clarke, D.D. 1986, Race-specific resistance in groundsel (*Senecio vulgaris*) to the powdery mildew, *Erysiphe fischeri*. *New Phytol.* 103: 167-175.

Harry, I.B. and Clarke, D.D. 1987, The genetics of race-specific resistance in groundsel (*Senecio vulgaris*) to the powdery mildew fungus, *Erysiphe fischeri*. *New Phytol.* 107: 715-723.

Harry, I.B. and Clarke, D.D, 1992, The effects of powdery mildew (*Erysiphe fischeri*) infection on the development and function of leaf tissue by *Senecio vulgaris*. *Physiol. Mol. Pl. Pathol.* 40: 211-224.

Junnnel, L. 1967, *Erysiphaceae of Sweden*. Symbolae Botanicae Upsaliensis XIX, 1.

Moseman, J.G. 1956, Physiological races of *Erysiphe graminis* f.sp. *hordei* in North America. *Phytopathology* 46: 318-322.

Pirozynski, K.A. and Hawksworth, D.L. 1988, Coevolution of fungi with plants and animals: Introduction and overview, pp. 1-29 In: Pirozynski, K.A. and Hawksworth, D.L. (eds.) *Coevolution of Fungi with Plants and Animals*, Academic Press, London.

Preece, T.F. and Francis, S.M. 1987, *Albugo on Senecio vulgaris*. *Mycologist* 21: 71.

Roberts, H.A. 1964, Emergence and longevity in cultivated soil of seeds of some annual weeds. *Weed Res.* 4:296-307.

Roelfs, A.P. 1988, Genetic control of phenotypes in wheat stem rust. *Annu. Rev. Phytopathol.* 26: 351-367.

Sabri, N. and Clarke, D.D. 1996, The relative tolerances of wild and cultivated oats to infection by *Erysiphe graminis* f.sp. *avenae*: 1. The effects of infection on vegetative growth and yield. *Physiol. Mol. Pl. Pathol.* (in press).

Shafer, J.F. 1971, Tolerance to plant disease. *Annu. Rev. Phytopathol.* 9: 235-252.

Smedegaard-Peterson, V. and Stølen, O. 1981, Effect of energy-requiring defence reactions on yield and grain quality in a powdery mildew resistant barley. *Phytopathology*, 71: 396-399.

Tarr, S.A.J. 1972, *Principles of Plant Pathology*. Macmillan, London.

Wahl, I., Eshed, N.S.A. and Sobel, Z. 1978, Significance of wild relatives of small grains and other wild grasses in cereal powdery mildews, pp. 84-100 In: Spencer, D.M. (ed.) *The Powdery Mildews*, Academic Press, London.

Wilson, M. and Henderson, D.M. 1966, *British Rust Fungi*. Cambridge University Press.

Wolfe, M.S. and McDermott, J.M. 1994, Population genetics of plant pathogen interactions: The example of the *Erysiphe graminis-Hordeum vulgare* pathosystem. *Annu. Rev. Phytopathol.* 32: 89-113.

7

GENE TRANSFER BETWEEN MICRO-ORGANISMS IN THE PHYLLOSPHERE

Mark J. Bailey, Andrew K. Lilley, and Julian P. Diaper

NERC, Institute of Virology and Environmental Microbiology
Molecular Microbial Ecology Group
Mansfield Road
Oxford
OX1 3SR
United Kingdom

INTRODUCTION

When considering the factors that influence the population biology of micro-organisms in natural environments, such as the surface of plants (the phytosphere), it is reasonable to assume that active gene transfer takes place. But, despite a large number of reports describing the isolation of mobile genetic elements from natural microbial communities no data is currently available that allows the quantification of such events. The ability to predict relative transfer frequencies and determine the ecological significance of changing genotypes is relevant to microbial ecologists and necessary for assessing the potential impact and risk associated with the environmental release of large quantities of bacterial inocula or genetically modified micro-organisms (GMMs) (Tiedje *et al.*, 1989). The growing interest in the population genetics of natural bacterial communities (see Baumberg *et al.*, 1995) has led to an appreciation of the central role that mobile genetic elements play in the dissemination of information necessary for the evolution of microbial communities. Although the rate of assimilation and adaptation of communities varies, variation can be linked to a range of factors which are influenced by immediate and diffuse changes in the local environmental conditions. Examples of immediate adaptation include the rapid rate at which the acquisition of antibiotic drug resistance spreads through microbial populations of medical, veterinary and agricultural importance. It is generally agreed that such resistances are sustained by the imposed selective pressures of chemotherapy or the widespread application of antibiotics and anti-microbial compounds. The rate of response can be attributed to the plethora of mechanisms that have evolved to disseminate genetic information within and between communities.

The study of acquired drug resistance underpins modern bacterial genetics and led to the identification of three mechanisms for the horizontal transfer of genetic

Aerial Plant Surface Microbiology, edited by Morris et al.
Plenum Press, New York, 1996

material: conjugation, transformation and transduction. For more detailed descriptions of the importance of the environmental transfer of genes by plasmids or transposons (conjugation), by bacteriophage (transduction) or by the active uptake of free DNA (transformation) we recommend the following: Levy and Miller, 1989; Fry and Day, 1992; Clewell, 1993; Lorenz and Wackernagel, 1994; Kokjohn and Miller, 1992). Of these three mechanisms of gene transfer, conjugation has been considered most relevant to the phyllosphere (Farrand, 1989, 1992). Despite the interest shown in the processes of transformation and transduction in soil, rhizosphere and aquatic environments (Fry and Day, 1992; Levy and Miller, 1989) only a single investigation of phage mediated transfer among phylloplane bacteria has been described (Kidambi *et al.*, 1994). The reason for this bias is unclear but may, in part, reflect the relative intensity of research effort in these different habitats.

In the following chapter we discuss the importance of mobile genetic elements in phyllosphere microbiology in the context of their ecological relevance and the experimental approaches taken to detect and monitor actual gene transfer. Due to the paucity of available information and the focus of our own studies, we will concentrate on the description of the plasmid biology of plant-associated microbes.

The dissemination of genetic information plays a central role in the adaptation and survival of populations. Reanny (1978) proposed that bacterial populations represent a commonwealth of clones that share a plasmid gene pool. Adaptations to changes in the environment might result in the selection of a plasmid containing strain or the mobilisation of a plasmid (carrying a beneficial trait) through the community. Such mechanisms provide bacteria with the genetic plasticity necessary to extend their niche range and survive. Many studies provide direct evidence for the stable maintenance of host and accessory elements (referred to here as co-evolution), although the ecological rational for sustaining the division of traits between plasmid and chromosome has not been satisfactorily resolved. Microcosm and laboratory studies into the adaptive evolution of host and plasmid illustrate the potential benefits that plasmid maintenance may have for the survival of bacteria in the environment. Furthermore, plasmids carrying a variety of ecologically important traits, are widely distributed in phyllosphere bacterial populations (Table 1). However, the majority of plasmid functions remain cryptic due to our limited understanding of the role that extrachromosomal DNA plays in bacterial survival. Identifying these traits will be fundamental for explaining the persistence of plasmids in bacteria. It is reasonable to assume that plasmids of plant-associated bacteria encode traits which facilitate colonisation or survival (Coplin, 1989; Hozore and Alexander, 1991). Such traits may include chemotaxis, adherence to plant cells, agglutination, tolerance of low osmotic potentials, resistance to antibiotics, symbiotic nitrogen fixation, synthesis of plant growth regulators, catabolism (of plant sugars, organic acids or hydrocarbons), etc. (Coplin, 1989; Table 1). The structural variation observed within and between plasmids has also been interpreted as an additional phenotype class conferring the selective advantage of genetic variability (Levy, 1985; Amabile-Cuevas and Chicurel, 1992). Periods of selection for plasmid-conferred traits have been shown to produce co-evolution of plasmid and chromosome, with recovery of fitness of plasmid carrying cells and even reduced fitness of plasmidless segregants (Lenski, 1995; Modi and Adams, 1991; Bouma and Lenski, 1988; Helling *et al.*, 1981). The stability of co-evolution may lead to a host reservoir in the environment from which transfer may be initiated when suitable conditions occur.

Three approaches have been taken to demonstrate that gene transfer is an active process in the environment of the plant surface.

Table 1. Examples of traits conferred by plasmids of plant-associated bacteria. (For an extensive review of this topic refer to Coplin, 1989.)

Plasmid conferred trait	Bacterial Source	Plasmid	Reference
Crown gall and hairy root disease, T-DNA transfer, synthesis of auxins, cytokinins and utilisation of opines.	*Agrobacterium tumefaciens.*	Ti and Ri plasmids.	Steck and Kado, 1990; Farrand, 1993.
Cytokinin biosynthesis, probably a pathogenicity factor.	*Pseudomonas syringae pv savastanoi.*	pCK2, 105 kbp.	MacDonald *et al.*, 1986.
Production of auxins and oleander knots.	*Pseudomonas syringae pv savastanoi.*		Comai and Kosuge, 1980.
Production of the phytotoxin coronatine, may improve fitness of pathogens *in planta*.	*Pseudomonas syringae pv. various.*	pPT23A and others tra^{+}	Sato *et al.*, 1983 cited in Bender *et al.*, 1989.
Avirulence.	*Pseudomonas syringae pv. pisi.*	pAV212, 54 kbp.	
Catechol dissimilation.	*Rhizobium* sp.	pAMG1 tra^{+}.	Gajendiran and Mahadevan *et al.*, 1990.
Degradation of herbicide 2,4-dichloro-phenoxy acetic acid, 2-methyl-4-chlorophenoxyacetic acid, and 3-chlorobenzoate.	*Alcaligenes eutrophus.*	pJP4, 80 kbp, tra^{+}	Don and Pemberton, 1985.
Mercuric resistance.	Phyllosphere bacteria and rhizosphere bacteria.	five genetically distinct groups identified, tra^{+} 60- 380 kbp	Lilley *et al.*, 1994. This chapter.
Cu^{++} resistance.	*Pseudomonas syringae.*	pCPP501, 61 kbp, tra^{+}.	Bender *et al.*, 1990; Cooksey, 1990; Sundin *et al.*, 1989.
Resistance to streptomycin.	*Pseudomonas syringae pv. papulans.*	146 kbp.	Burr *et al.*, 1988.
Exopolysaccharide production, cell motility and acid tolerance	*Rhizobium leguminosarum.*		
Utilisation of adonitol, arabinose, catechol, inositol, lactose, malaye, rhamnose, sorbitol and nitrate.	*Rhizobium leguminosarum.*		Baldani *et al.*, 1992.
Tannin utilisation.	*Pseudomonas solanacearum.*	pAMB1.	Boominathan and Mahadevan, 1987.
Confers motility (chemotaxis) resistance to ampicillin, rhizocoenosis and adsorption to roots (exopolysaccharide production).	*Azospirillum brasilense.*	p90 139 kbp, tra^{+}.	Croes *et al.*, 1991.
Nitrogen fixation genes.	*Enterobacter agglomerans.*	pEA9 tra^{+}, 200 kbp.	Klingmüller *et al.*, 1990.
Root nodulation, nitrogen fixation.	*Rhizobium* spp	Sym plasmids tra^{+},	" "
carotenoid pigment production conferring UV resistance.	*Erwinia herbicola.*		Gantotti and Beer, 1982.
Cell motility.	*Rhizobium leguminosarum.*		Chen *et al.*, 1993; Baldani *et al.*, 1992.
Ice nucleation-active protein.	*Pseudomonas*		Lindow *et al* 1978.
Bacteriocin production.	*Erwinia herbicola.*		Jones *et al.*, 1988.
Production of herbicolin antifungal agent.	*Erwinia herbicola.*	pHER1065, 170 kbp.	Tenning *et al.*, 1993.
Entomopathogenic toxin production	*Bacillus thuringiensis*		Lereclus *et al.*, 1993

INDIRECT EVIDENCE

Plasmid Distribution, Phylogeny, and Interaction with Host Bacteria

Studies into plasmid distribution in the phyllosphere have utilised a number of techniques for their detection and characterisation, including the screening of isolates for a presumed phenotype, the physical disruption of cells and assessment of plasmid content by agarose gel electrophoresis. Plasmids isolated in this way may be further grouped in terms of total mass, or the number of different plasmids in an isolate, the restriction endonuclease pattern of plasmids and the extent of shared homology with known or presumed genes (Bender et al., 1990; Lim and Cooksey, 1993; Bailey et al., 1994; Kobayashi and Bailey, 1994; Sundin et al., 1994; Lilley et al., 1996). Such analyses confirm the genetic diversity of both plasmid and host genomes and their assortment in natural populations of phytosphere bacteria. Laguerre and co-workers (Laguerre et al., 1993) compared the distribution of Sym plasmids in populations of Rhizobium leguminosarum isolated from nodules of a variety of plants grown in the same soil type. This detailed study confirmed that recombination occurred, although specific plasmid-types and host-types were more common in isolates taken from a particular host plant species. However, the different Sym plasmids were found to be more host-plant specific than the bacteria indicating that, unlike some of the associated cryptic plasmids in these strains, the Sym plasmids were not tightly linked to the chromosomal types of R. leguminosarum and must have transferred at some point in time. Indeed, similar conclusions have been drawn from the description of a highly conserved plasmid (pTi2516) within a background of Agrobacterium tumefaciens strains (Nesme et al., 1992), and within Xanthomonas campestris and Pseudomonas syringae pathovars (Lazo and Gabriel, 1987; Curiale and Mills, 1983; Bender and Cooksey, 1986; Sundin and Bender, 1993) and copper and streptomycin resistant strains isolated from bactericide-treated orchards (Sundin et al., 1994).

The detection of the same or similar plasmids in different strains has generally been regarded as sufficient evidence that transfer of plasmid DNA between bacteria takes place. The most extensive studies of the distribution of related plasmids have concentrated on Rhizobium Sym plasmids (Young and Wexler, 1988) or the Ti plasmids found in Agrobacterium spp. (Paulus et al., 1989; Michel et al., 1990; Bouzar et al., 1993). Several groups have also investigated plasmids found in other phytosphere bacteria. For example Coplin et al. (1981), observed that 87% of Erwinia stewartii strains isolated from the phyllosphere contained common electrophoretic profiles of total undigested plasmid DNA, whereas Falkenstein et al. (1988) reported the presence of a 30 kbp plasmid in all isolates of Erwinia amylovora. Similarly Steinberger (1990) found a common 30 kbp plasmid in all isolates of E. amylovora as well as an additional 56 kbp plasmid in some isolates. The 30 kbp plasmid may reflect stable adaptation and maintenance with the chromosome whereas the 56 kbp plasmid in some E. amylovora isolates may indicate more recent plasmid transfer. Related plasmids have also been found within several pathovars of Pseudomonas syringae (Curiale and Mills, 1983; Quant and Mills, 1984; Bender and Cooksey, 1986) and other bacteria (Mogen et al., 1988). These reports suggest that related plasmids can be commonly found in members of the same bacterial group. Whether this commonality is a consequence of co-evolution between plasmid and chromosome or evidence of recent plasmid transfer is unclear. To address this point we have adopted a phylogenetic approach, illustrated below and proposed by Coplin (1989), to compare genotypes by cross-hybridisation and RFLP analyses of chromosomal and plasmid DNA isolated from the populations of interest.

In our laboratory we have established a collection of bacteria taken from the phyllosphere of sugar beet and grasses grown at the University Field Station, Wytham

(Oxfordshire) and from commercial fields in Northamptonshire (Powell *et al.*, 1993; Kobayashi and Bailey, 1994). The diversity of chromosomal and plasmid content has been examined over a 5 year period. These investigations have focused on *Erwinia* spp. as they are the second most abundant bacterial group in the phylloplane of sugar beet (Thompson *et al.*, 1993) and because all of the *Erwinia* isolates examined contain plasmid DNA. Individual isolates may contain between 1 and 6 different plasmids ranging from 10 to 150 kbp in size (unpublished). The distribution of plasmids and conserved regions of plasmid DNA within this collection has been determined by cross-hybridisation using labelled individual plasmids as probes against undigested and restriction enzyme digested plasmid DNA. By RFLP analysis the distribution of several unrelated plasmids within this collection has now been studied. Typically, the structure of the sampled community appears clonal, where identical strains were only isolated on a single sampling occasion. However, by grouping bacterial isolates according to ribotype, based on chromosomal RFLP patterns generated with different enzyme digests and probed with the 16s rRNA operon, a more predictable pattern of community succession was observed. A limited number of identical or closely related bacterial genotypes not only predominated on different plant species growing in the same experimental field plot but they also reappeared season to season, and were detected on sugar beet grown in geographically distant field sites. A comparison of plasmid and genome DNA-RFLP patterns, produced for a number of isolates that hybridised to one particular plasmid (pSF005, isolated from the Wytham site), revealed that they all belonged to closely related ribotypes. These isolates were collected each year of sampling at the Oxfordshire site and also from a Northamptonshire site, and formed a particular "ribogroup", ribogroup-EHPrGpI (Diaper and Bailey, in preparation). However, not all the members of this "ribogroup" contained related plasmid DNA. This observation was supported by similar results obtained using another plasmid as a probe (pSN3178, isolated from Northamptonshire). When the distribution of a third plasmid (pSB407, isolated from Wytham) was assessed it hybridised to a larger number of isolates which belonged to a wider variety of ribogroups (unpublished data). These observations illustrate that gene flux takes place and that the genetic pool is varied in both its content and apparent transfer activity. The diversity in the bacterial genome (plasmid and chromosome) may facilitate the persistence and survival of this group of bacteria in the extreme habitats provided by leaf surfaces. The close association of host and plasmid and their stability even under laboratory conditions may indicate the development of mutual tolerance, interdependence or a symbiotic relationship.

The distribution of related plasmids in different chromosomal backgrounds has led to the suggestion that the transfer may be limited to certain sub-populations where the transfer of plasmids between strains is compartmentalised allowing plasmids to move freely between some isolates but rarely between others (Young and Wexley, 1988). Our experiments with the *Erwinia* populations of the sugar beet phyllosphere support this view, although in contrast to the *Rhizobium* spp. Sym plasmids, the exact nature of the selection for the *Erwinia*-associated plasmids is unclear.

ISOLATION AND CHARACTERISATION OF SELF-TRANSMISSIBLE PLASMIDS

Isolation of Self-Conjugative Plasmids

In the previous section data were presented which provide strong indirect evidence for gene transfer *in planta*. However more direct studies are necessary if the mechanisms and frequencies of gene transfer are to be resolved. Self-conjugative plasmids have been isolated from bacteria associated with plants and rhizosphere soils. These plasmids confer a wide range of traits which, while not essential for bacterial growth and replication, may facilitate survival in the environment (Table 1).

The demonstration that a presumed plasmid-encoded phenotype can be transferred to suitable recipients - in the laboratory or in the plant - has been the basic approach to the study of plasmid-mediated gene transfer on plant surfaces. For example, this method has been used to isolate plasmids conferring mercury resistance (Kelly and Reanney, 1988) or copper resistance (Bender and Cooksey, 1986). Transfer proficient, cryptic plasmids have also been detected by the introduction of selectable marker genes (e.g., antibiotic resistance) carried on transposons (Obukowicz and Shaw 1985; Coplin and Rowen, 1978; Coplin, 1985). Alternatively, assays can be applied to determine whether natural isolates (containing a transposon-marked plasmid, or a plasmid carrying a known phenotype) can be mobilised by a second self-transferable plasmid (e.g., RP4) or are able to mediate the transfer of broad host range transfer negative (tra⁻) mobilisable (mob⁺) plasmids (i.e., IncQ, R300B) to suitable recipients. In one study one third of the *Erwinia* isolates collected from the sugar beet phyllosphere were shown to be transfer proficient and mobilised R300B to a fluorescent pseudomonad also isolated from the leaf surface (Powell *et al.*, 1993). Isolates may also be assessed for their ability to back-mobilise plasmids or retrotransfer plasmids from recipients, containing a transfer deficient plasmid, to conjugative donors (Mergeay *et al.*, 1987). Although this mechanism has not yet been reported for phyllosphere isolates it should be considered as a viable means of disseminating genetic information.

A more direct, one-step method has been developed for the isolation of tra⁺ plasmids encoding selectable traits from phytosphere bacteria (Lilley *et al.*, 1994). Transconjugant recipients, which had acquired resistance to a known (or assumed) natural plasmid-associated marker, were selected following the incubation of a suitable pseudomonad recipient in a plant tissue suspension. In our studies of sugar beet phytosphere microflora, resistance to mercury was adopted as the selectable marker because it is commonly associated with mobile genetic elements, (plasmids, transposons) (Silver and Walderhaug, 1992). The field site used for our investigations was, prior to the planting of small plots of sugar beet in 1989, pristine pasture with no history of mercury exposure. And although mercury resistance determinants are more common in bacteria indigenous to mercury-contaminated environments, they have been detected in bacteria of other unpolluted environments (Osborn *et al.*, 1993; Kelly and Reanney, 1984).

In the first reported use in the terrestrial environment of this one-step method for the isolation of tra⁺ plasmids we were able to isolate, conjugative plasmids encoding mercury resistance from the microflora of sugar beet phyllosphere and rhizosphere using rifampicin resistant *Pseudomonas putida* UWC1 as recipients (Lilley *et al.*, 1994). These conjugative mercury resistance plasmids (p*Mer* plasmids) were extracted from the UWC1 transconjugants and characterised, by numerical analyses of their restriction endonuclease digest patterns, into 5 genetically distinct groups (groups I to V) of 60 kbp to 383 kbp in size (Lilley *et al.*, 1996). All 100 isolates so far characterised confer the capacity for reductase detoxification of mercury. However, none of these plasmids conferred resistance to a range of antibiotics, to other heavy metals or to UV radiation. Plasmids of different sizes with minor differences in their REN fingerprint were identified in each of the five groups, indicating the presence of structural variants. Plasmid instability resulting in deletion, rearrangement and recombination has been noted in many genera including *E. coli* (Davies *et al.*, 1982), *Pseudomonas cepacia* (Gaffney and Lessie, 1987; Rochelle *et al.*, 1988), *Shigella flexneri* (Sasakawa *et al.*, 1986) and *Rhizobium leguminosarum* (Schofield *et al.*, 1987). These mechanisms of recombination and rearrangement are recognised as having adaptive value in plasmid evolution (Sykora, 1992; Lee *et al.*, 1984) as observed with the reassortment of genes which extend the catabolic capacity of bacteria (Top *et al.*, 1995). Increased mutability from transposon and insertion sequence activity has also been associated with plasmid adaptation, where established replicons are more likely to acquire new resistance determinants than previously rare plasmids are to proliferate (Datta and Hughes, 1983; Jones and Stanley, 1992). Although the function of the p*Mer* plasmids in the microflora from sugar beet phytosphere has not been resolved, we have been able to demonstrate that plasmids from three of the five groups identified could be isolated from the leaves and roots of sugar beet grown at the same field site for 4 consecutive years. The recurrence of these distinct plasmid groups demonstrates their established presence, suggesting that they play an ecologically significant role in this unpolluted habitat and that their natural hosts are transfer competent at the time of isolation. The availability and demonstration of the ubiquity of these indigenous mobile genetic elements facilitated the investigations into *in situ* gene transfer described in the following sections.

IN SITU EVALUATION OF CONJUGATIVE GENE TRANSFER IN THE PHYTOSPHERE

Overview

Despite the supporting data that comes from determining the extent of homology within and between isolates, the most convincing evidence for the occurrence of actual gene transfer on the plant surface can only be provided by actual experiments performed *in planta*. Surprisingly few evaluations have been undertaken in the phyllosphere. These assessments have been confined to observations of transfer between donors and recipients introduced, often at high densities, by spray misting, direct injection, infiltration or wounding. Often these experiments have been conducted with glasshouse- or growth chamber-propagated plants or on leaves, blossoms or seed pods removed from the plants. Although useful data can be collected from these careful studies, investigations have not been performed with combinations of indigenous bacteria and plasmids in a natural state. Many published investigations make use of the transfer constitutive, broad host range IncPα plasmid RP4, (isolated from a burns victim in a Birmingham hospital). RP4 or related plasmids, have been used to study conjugation in the phyllosphere (Lacy, 1978; Lacy and Leary, 1975; Lacy *et al.*, 1984; Manceau *et al.*, 1986; Björklöf *et al.*, 1995), the rhizosphere, soil and other environments. One criticism of these studies, where indigenous donor and recipients have been used in preference to exotic inocula such as *E. coli*, is the lack of evidence to suggest that RP4, like plasmids, are indigenous to unpolluted environments. Despite this limitation Manceau and his colleagues (Manceau *et al.*, 1986) undertook perhaps the first *in planta* investigation of conjugative plasmid transfer under field conditions. Bacteria were introduced to wound sites on year old shoots of hazelnut trees. The intraspecific transfer of RP4 among cells of the phytopathogenic strain of *Xanthomonas campestris* pv. *corylina*, K100.1 was determined for high, medium and low inoculation densities. Transfer of RP4 to the rifampicin resistant recipient was detected within 24 hours in wounds inoculated at high density; for intermediate or low concentrations of bacteria the appearance of transconjugants (transfer) was delayed by 28 and 116 days, respectively. The inocula persisted over the winter and recipient and transconjugant xanthomonad populations increased in the spring. Donor and transconjugant counts were always lower than those of the plasmid free recipient implying that either a small fraction of the community was active as recipients or the presence of the plasmid was a disadvantage for the host bacteria (Manceau *et al.*, 1986). Transconjugants were isolated from wound sites inoculated in the winter, spring and summer indicating that the physiological state of the host tissue was not important. However, transfer correlated with the actual density of bacterial populations and appeared to be governed, to some extent, by the frequency of contact between donor and recipient. Conjugation was less efficient when interspecific transfers were attempted: RP4 transferred from *Erwinia herbicola*, C1111b to *X. campestris* pv. *corylina*, K100.1 but not in the reverse cross from *X. campestris* to *E. herbicola*. During the investigation two RP4-containing natural epiphytic residents, *Pseudomonas fluorescens* and *Erwinia herbicola* were isolated on a single occasion in early December, some four months after the introduction of inocula.

Similarly, the transfer of RP1, an RP4-like plasmid, *in planta* was confirmed when 41 indigenous bacteria were isolated following the inoculation of 6 week old plants with *P. syringae* (RP1) donors (Björklöf *et al.*, 1995). These natural transconjugants represented 10 species from 5 genera. Under artificial conditions in growth chambers a number of investigations have also modelled the transfer of RP1 (RP4-like) between phyllosphere pseudomonads in excised bean pods and leaves (Lacy and Leary, 1975), between *E. chrysanthemi* in maize seedlings (Lacy, 1978) and between *E. herbicola* or *P. syringae* donors and *E. amylovora* recipients on pear blossom (Lacy *et al.*, 1984). In some instances transconjugants accounted for as many as 10% of potential recipients sampled from injected or mist sprayed bean pods (Lacy and Leary, 1975) or pear blossoms (Lacy *et al.*, 1984). The high transfer frequencies observed were similar to those recorded in filter mating controls performed on bacteriological agars. In another investigation, using excised radish and bean leaf discs, Knudsen *et al.* (1988) observed a transfer rate of approximately 10^{-4} transconjugants per recipient after 24 hours for the conjugative transfer of plasmid R388::Tn1721 between strains of *Pseudomonas cepacia*. They concluded that this rate of transfer correlated with cell density. In a recent study, RP1 transfer between populations of *P. syringae* growing epiphytically on beans was higher on the leaf surface under humid conditions, than in the soil or rhizosphere (Björklöf *et al.*, 1995). Factors that influence conjugation frequencies on the leaf included the availability of nutrients and the physical presence of the leaf surface.

Further direct evidence for *in planta* transfer comes from investigations using indigenous bacteria and naturally isolated, conjugative plasmids. Sundin *et al.* (1989) observed that the transfer frequency of a copper resistant 61 kbp plasmid between pseudomonad isolates collected from different cherry orchards on excised bean leaves was as low as 0.01% of the transfer frequency on bacteriological media. This confirmed that transfer is neither dependent on the selective pressure of copper nor is entirely dependent on the host plant type. The transferred plasmids remained stable in recipients, in the absence of selection, and transferred at higher frequencies to isolates taken from the same orchard, indicating a local adaptation between host bacteria and plasmid.

Other studies have also demonstrated that the root surface is a site where conjugative transfer of RP4 from pseudomonads can be promoted when compared to frequencies recorded in bulk soils (van Elsas *et al.*, 1988, 1989, 1990; Smit *et al.*, 1991; Richaume *et al.*, 1992a,b). Transfer in soils can also be stimulated by nutrient amendment or soil sterilisation (van Elsas *et al.*, 1989, 1990). However, not all plasmid transfers are promoted similarly. Klingmüller *et al.* (1991) and Klingmüller (1993) failed to detect transfer between *Escherichia coli* and *Alcaligenes eutrophus* on wheat roots, but were able to detect transfer of a transposon-labelled nitrogen fixation (*nif*) plasmid between homologous rhizosphere strains of *Enterobacter agglomerans* when soils were amended with sucrose, Luria broth, ground sugar beet or when the soil was steam-sterilised. In conclusion, these studies demonstrate that successful plasmid transfer in the phyllosphere, rhizosphere or soil is dependent on cell density, activity and nutrient availability. Transfer can be promoted by nutrient amendment, changes in the micro-climate (pH temperature, moisture), surface contact with clays or plant tissues, soil sterility and the size of inocula. All these factors correlate with improved survival or growth of donors and recipients.

In situ Investigations of p*Mer* Plasmid Transfer on the Surface of Sugar Beet Plants

To undertake ecologically relevant investigations of the factors that affect the incidence of conjugative gene transfer in any natural environment it is essential that investigations be conducted between recipient and donors indigenous to the selected habitat with a similarly natural plasmid. The donor (*Pseudomonas marginalis* 376N) and recipient (*Pseudomonas aureofaciens* 381R) bacteria, and the group I conjugative plasmid conferring mercury resistance (pQBR11, type plasmid for p*Mer*I group) used in the *in situ* studies described below were all isolated from the bacterial community of the sugar beet rhizosphere. Spontaneous nalidixic acid and rifampicin resistant mutants of these bacteria were used as donors and recipients of pQBR11 for *in situ* matings (Fig 1).

Fresh field soil was mixed with washed donor and recipient cells to give a soil mating mix (SMM) which was placed underground on the surface of a sugar beet root storage organ. Plasmid transfer in the SMM was determined after 24 h. In our studies transfer was demonstrated in 10 out of 12 occasions at frequencies between 1.3×10^{-8} and 5.1×10^{-5} transconjugants per recipient. Notably higher transfer frequencies (1.7×10^{-6} to 1.3×10^{-2})

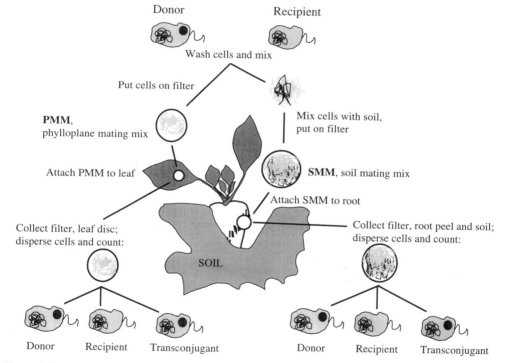

Figure 1. Protocol for *in situ* study of conjugative transfer on leaves and roots. Schematic diagram of the protocol adopted to evaluate plasmid transfer between bacteria introduced to the surface of mature sugar beet plants. Plants were grown in field soils and bacterial inocula prepared by washing saturated 16h broth cultures in Ringer's solution. Bacteria (donor and recipient) were mixed and applied to filters and attached to leaves, or mixed with soil, applied to filters and attached to the root surface (Lilley *et al.*, 1994). The illustration shows a cross section through the soil and the positioning of the filter on the surface of the sugar beet storage root. Filters were left in place for specified periods, typically 24h before collection and removal of bacteria. In all instances controls included SMMs or PMMs inoculated with only donor or recipient, and estimates of transfer frequencies between bacteria on filters applied to nutrient agar.

were recorded on the peel (the root surface) adjacent to the SMM. No transfer of mercury resistance was detected in SMM controls incubated *in vitro* or placed in soil at distances of more than 5 cm from plants (Lilley *et al.*, 1994). Transfer frequencies of 3.3 x 10^{-3} transconjugants per recipient were typical between these donors and recipients for filter matings on tryptic soy broth agar (TSBA). The difference in transconjugant frequencies observed between the peel and rhizosphere soil may have been related to a declining concentration of nutrient exudates away from the root. Filters imbibed with donors and recipients were also placed on leaves of different ages, (a phyllosphere mating mix, PMM). No transfer to recipient bacteria on the leaf was observed, although coincident transfer of the p*Mer* plasmid was recorded on the rhizoplane using SMMs (Diaper and Bailey, unpublished observations). Inocula applied to the leaves (approximately 10^5 colony forming units (cfu) cm^{-2}) survived, without transfer being detected, for the duration of the 6-day experiment (7.8 x 10^3 cfu cm^{-2}). Altering the position of donors and recipients had no effect; transfer was not observed after inoculation onto either the abaxial or adaxial leaf surface or on top of leaf veins. We have not been able to establish which factors are required to stimulate transfer between introduced inocula on the leaf surface. Nutrient deficiency and subsequent low activity of the inoculated cells may play an important role in the regulation of transfer, as the use of unwashed cells from overnight (16h) cultures results in some transfer.

In a second assay developed for evaluating conjugation in the phytosphere, equal numbers of washed donor and recipient bacteria were introduced as a dressing to pelleted sugar beet seed before they were planted in moistened vermiculite. The donor and recipient strains were the same as those used above in the SMM and PMM investigations. Triplicate samples were collected over a 21 day period (Fig 2).

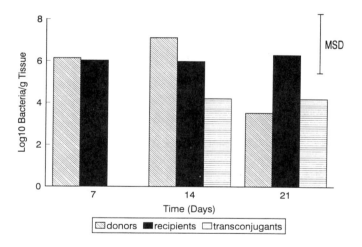

Figure 2. Gene transfer between introduced donor and recipient bacteria on sugar beet seedlings. Equal numbers (ca. 1 x 10^7 cfu/seed) of the donors, *Pseudomonas marginalis* 376N(pQBR11) and recipient *Pseudomonas aureofaciens* 381R were prepared from washed broth cultures, (16h, 28°C) were mixed briefly and applied to pelleted sugar beet seeds. Treated seeds were planted into moistened vermiculite held in large boiling tubes and incubated at 20°C under an 18h photoperiod. Plants were destructively sampled and the numbers of donors, recipients and transconjugants determined on plate count agar (PCA) supplemented with nalidixic acid -HgCl$_2$, rifampicin and rifampicin - HgCl$_2$ respectively. The combination of donor and recipient used in the seedling assay were the same as those used in the *in situ* rhizosphere transfer experiments described by Lilley *et al. (1994). Plasmid pQBR11 is the type isolate of the pMer group I plasmids that were isolated from the indigenous microflora of sugar beet, described above.

Both donors and recipients colonised the developing seedling. Transconjugants were not detected on germinating seedlings sampled 7 days after inoculation, but were found in the leaf and root samples collected on day 14 (limit of detection = 30 cfu/ seedling) at frequencies of 2.1×10^{-3} transconjugants/recipient. Experiments in vermiculite were not continued beyond 21 days. When the seedling assay was repeated with field soil similar results were obtained. The recipient persisted in the phyllosphere (1.5×10^5 cfu/g leaf) whereas donor levels fell below the limits of detection 35 days after sampling. Transconjugants were again detected in the developing phyllosphere (<28 days), but were absent from plants sampled after 35 days coincident with the development of true leaf sets. These data indicate that the seedling can be used as a suitable model for evaluating the transfer of indigenous plasmids between donor and recipient bacteria native to the studied habitat. The poor survival of transconjugants over time may reflect their failure to compete with the plasmid-free recipient. Furthermore, the limited period where transfer was observed in either the roots or leaves of the seedling may also reflect a defined period in the development of the plant that provides either the stimuli or conditions which favour the exchange of genetic material.

Although these *in situ* methods did not use donor and recipient populations in a natural physiological state, the results indicate that environmental plasmid transfer may not be an uncommon event provided bacteria are present in appropriate habitats.

TRANSFER OF CONJUGATIVE PLASMIDS BETWEEN ESTABLISHED BACTERIAL POPULATIONS IN THE PHYLLOSPHERE OF SUGAR BEET

The mechanisms by which plant surfaces may promote plasmid transfer between epiphytic populations of bacteria have not been resolved, and may involve exudates, high bacterial densities, chemotactic migration of bacteria, the provision of surfaces and rhizoplane biofilms. Despite these studies none of the experiments described above have established whether or not the plant itself has any direct effect on conjugative transfer. In this context, the original, and still the most pertinent, investigations into the *in planta* transfer of genetic traits were performed by Kerr (Kerr, 1969). Kerr demonstrated that the plasmid conferring the ability to induce tumours could be passed to avirulent strains of *Agrobacterium* introduced to crown-gall tumours containing virulent *A. tumefaciens*. Further work has now established that the conjugative transfer of Ti plasmids, which occurs at very low frequencies in filter matings, is specifically induced in the presence of opines. In the absence of opines transfer frequencies of less than 1×10^{-8} per donor are typical. However, after growth of donors in the presence of opines the frequency of transfer increased to 1×10^{-2} per donor (Kerr *et al.*, 1977). The molecular basis of the mechanisms and regulation of transfer has recently been described (Farrand, 1993; Piper *et al.*, 1993; Zhang *et al.*, 1993; Hwang *et al.*, 1995), and will not be discussed here.

In the following section we describe some of our studies of conjugation on the leaves and roots of plants. We approached these investigations with the intention of establishing whether transfer actually takes place at frequencies that can be measured.

Construction and Release of Genetically Modified Micro-Organisms (GMMs) Suitable for Monitoring Gene Transfer in the Phyllosphere

In order to clarify the risks associated with the release of GMMs and determine whether gene transfer occurs at ecologically significant rates, information on the survival, establishment and dissemination of bacteria in the environment is needed. To do this effectively, studies must be undertaken in natural habitats - rather than in microcosms - using indigenous micro-organisms as donors and recipients of transferable agents also isolated from the study site. In this context we have undertaken a series of experiments following the field release of a genetically modified pseudomonad. The candidate bacterium was selected from the natural community of fluorescent pseudomonads associated with the phyllosphere of sugar beet plants grown at the field site proposed for the release experiments. Studies on the microbial ecology and community succession over several consecutive seasons demonstrated that fluorescent pseudomonads were the most abundant population colonising sugar beet (Thompson *et al.,* 1993; Thompson *et al.* 1995a; Rainey *et al.,* 1994). A plasmid free strain, *Pseudomonas fluorescens* SBW25, non-pathogenic to plants and animals, was selected and genetically modified by the insertion of two marker gene cassettes (*lac*ZY and kan^r-*xyl*E) into two non-essential sites of the chromosome of SBW25 (Bailey *et al.,* 1995). The chosen markers allow unequivocal detection of the GMM, *P. fluorescens* SBW25EeZY6KX, on media containing selective substrates (*lac*ZY confers the ability to utilise lactose and *kan^r* confers resistance to kanamycin) at sensitivities of a single cell per gram of plant tissue or soil (Bailey *et al.,* 1996). These marker gene cassettes were introduced into two sites approximately 1 Mbp apart on the physically mapped 6.65 Mbp bacterial chromosome (Rainey and Bailey, 1995). The location of the different marker cassettes at different chromosomal sites ensured genetic stability and provided methods for detecting chromosomal gene transfer.

The fate of the inocula, introduced as a seed dressing to sugar beet, was assessed on two separate occasions at the same field site on consecutive growing seasons, the first in 1993 and again in 1994. The GMM successfully colonised the leaves and roots of field-grown plants where populations established in immature, mature and senescent leaves throughout the growing season (Thompson *et al.,* 1995b). Greatest population densities of the GMM were found on the emerging leaves located at the growing point of the plant where new leaf sets develop every 14-20 days. As leaves matured and expanded fewer GMMs could be isolated, although towards the end of the season an increase in the population size of the GMM was recorded in samples of senescing leaves (Thompson *et al.,* 1995b), suggestive of a mechanism of phyllosphere bacterial survival where population size increases in the soil in association with decaying plant material. GMMs could not be detected in the phytosphere of over-wintering plants although the density of total pseudomonads did not change significantly. When data from the two separate releases (1993 and 1994) were compared, it was apparent that the inocula became established at significantly higher densities in 1994 in both the rhizosphere and phyllosphere (Bailey *et al.,* 1996). In 1993 the GMM constituted a maximum of 6% of pseudomonads isolated from the phyllosphere of seedlings. In 1994 population densities for the GMM were up to 10,000 fold greater, and on average they represented 62% of the total pseudomonads isolated from emerging leaves sampled 100 days after planting. Later in the 1994 season (five months after planting) the density of GMMs was similar to that observed in 1993 for plants of a similar age. The reasons for the improved phytosphere colonisation by the GMM during the repeated field investigations in 1994 have not been clearly resolved. The inoculated seeds germinated at the same rate each year, but as a consequence of a prolonged hot, dry period in May 1994, the emergence of the first true leaf set was delayed by approximately 14 days compared to that observed in 1993. This delay may have been sufficient to allow pre-emptive colonisation of the seedlings by the GMM. The ability of the marked inoculum, *P. fluorescens* SBW25EeZY6KX, to effectively compete with other indigenous populations and colonise the phytosphere of sugar beet and the ease with which it and the inserted markers could be isolated and identified provided an ideal opportunity to study gene transfer in the phytosphere. All attempts to determine whether either of the marker gene cassettes had been transferred by conjugation, transduction or transformation to other resident pseudomonads were unsuccessful. However, gene transfer between the indigenous microflora and the GMM was observed as described below.

Conjugative Plasmid Transfer in the Phytosphere of Field Grown Sugar Beet

We have demonstrated the presence of a family of genetically diverse, conjugative mercury resistance plasmids in the sugar beet phytosphere that transfer on the plant surface (Lilley, 1994; Lilley et al., 1994, 1996). Therefore an essential consideration in the experimental design for monitoring the fate and impact of the released GMM was to evaluate whether it was able to function as a recipient of these plasmids.

Mercury resistant *P. fluorescens* SBW25EeZY6KX were isolated from the leaves and roots of mature sugar beet plants sampled after the release of this bacterium as a seed dressing in 1993 and again in the second release undertaken in 1994. All of the isolates examined contained conjugative plasmids carrying resistance to inorganic mercury (Bailey and Lilley, in preparation). The isolation of transconjugants appeared to be limited to a specific period in the development of the plants as mercury resistant GMMs were only isolated between 50 and 100 days after the appearance of the first true leaf set. The apparent absence of transconjugant GMMs (limit of detection = 20 cfu / g) in seedlings and immature plants may reflect the ecological status of the indigenous donor population. In studies where the transfer proficiency of the microflora of sugar beet has been monitored over entire seasons p*Mer* plasmids were only isolated from mature plants sampled up to 180 days after germination (Lilley and Bailey, in preparation). Another important observation was that the actual number of transconjugant GMMs isolated each year was not significantly different, although considerably higher population densities of the GMM (recipient) were recorded in the 1994 release. From these findings we infer that the spatial density and spatial distribution of donors (and recipients) strongly influence the transconjugant frequency. This will be discussed towards the end of this chapter. The greatest number of transconjugants was isolated from immature leaves although transconjugants were also detected in the rhizosphere. In one sampling occasion in 1994, transconjugants were isolated from seven out of the nine plants sampled on a single occasion (102 days after release), demonstrating the ubiquity of these mobile elements within the microbial community at the field site. Molecular characterisation of the isolated plasmids, by comparing restriction enzyme digestion patterns on agarose gels, demonstrated that they belong to three of the five p*Mer* groups identified at the field site as described above (see Lilley et al., 1996). Cross-hybridisation studies with labelled whole plasmid DNA also confirmed the separation of these groups and revealed, despite their size of <350 kbp, that these plasmids share little or no homology. The lack of homology, and the differences observed in the restriction fragment patterns for many of the isolates sampled from the same plant (Lilley and Bailey, manuscript in preparation) was taken to indicate that numerous individual transfer events had taken place.

The detection of these transfer events demonstrates the value of GMMs for "mark and recapture" type experiments in the natural environment. The absence of known antibiotic selection (lack of detectable mercury pollution) allows the speculation that the plant habitat influences the movement of these p*Mer* plasmids and provides some unresolved advantage to the recipient. Whatever the "advantage" they seem to have, transconjugants do not appear to proliferate and do not persist as well as recipients. Nevertheless, actual gene transfer to a natural community of bacteria has been demonstrated. This is an exciting observation as it demonstrates further the genetic plasticity of natural microbial communities and that predictions of the fate and genetic stability of inocula cannot be guaranteed if, as appears in our studies, gene transfer is in fact a common occurrence.

FACTORS OF THE PHYSICAL AND BIOLOGICAL ENVIRONMENT IMPORTANT FOR PLASMID TRANSFER

Nutrients and the metabolic condition of bacteria have been identified as factors that influence plasmid transfer. Another important factor influencing transfer, which we wish to consider here, is the spatial distribution of micro-organisms in the phytosphere. The physical distribution of bacteria in the environment is important as conjugative plasmid transfer requires that competent donors and recipients come into physical contact. Rovira *et al.* (1974) using direct microscopy found for 8 plant species that bacteria covered only 4% to 10% of the root surface. This spatial distribution was shown statistically, by Newman and Bowen (1974), not to be random, but to be aggregated with bacteria probably located at sites of nutrient exudation. The aggregation of bacteria on root and leaf surfaces may facilitate local plasmid transfer, though any constraint on mobility between these habitats (such as low moisture levels) could contain the dispersal of bacteria and hence the spread of a plasmid. Donor and recipient densities are known to influence plasmid transfer on plates (Rochelle *et al.*, 1988), in soil and rhizosphere microcosms (van Elsas *et al.*, 1988; Kinkle and Schmidt, 1991; Ramos-Gonzalez *et al.*, 1991) and in hazelnut tissues (Manceau *et al.*, 1986). It is common for these and other studies to report no transfer or reduced transfer at low donor and recipient densities. Clearly mechanisms that influence dispersal and the distribution of bacteria in the environment contribute significantly to net gene flow.

Understanding the distribution and diversity of bacteria, particularly on the leaf surface, is vital to our understanding of gene mobilisation. The existence of aggregates, or biofilms, of micro-organisms on plant leaf surfaces has been confirmed by direct microscopic examination (Gras *et al.*, 1994). It is probable that, within such aggregates, the rate of active gene transfer would be dictated not only by the relative distribution and density of cells but also by the frequency that micro-colonies coincide or exchange immigrants. If it is more normal for phyllosphere bacteria to exist as diverse aggregates then the content, position and relative activity of the component species is as important as the rate of immigration. A microbial aggregate from the leaf surface of broad-leaved endive has been isolated and was found to be composed of yeast, bacteria and fungi (Morris *et al.*, 1994). A detailed analysis of the 139 bacterial strains of this aggregate demonstrated genetic and plasmid diversity within the bacterial genera (Jacques *et al.*, unpublished). When the phylogeny of component fluorescent pseudomonads was compared, by standard ribotyping methods, a number of related but distinctive genotypes were identified. Particular pseudomonad isolates from the sample, resistant to inorganic mercury, contained an identical plasmid of approximately 60 kbp. The plasmid was present in 50% of the pseudomonads isolated from the micro-aggregate, and was detected in at least one representative of the six different pseudomonad ribotypes identified (Jacques *et al.*, unpublished). Because of the distribution of this identical plasmid within different host genomic backgrounds we presume that plasmid transfer has taken place in the phytosphere and interpret the findings as supportive evidence that microbial aggregates are a hot spot for gene transfer on the plant surface.

Our own results indicate that the extent of spread of plasmids within the GMM population on a plant is related to the type and age of a plant surface. A possible factor is that transconjugants occur in some tissues within discrete aggregates and cannot transfer plasmids to other similarly isolated groups (islands) of cells. The immature leaves and apical bud of sugar beet, for example, are noticeably sticky with "exudates" which may provide a continuous fluid film to support bacteria. This film may then fracture as the leaf expands producing islands of discrete aggregates which become the dominant form of microbial distribution on mature leaves.

Bacterial densities are significant not only for the likelihood of cell-to-cell contact but also because gene expression can be influenced by cell density dependent factors. Quorum sensing has been clearly shown to regulate *Agrobacterium* Ti plasmid transfer. The conjugation factor (CF) identified in the culture supernatant of transfer constitutive strains of *A. tumefaciens* (Zhang and Kerr, 1991) is required for activation of Ti plasmid *tra* gene transcription. CF is an autoinducer and has been identified as a homoserine lactone, HSL (Zhang *et al.*, 1993; Piper *et al.*, 1993). Autoinduction mediated by an activator and a cognate substituted HSL is recognised as a common environmental sensor that activates transcription under conditions of high cell density (Hwang *et al.*, 1995).

CONCLUDING REMARKS

The presence of transferable plasmids in the phytosphere conferring ecologically significant traits is well established. Microcosm and field experiments have shown plant-associated plasmid transfer between specified strains and that roots promote transfer in soil. We have shown that three groups of plasmids conferring resistance to mercury persist in the phytosphere of 4 annual sugar beet crops and that roots and seedlings actively promote transfer of a group I p*Mer* plasmid. Acquisition of p*Mer* plasmids from all 3 groups by the GMM (fluorescent pseudomonad) colonising 2 successive crops has further shown that transconjugants do occur "normally" *in planta* and at detectable frequencies.

While *in situ* and microcosm experiments have shown that conjugation can occur at quite high frequencies, epidemiological data indicates that new stable associations of plasmids and hosts may be somewhat rarer. Many plasmids in phytosphere bacteria have been observed to be highly stable (Coplin 1989; Coplin *et al.*, 1981; Mogen *et al.*, 1988). These persistent plasmid-host associations imply either that the advantage of carrying a plasmid outweighs the costs, or that co-evolution has integrated plasmid regulation and expression of traits into the overall harmony of cell regulation (Modi and Adams, 1991; Bouma and Lenski, 1988). Alternatively, it has been proposed that tra⁺ plasmids encoding no selective advantage may persist in a community if the cell density and the conjugative transfer rate are sufficient for infectious transmission to outweigh losses through segregation and selection (Stewart and Levin, 1977; Simonsen, 1991). Though Smets *et al.* (1993) have suggested that these conditions apply to the TOL plasmid, there is little evidence from the phytosphere regarding this possibility.

Some plasmids of phytosphere bacteria have been observed to be less stable (Albiach and Lopez 1992; Gaffney and Lessie, 1987). Plasmids will, of course, occur in opportunistic relationships with relatively new hosts following conjugation or mobilisation. Many such transconjugants will not persist without specific selective conditions. A similar result may have occurred in our field experiments where p*Mer* plasmid transconjugants only appeared on roots and leaves of mature, mid-season plants. To understand the persistence of plasmids and their fixation in communities it is essential to identify what traits they confer on their hosts. It is plausible that traits, such as the ability to fix nitrogen, synthesise plant growth regulators, confer resistance to antibiotics, catabolise (plant sugars, organic acids, hydrocarbons) and direct cellular chemotaxis have clear adaptive advantage to bacteria of the phytosphere. These traits may be of only occasional value and sometimes may be adaptive in other environments. Though many traits are known, the size of environmental plasmids often indicates a considerable coding capacity to which functions have not been assigned and therefore the significance and behaviour of these plasmids is not well understood. Plasmids also confer the less easily assessed trait of increased phenotypic and geneotypic plasticity on bacterial communities. They do this by acting as carriers of an extended gene pool and as sources of variability providing a mobile self replicating extra-chromosomal site for recombination.

Many factors affect the frequency with which transconjugants appear, their persistence and significance to host ecology. Progress with the development of techniques in microbial ecology, bacterial population genetics and studies in the organisation of the micro-niche will all continue to contribute to our understanding of these factors. This will assist in explaining the high diversity of plasmids whose various behaviours may be judged to be somewhere from the pathogenic or parasitic (such as phage) to mutualistic (as in extra-chromosomal element) or symbiotic. Important aspects of a plasmid's characterisation include not only its interactions with its host but also its regulation of conjugation. In one succinct description, Wilkins (1995) observed that "there is increasing evidence that conjugation systems are adapted through their mating apparatus and regulatory controls to promoting gene transfer between particular groups of organisms in preferred ecological conditions".

THE FUTURE, WHERE DO WE GO FROM HERE?

There is an urgent need to develop our understanding of the diversity and distribution of plasmids in the context of the host cell background as well as establishing the significance of plasmid function in bacterial and plant ecology. Such approaches will require the co-operation of scientists from a broad range of biological disciplines including both microbial ecologists and molecular geneticists. Key features of any such efforts should consider the following:

- The significance of plasmids to host ecology.
- Integration of recent advances in population genetics and studies of the spatial distribution of phytobacteria.
- Development of molecular tools to define the functional backbone of these plasmids so as to create a phylogeny of replicons.
- Elaboration of the conjugative apparatus and its regulation for a range of plasmids.
- Improved recognition of plasmid-conferred traits.
- Studies of the molecular signalling between bacteria and plants.

ACKNOWLEDGMENTS

This work has been undertaken following support by NERC core funding, EU-Biotechnology awards (AKL, Biot2 - CT92-0491), Ministry of Agriculture Fisheries and Food (JPD, RG 0101) and Department of the Environment research contracts (PECD 6/8/143 and PECD 7/8/61). The views expressed in this review are those of the authors and not necessarily those funding agencies or NERC.

REFERENCES

Albiach, M. R. and Lopez, M. M. 1992, Plasmid heterogeneity in Spanish isolates of *Agrobacterium tumefaciens* from 13 different hosts. *Appl. Environ. Microbiol.* 58:2683-2687.
Amabile-Cuevas, C. F. and Chicurel, M. E. 1992, Bacterial plasmids and gene flux. *Cell.* 7:189-199.

Bailey, M. J., Kobayashi, N., Lilley, A. K., Powell, B. J. and Thompson, I. P. 1994, Potential for gene transfer in the phytosphere: Isolation and characterisation of naturally occurring plasmids. pp. 77-98. In: Bazin, M. J. and Lynch, J. M., (eds.). *Environmental Gene Release*. Chapman Hall, London.

Bailey, M. J., Lilley, A. K., Thompson, I. P., Rainey, P. B. and Ellis, R. J. 1995, Site directed chromosomal marking of a fluorescent pseudomonad isolated from the phytosphere of sugar beet; stability and potential for marker gene transfer. *Molec. Ecol.*. 4:755-764.

Bailey, M. J., Lilley, A. K., Ellis, R. J. Bramwell, P. A., Thompson, I. P. 1996, Microbial ecology, inoclula distribution and gene flux within populations of bacteria colonising the surface of plants: Case study of a GMM field release in the UK. In: Van Elsas, J. D., Wellington, E. M. and Trevors, J. T. (eds.). *Modern Soil Microbiology*. Marcel Dekker, New York. (in press).

Baldani, J. I., Weaver, R. W., Hynes, M. F. and Eardly, B. D. 1992, Utilization of carbon substrates, electrophoretic enzyme patterns and symbiotic performance of a plasmid-cured clover rhizobia. *Appl. Environ. Microbiol.* 58:2308-2314.

Baumberg, S., Young, J. P. W., Wellington, E. M. H. and Saunders, J. R. (eds.) 1995, *Population Genetics of Bacteria*. Society for General Microbiology Symposium 52. Cambridge University Press, Cambridge. 348 pp.

Bender, C. L., Malvick, D. K. and Mitchell, E. R. 1989, Plasmid mediated production of the phytotoxin coroatine in *Pseudomonas syringae* pv. *tomato*. *J. Bacteriol*. 171:807-812.

Bender, C. L. and Cooksey, D. A. 1986, Indigenous plasmids in *Pseudomonas syringae* pv. *tomato*: Conjugative transfer and role on copper resistance. *J. Bacteriol*. 165:534-541.

Bender, C. L., Malvick, D. K., Conway, K. E., George, S. and Pratt, P. 1990, Characterization of pXv10A, a copper resistance plasmid in *Xanthomonas campestris* pv. *vesictoria*. *Appl. Environ. Microbiol.* 56:170-175.

Björklöf, K., Suoniemi, A., Haahtela, H. and Romantschuk, M. 1995, High frequency of conjugation verses plasmid segregation of RP1 in epiphytic *Pseudomonas syringae* populations. *Microbiol*. 141. 2719-2727.

Boominathan, K. and Mahadevan, A. 1987, Plasmid encoded dissimilation of condensed tannin in *Pseudomonas solanacearum*. *FEMS Microbiol. Ecol.* 40:147-150.

Bouma, J. E. and Lenski, R. E. 1988, Evolution of a bacteria plasmid association. *Nature*. 335:351-352.

Bouzar, H., Ouadah, D., Krimi, K., Jones, J. F., Trovato, M., Petit, A. and Dessaux, Y. 1993, Corretative association between resident plasmids and the host chromosome in a diverse *Agrobacterium* soil population. *Appl. Environ. Microbiol.* 59:1310-1317.

Burr, T. J., Katz, B., Hoying, S. A., Wilcox, W. F. and Norelli, J. L. 1988, Streptomycin resistance of *Pseudomonas syringae* pv. *papulans* in apple orchards and its association with a conjugative plasmid. *Phytopathology*. 78:410-413

Chen, H., Gartner, E. and Rolfe, B. G. 1993, Involvement of genes on a mega plasmid in the acid-tolerant phenotype of *Rhizobium leguminosarum* biovar *trifolii*. *Appl. Environ. Microbiol*. 59:1058-1064.

Clewell, D. B. (ed.) 1993, *Bacterial Conjugation*. Plenum Press, New York. 413 pp.

Comai, L. and Kosuge, T. 1980, Involvement of plasmid deoxyribonucleic acid in indolacetic acid synthesis in *Pseudomonas savastanoi*. *J. Bacteriol*. 143:950-57.

Coplin, D. L. 1985, Characterisation of a conjugative plasmid from *Erwinia stewertii*. *J. Gen. Microbiol.* 131:2985-2991.

Coplin, D. L. 1989, Plasmids and their role in the evolution of plant pathogenic bacteria. *Annu. Rev. Phytopathol*. 27:187-212.

Coplin, D. L., Chisholm, D. A., Rowan, R. G. and Whitmoyer, R. E. 1981, Characterization of plasmids in *Erwinia-stewartii*. *Appl. Environ. Microbiol*. 42:599-604.

Croes, C., van Bastelaere, E., Declercq, E., Eyers, M., van Derleyden, J. and Michiels, K. 1991, Identification and mapping of loci involved in mobility, adsorption to wheat roots, colony morphology, growth in minimal medium on the *Azospirillum brasilense* Sp7 90-MDa plasmid. *Plasmid* 26:83-93.

Curiale, M. S. and Mills, D. 1983, Molecular relatedness among cryptic plasmids in *Pseudomonas syringae* pv. *glycinea*. *Genetics* 73:1440-1444.

Datta, N. and Hughes, V. M. 1983, Plasmids in the same Inc groups in Enterobacteria before and after the medical use of antibiotics. *Nature*. 306:616-617.

Davies, D. L., Binns, M. M. and Hardy, K. G. 1982, Sequence rearrangements in the plasmid colv, I-K94. *Plasmid*. 8:55-72.

Don, R. H. and Pemberton, J. M. 1985, Genetic and physical map of the 2-4-dichlorophenoxyacetic acid degradative plasmid pjp4. *J. Bacteriol*. 16:466-468.

Falkenstein, C., Bellemann, P., Walter, S., Zeller, W. and Geider, K. 1988, Identification of *Erwinia amylovora*, the fireblight pathogen, by colony hybridization with DNA from plasmid pEA29. *Appl. Environ. Microbiol.* 54:2798-2802.

Farrand, S. K. 1989, Conjugal transfer of bacterial genes on plant, pp.261-285 In: Levy, S.B. and Miller, R.V. (eds.), *Gene Transfer in the Environment*. McGraw-Hill, New York.

Farrand, S.K. 1992, Conjugal gene transfer on plants, pp. 345-362 In: Levin, M. A. Seidler, R. J. and Gogul, R. (eds.), *Microbial Ecology: Principles, Methods and Applications*. McGraw-Hill, New York.

Farrand, S. K. 1993, Conjugal transfer of agrobacterium plasmids. pp. 255-285 In: Clewell, D.R. (ed.), *Bacterial Conjugation*, Plenum Press, New York.

Fry, J. C. and Day, M. J. (eds.) 1992, *Release of Genetically Engineered and Other Micro-organisms*. Cambridge University Press. Cambridge. 178 pp.

Gaffney, T. D. and Lessie, T .G. 1987, Insertion-sequence-dependent rearrangements of *Pseudomonas cepacia* plasmid pTGL1. *J. Bacteriol.* 169:224-230.

Gajendiran N. and Mahadevan, A. 1990, Plasmid borne catechol dissimilation in *Rhizobium* sp. *FEMS Microbiol. Ecol.* 73:125-130.

Gantotti, B. V. and Beer, S. V. 1982, Plasmid borne determinants of pigmentation and thiamine prototrophy in *Erwinia herbicola*. *J. Bacteriol.* 151:1627-1629.

Gras, M.H., Druet-Michaud, C. and Cerf, O. 1994, La flore bactérienne des feuilles des salades fraîches. *Science des Aliments* 14:173-188.

Helling, R. B., Adams, J. and Kinney, T. 1981, The maintenance of plasmid-containing organisms in populations of *Escherichia coli*. *J. Gen. Microbiol.* 123:129-141.

Hozore, E. and Alexander, M. 1991, Bacterial characteristics important to rhizosphere competence. *Soil Biol Biochem.* 23:717-723.

Hwang, I., Cook, D. M. and Farrand, S. K. 1995, A new regulatory element modulates homoserine lactone-mediated autoinduction of Ti plasmid conjugal transfer. *J. Bacteriol.* 177: 449-458.

Jones, C. and Stanley, J. 1992, *Salmonella* plasmids of the pre-antibiotic era. *J. Gen. Microbiol.* 138:189-197.

Jones, D. A., Ryder, M. H., Clare, B. G., Farrand, S. K. and Kerr, A. 1988, Construction of a tra deletion mutant of pAgK84 to safeguard the biological control of crown gall. *Mol. Gen. Genet.* 212:207-14.

Kelly, W. J. and Reanney, D. C. 1984, Mercury resistance among soil bacteria: ecology and transferability of genes encoding resistance. *Soil. Biol. Biochem.* 16:1-8.

Kerr, A. 1969, Transfer of virulence by non-pathogenic isolates of *Agrobacterium*. *Nature* (London) 223:1175-1176.

Kerr, A., Manigault, P. and Tempé, J. 1977, Transfer of virulence *in vivo* and *in vitro* in *Agrobacterium*. *Nature* (London) 265:560-561.

Kidambi, S. P., Ripp, S. and Miller, R. V. 1994, Evidence for phage-mediated gene transfer among *Pseudomonas aerugenosa* strains on the phylloplane. *Appl. Environ Microbiol.* 60:496-500.

Kinkle, B. K. and Schmidt, E. L. 1991, Transfer of a pea symbiotic plasmid pJB5JI in nonsterile soil. *Appl. Environ. Microbiol.* 57:3264-3269

Klingmüller, W. 1991, Plasmid transfer in natural soil - a case by case-study with nitrogen-fixing *Enterobacter*. *FEMS Microbiol. Letts.* 85:107-115.

Klingmüller, W., Dally, A., Fentner, C. and Stienlein, M. 1990, Plasmid transfer between soil bacteria. pp 133-151 In: Fry, J.C. and Day, M.J. (eds.), *Bacterial Genetics in Natural Environments*. Chapman and Hall, London.

Klingmüller, W. 1993, Plasmid transfer in a natural soil, as stimulated by sucrose, wheat rhizosphere and ground sugar beets: a study with nitrogen fixing *Enterobacter*. *Microb. Rel.* 1:229-235.

Knudsen, G.R., Armstrong, J. L., Porteous, L. A., Prince, V. J. Seidler, R. J. and Walter, M. V. 1988, Predictive model of conjugative plasmid transfer in the rhizosphere and phyllosphere. *App. Environ. Microbiol.* 52:343-347.

Kobayashi, N. and Bailey, M. J. 1994, Plasmids isolated from the sugar beet phyllosphere show little or no homology to molecular probes currently available for plasmid typing. *Microbiol.* 140:289-296.

Kokjohn, T. A. and Miller, R. V. 1992, Gene transfer in the environment: transduction. pp. 54-81 In: Fry, J. C. and Day, M. J., (eds.), *Release of Genetically Engineered and Other Micro-organisms*. Cambridge University Press, Cambridge.

Lacy, G. H. 1978, Genetic studies with plasmid RP1 in *Erwinia chrysanthemi* strains pathogenic on maize. *Phytopathology.* 68:1323-1330.

Lacy, G. H. and Leary, J. V. 1975, Transfer of antibiotic resistance plasmid RP1 into *Pseudomonas glycinea* and *Pseudomonas phaseolicola in vitro* and *in planta*. *J. Gen. Microbiol.* 88:49-57.

Lacy, G. H., Stromberg, V. K. and Cannon, M. P. 1984, *Erwinia amylovora* mutants and *in planta* derived transconjugants resistant to oxytetracyclin. *Can J. Plant Path.* 6:33-39.

Laguerre, G., Mazurier, S. I. and Amarger, N. 1993, Plasmid profiles and restriction-fragment-length-polymorphism of *Rhizobium-leguminosarum* bv. *viciae* in field populations. *FEMS Microbiol. Letts.* 101:17-26.

Lazo, G. R. and Gabriel, D. W. 1987, Conservation of plasmid DNA sequences and pathovar indentification of strains of *Xanthomonas campestris*. *Phytopathology* 77:448-453.

Lereclus, D., Delecluse, A. and Lecadet, M. M. 1993, Diversity of *Bacillus thuringiensis* toxins and genes, pp. 37-70 In: Entwistle, P., Cory, J. S., Bailey, M. J., Higgs, S. (eds.), *Bacillus thuringiensis, an Environmental Biopesticide: Theory and Practice.* Wiley and Sons, New York.

Lee, S. C., Cleary, P. P. and Gerding, D. N. 1984, Plasmid macroevolution in a nosocomial environment - demonstration of a persistent molecular polymorphism and construction of a cladistic phylogeny on the basis of restriction data. *Mol. Gen. Genet.* 194:173-178.

Lenski, R. E. 1994, Evolution in experimental populations of bacteria. pp. 193-215 In: Baumberg, S., Young, J.P.W., Wellington, E.M.H and Saunders, J.R. (eds.), *Population Genetics of Bacteria.* Society for General Microbiology Symposium 52. Cambridge University Press, Cambridge.

Levy, S. B. 1985, Ecology of plasmids and unique DNA sequences. pp 56-88 In: Halvorson, H.O., Pramer, D. and Rogul, M. (eds), *Engineered Organisms in the Environment: Scientific Issues.* American Society for Microbiology. Washington D.C.

Levy, S. B. and Miller, R. V. (eds). 1989, *Gene Transfer in the Environment.* McGraw-Hill Publishing Company, New York. 434 pp.

Lilley, A. K., Fry, J. C., Day, M. J. and Bailey, M. J. 1994, *In situ* transfer of an exogenously isolated plasmid between *Pseudomonas* spp. in sugar beet rhizosphere. *Microbiol.* 140:27-33.

Lilley, A. K., Bailey, M. J., Day, M. J., and Fry, J. C. 1996, Diversity of mercury resistance plasmids obtained by exogenous isolation from the bacteria of sugar beet in three successive seasons. *FEMS Microbiol. Ecol.* 20: 211-228.

Lim, C. K. and Cooksey, D. A. 1993, Characterisation of chromosomal homologs of plasmid borne copper resistance in *Pseudomonas syringae*. *J. Bacteriol.* 175:4492-4498.

Lindow, S. E. and Panopoulos, N. J. 1988, Field tests of recombinant ice⁻*Pseudomonas syringae* for biological frost control in potato. pp. 121-138 In: Sussman, M., Collins, C.H., Skinner, F.A. and Stewart-Tull, D.E.S. (eds.), *The Release of Genetically Engineered Micro-organisms.* Academic Press, London.

Lorenz, M. G. and Wackernagel W. 1994, Bacterial gene transfer by natural genetic transformation in the environment. *Microbiol. Rev.* 58:563-602.

Macdonald, R. M. 1986, Sampling soil microfloras- problems in estimating concentration and activity of suspensions of mixed populations of soil-micro-organisms. *Soil Boil. Biochem.* 18:411-416.

Manceau, C., Devaux, M. and Gardan, L. 1986, Dynamics of RP4 plasmid transfer between *Xanthomonas campestris* pv. *corylina* and *Erwinia herbicola* in hazelnut tissues, *in planta. Can. J. Microbiol.* 32:835-841.

Mergeay, M, Lejeune, P., Sadouk, A., Gerits, J. and Fabry, L. 1987, Shuttle transfer (or retrotransfer) of chromosomal markers mediated by plasmid pULB113. *Mol. Gen. Genet.* 209: 61-70.

Michel, M-F., Brasileiro, A. C. M., Depierreux, C., Otten, l., Delmotte F. and Jouanin, L. 1990, Identification of different *Agrobacterium* strains isolated from the same forest nursery. *Appl. Environ. Microbiol.* 56:3537-3545.

Modi, R. I. and Adams, J. 1991, Coevolution in bacterial-plasmid populations. *Evol.* 45:656-667.

Mogen B. D., Oleson, A. F., Sparks, R. B., Gudmestad, N. C. and Secor, G. A. 1988, Distribution and partial characterisation of pCS1, a highly conserved plasmid present in *Clavibacter michiganense* subsp. *sepedonicum*. *Phytopathology* 78:1381- 1386.

Morris, C. E., Jacques, M-A. and Nicot, P. 1994, Microbial aggregates on leaf surfaces: Characterisation and implications for the ecology of epiphytic bacteria. *Mol. Ecol.* 6:613.

Nesme, N., Ponsonnet, C., Picard, C. and Normand, P. 1992, Chromosomal and pTi genotypes of *Agrobacterium* strains isolated from *Populus* tumors in two nurseries. *FEMS Microbiol. Ecol.* 101:189-196.

Newman, E. I. and Bowen. H. J. 1974, Patterns of distrbution of bacteria on root surfaces. *Soil Biol. Biochem.* 6:205-209.

Obukowicz, M. and Shaw, P. D. 1985, Construction of Tn3- containing plasmids from plant-pathogenic pseudomonads and an examination of their biological properties. *Appl. Environ. Microbiol.* 49:468-473.

Osborn, A. M., Bruce, K. D., Strike, P. and Ritchie, D. A. 1993, Polymerase chain reaction-restriction fragment length polymorphism analysis shows diversity among *mer* determinants from Gram-negative soil bacteria indistinguishable by DNA-DNA hybridisation. *Appl. Environ. Microbiol.* 59:4024-4030.

Paulus, F., Ride, M. and Otten, L. 1989, Distribution of two *Agrobacterium tumefaciens* insertion elements in natural isolates: Evidence for stable association between Ti plasmids and their bacterial hosts. *Mol. Gen. Genet.* 219:145-152.

Piper, K. R., Beck von Bodman, S. and Farrand, S. K. 1993, Conjugation factor of *Agrobacterium tumefaciens* regulates Ti plasmid transfer by auto induction. *Nature.* 362:448-450.

Powell, B. J., Purdy, K. J., Thompson, I. P. and Bailey, M. J. 1993, Demonstration of tra⁺ plasmid activity in bacteria indigenous to the phyllosphere of sugar beet; gene transfer to a genetically modified pseudomonad. *FEMS Microbiol. Ecol.* 12:195-207.

Quant, R. L. and Mills, D. 1984, An integrative plasmid and multiple-sized plasmids of *Pseudomonas syringae* pv. *phaseolicola* have extensive homology. *Mol. Gen. Genet.* 193:459-466.

Rainey, P. B., Bailey, M. J. and Thompson, I. P. 1994, Phenotypic and genotypic diversity of fluorescent pseudomonads isolated from field grown sugar beet. *Microbiol.* 140: 2315- 2331.

Rainey, P. B. and Bailey, M. J. 1996, Physical and genetic map of the *Pseudomonas fluorescens* SBW25 chromosome. *Molec. Microbiol.* 19:521-533.

Ramos-Gonzalez, M. I., Duque, E. and Ramos, J. L. 1991, Conjugational transfer of recombinant-DNA in cultures and in soils - host range of *Pseudomonas-putida* tol plasmids. *Appl. Environ. Microbiol.* 57:3020-3027.

Reanney, D. C. 1978, Coupled evolution: adaptive interactions among the genomes of plasmids, viruses and cells. *Int. Rev. Cytol. (Suppl.)* 8:1-68.

Richaume, A., Smit, E., Faurie, G. and van Elsas, J. D. 1992a, Influence of soil type on the transfer of plasmid-Rp4(p) from *Pseudomonas fluorescens* to introduced recipient and to indigenous bacteria. *FEMS Microbiol. Ecol.* 101:281-292.

Richaume, A., Bernillon, D. and Faurie, G. 1992b, Role of the intraspecific competition in the regulation of *Agrobacterium tumefaciens* transconjugant population-level in soil experiments. *FEMS Microbiol. Ecol.* 86:321-329.

Rochelle, P. A., Fry, J. C. and Day, M. J. 1988, Structural rearrangements of a broad host range plasmid encoding mercury resistance from an epilithic isolate of *Pseudomonas cepacia. FEMS Microbiol. Letts.* 52:245-250.

Rovira, A. D., Newman, E. I., Bowen, H. J. and Campbell, R. 1974, Quantitative assessment of the rhizoplane microflora by direct microscopy. *Soil Biol. Biochem.* 6:211-216

Sasakawa, C., Kamata, K., Makino, S., Murayama, S. Y., Sakai, T. and Yoshikawa, M. 1986, Molecular alteration of the 140- megadalton plasmid associated with loss of virulence and congo-red binding-activity in *Shigella flexneri. Infect. Immun.* 51:470-475.

Schofield, P. R., Dudman, W. F., Gibson, A. H. and Watson, J. M. 1987, Evidence for genetic exchange and recombination of *Rhizobium* symbiotic plasmids in a soil population. *Appl. Environ, Microbiol.* 53:2942-2947.

Silver, S. and Walderhaug, M. 1992, Gene regulation of plasmid determined and chromosome determined inorganic ion transport in bacteria. *Microbiol. Rev.* 56:195-228.

Simonsen, L. 1991, The existence conditions for bacterial plasmids: Theory and reality. *Microb. Ecol.* 22:187-205.

Smets, B. F., Rittmann, B. E. and Stahl, D. A. 1993, The specific growth rate of *Pseudomonas putida* PAW1 influences the conjugal transfer rate of the TOL plasmid. *Appl. Environ. Microbiol.* 59:3430-3437.

Smit, E., van Elsas, J. D., van Veen, J. A. and Devos, W. M. 1991, Detection of plasmid transfer from *Pseudomonas fluorescens* to indigenous bacteria in soil by using bacteriophage Phi-R2F for donor counter selection. *Appl. Environ. Microbiol.* 57:3482-3488.

Steck, T. R. and Kado, C. I. 1990, Virulence genes promote conjugative transfer of the Ti plasmid between *Agrobacterium* strains. *J. Bact.* 172:2191-2193.

Steinberger, E. M., Cheng, G. Y. and Beer, S. V. 1990, Characterisation of a 56 Kb plasmid of *Erwinia amylovora* Ea322: its non-involvement in pathogenicity. *Plasmid.* 24:12-24.

Sundin, G. W., Jones, A. L. and Fulbright, D. W. 1989, Copper resistance in *Pseudomonas syringae* pv. *syringae* from cherry orchards and its associated transfer *in vitro* and *in planta* with a plasmid. *Phytopathology* 79:861-865.

Sundin, G. W., Demezas, D. H. and Bender, C. L. 1994, Genetic and plasmid diversity within natural populations of *Pseudomonas syringae* with various exposures to copper and streptomycin bactericides. *Appl. Environ. Microbiol.* 60:4421-4431.

Sundin, G. W. and Bender, C. L. 1993, Ecological and genetic analysis of copper and streptomycin resistance in *Pseudomonas syringae* pv. *syringae. Appl. Environ. Microbiol.* 59:1018-1024.

Sykora, P. 1992, Macroevolution of plasmids: A model for plasmid speciation. *J. Theor. Biol.* 159:53-65.

Tenning, P., Vanrijsbergen, R., Zhao, Y. and Joos, H. 1993, Cloning and transfer of genes for antifungal compounds from *Erwinia herbicola* to *Escherichia coli. Molec. Plant-Microb. Interact.* 6:474-480.

Thompson, I. P., Bailey, M. J., Fenlon, J. S., Fermor, T. R., Lilley, A. K., Lynch, J. M., McCormack, P. J., McQuilken, M., Purdy, K. J., Rainey, P. B. and Whipps, J. M. 1993, Quantitative and qualitative seasonal changes in the microbial community from the phyllosphere of sugar beet (*Beta vulgaris*). *Plant. Soil.* 150:177-191.

Thompson, I. P., Bailey, M. J., Ellis, R. J., Lilley, A. K., McCormack, P. J. Purdy and Rainey, P. B. 1995a, Short term community dynamics in the phyllosphere microbiology of field grown sugar beet. *FEMS Microbiol. Ecol..* 16:205-211.

Thompson, I. P., Lilley, A. K., Ellis, R. J., Bramwell, P. A. and Bailey, M. J. 1995b, Survival, colonisation and dispersal of genetically modified *Pseudomonas fluorescens* SBW25 in the phytosphere of field grown sugar beet. *Biotechnology* 13:1493-1497.

Tiedje, J. M., Colwell, R. K., Grossman, Y. L., Hodgson, R. E., Lenski, R. E., Mack, R. N. and Regal, P. J. 1989, The planned introduction of genetically modified organisms: ecological considerations and recommendations. *Ecology* 70:298-315.

Top, E. M., Holben, W. E. and Forney, L. J. 1995, Characterisations of diverse 2,4-dichlorophenoxyacetic acid degradative plasmids isolated from soil by complementation. *Appl. Environ. Microbiol.* 61:1691-1698.

van Elsas, J. D., Starodub, M. E. and Trevors, J. T. 1988, Bacterial conjugation between pseudomonads in the rhizosphere of wheat. *FEMS Microbiol. Ecol.* 53:299-306.

van Elsas, J. D., van Overbeek, L. S. and Nikkel, M. 1989, Detection of plasmid RP4 transfer in soil and rhizosphere and the occurrence of homology to RP4 in soil bacteria. *Curr. Microbiol.* 19:375-381.

van Elsas, J. D., Trevors, J. T., Starodub, M. E. and van Overbeek, L. S. 1990, Transfer of plasmid RP4 between pseudomonads after introduction into soil - influence of spatial and temporal aspects of inoculation. *FEMS Microbiol. Ecol.* 73:1-11.

Weatcroft, R. and Williams, P. A. 1981, Rapid method for the study of both stable and unstable plasmids in *Pseudomonas. J. Gen. Microbiol.* 12:4433-437.

Wilkins, B. M. 1995, Gene transfer by bacterial conjugation: Diversity of systems and functional specialisations. pp. 59-88 In: Baumberg, S., Young, J.P.W., Wellington, E. M. H. and Saunders, J. R. (eds.), *Population Genetics of Bacteria.* Society for General Microbiology Symposium 52. Cambridge University Press, Cambridge.

Young J. P. W. and Wexler, M. 1988, Sym plasmid and chromosomal genotypes are correlated in field populations of *Rhizobium leguminosarum. J. Gen. Microbiol.* 134:2731-2739.

Zhang, L., Murphy, P. J., Kerr, A. and Tate, M. E. 1993, *Agrobacterium* conjugaton and gene regulation by N-acyl-homoserine lactones. *Nature.* 362:446-448.

Zhang, L. and Kerr, A. 1991, A diffusible compound can enhance conjugal transfer of the Ti plasmid in *Agrobacterium tumefaciens. J. Bacteriol.* 173:1867-1872.

8

MICROBIAL INTERACTIONS PREVENTING FUNGAL GROWTH ON SENESCENT AND NECROTIC AERIAL PLANT SURFACES

William F. Pfender

Department of Plant Pathology
4024 Throckmorton Plant Science Center
Kansas State University
Manhattan, Kansas 66506-5502

INTRODUCTION

Although there is an extensive literature concerning many aspects of phyllosphere microbial ecology, only a small proportion addresses senescent and necrotic aerial plant tissues. But microbial activity on these surfaces is important, in part because the microflora there can interact with microbes on living plant surfaces (see Köhl and Fokkema, 1994; Fokkema, 1993). Furthermore, a consideration of the environment on senescent or non-living plant surfaces can bring forward, through contrast with the living phyllosphere, general principles of microbial ecology and interactions. In another chapter of this volume, Alan Rayner discusses fungal interactions on bark, a non-living layer of aerial tissue in woody plants. In contrast, this chapter addresses senescent and non-living aerial tissues of herbaceous plants - specifically, the physical and nutritional environment of these surfaces as it affects the microbial community, the microbial interactions that can limit the survival or activity of fungi there, and how these effects are relevant to microbial activity in phyllospheres of living plants.

As a habitat for microbes, senescent and necrotic plant tissue is a non-equilibrium environment in which several important, directional transitions define the possibilities for microbial life and interactions. The underlying transition is the change through time from living through senescent to necrotic tissue. It is a gradual change, with no sharp demarcation between living and senescent, nor between senescent and dead - one merges into the other. And because the nature of the plant-surface environment changes during this process, the microbial communities of these surfaces also change by degree to eventually become qualitatively different from what they initially were.

One of the important transitions generated by the change in plant vitality is a change in integrity of the plant-surface boundary. It becomes increasingly indistinct as senescence proceeds. On the living plant, the distinction between what is inside and what is outside the plant is relatively clear. The plant maintains most of its biochemicals inside the leaf (although

Aerial Plant Surface Microbiology, edited by Morris et al.
Plenum Press, New York, 1996

a certain amount of these materials exude). Plant responses to attempted microbial invasion, and pre-existing physical and chemical boundaries, act to separate the inside from the outside of the plant with respect to other organisms, as noted by Rayner (this volume). But even on the living plant, the distinction is not entirely clear - some of the best "epiphytic colonisers", as defined operationally by the technique of leaf washing used to recover these organisms for isolation, may occupy stomatal openings and thus be in a transitional region between inside and outside the plant (Beattie and Lindow, 1994). And during the change of tissues from living to senescent to necrotic, the boundaries of the plant surface become increasingly indistinct as the vigour of the plant's "living field" (see Rayner, this volume) declines and microbes begin to penetrate more and more into the interior of the tissue. So the microbial communities, and the boundary between inside and outside, exist on a continuum as plant tissue proceeds through time from living to senescent to necrotic.

In a manner that is perhaps analogous to these transitional features of the microbial habitat, the concept itself of microbial ecology on senescent and necrotic aerial plant tissue has indistinct boundaries. In considering early stages of senescence, the investigations are a minor variation on microbial ecology of the living phyllosphere; but as one considers environmental conditions and microbial events on increasingly senescent and then necrotic tissue, the subject gradually but surely becomes qualitatively different from the study of living phyllospheres. Again, the point of demarcation is indistinct. But there is nonetheless a strong relationship among microbial communities all along this continuum, so that phenomena occurring even on dead plant tissues may be of interest to the study of microflora on living phyllospheres.

MICROBIAL ENVIRONMENT OF SENESCENT AND NECROTIC AERIAL PLANT TISSUES

Two major aspects of the microbial environment are affected by the plant's transition from living to dead (Figure 1). These are the nutritional environment and the physical conditions (temperature and moisture) of the habitat.

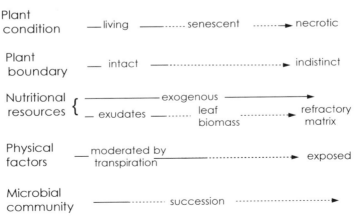

Figure 1. Transitions in the microbial environment of aerial plant surfaces as plant tissue proceeds from living to necrotic.

Nutritional Environment

Most of the microbes found on herbaceous plant surfaces, and particularly those of concern for plant disease and its control, are heterotrophs. They obtain their energy from carbon that has been fixed by primary producers. On any aerial plant surface, some components of this primary production are exogenous - their origin is remote from the plant tissue on which they are utilised. These resources, which include pollen for example, are deposited on the leaf and may be a major nutrient source for epiphytic microbes. And on the living leaf, there is also an ongoing supply of plant-produced nutrients from the very leaf on which the microbe is located; it is available as exudates and insect-transformed plant materials such as aphid honeydew (Dik *et al.*, 1992) and it is provided while the plant tissue is alive. Although the exudates are relatively scant (Fiala *et al.*, 1990), rendering the phyllosphere a carbon-limited environment (Wilson *et al.*, 1995), they contain a particularly assimilable form of nutrients for microbes (Morgan and Tukey, 1964).

But as the leaf senesces and dies, loss of the distinction between surface and interior has important implications for the nutrient resources available to the microbes. Although the senescing leaf depletes some proportion of its soluble nutrients by export to other parts of the plant (Baddeley, 1971), the loss of surface integrity nonetheless means that there will be a greater amount of the leaf's resource available to the microbes associated with it than was available on the living leaf. But this resource, unlike the exudate from living leaves, will change in composition as the easily-metabolised components are used and the more refractory components predominate in the declining nutrient source that remains.

The changing nature of this nutrient source, and the various strategies of microbes to exploit it, are determinative factors in the succession of the microbial community. Specifically, senescing tissue becomes increasingly less resistant to invasion by microbes, and mycelial forms are better able to extend into and exploit the resource than are unicellular forms. Although there is a lack of experimental data, it is plausible that in this succession competitive interactions among microbes become more and more important as the resource, now released from the plant's protective processes, becomes available, and then becomes progressively less in amount and more refractory in composition (Figure 1).

Physical Environment

In addition to changes in the nutritional nature of the environment as the leaf senesces and dies, there are changes in the physical aspects of the leaf-surface environment (Figure 1). Transpiration from the living leaf has a moderating effect on the leaf surface: the relative humidity is generally higher in the boundary layer of a transpiring leaf than in the ambient air, and heat transfer with evaporating water cools the temperature (Burrage, 1976). Because the boundary layer thins as leaf trichomes shrivel, and then the supply of transpirational water vapour stops, these moderating influences decline and then cease in senescent and dead plant tissue. Microbes in these communities thus will be exposed to greater extremes of humidity and temperature than on a similarly-exposed living plant. Available liquid water as well will probably fluctuate more widely on dead tissue than on living surfaces. Dry necrotic tissue will dry out more completely, and may take longer to become wet again (James *et al.*, 1984), than living tissue. Köhl and Fokkema (1994) suggest, however, that necrotic tissue, once wet, may remain wet longer than the surface of living tissue.

Considering the level of microbial competition for an initially substantial but diminishing resource base, and the extremes of heat and desiccation, it is likely that microbes will experience more stress, or certainly different stress, on senescent and dead tissue than on living phyllosphere environments. It should be noted, however, that the physical and nutritional aspects of this environment have been very little studied compared to those of the living plant surface. A few studies have examined the water status of necrotic tissue under field conditions (Fernandes *et al.*, 1991; Zhang and Pfender, 1992), but the measurement is technically difficult and has not often been done. Additional research would be beneficial to our understanding of microbial development in this habitat.

This brief description of the environmental conditions on senescent and dead aerial plant tissues may be used as a context in which to consider the relationships between microflora on these surfaces and microflora on living surfaces.

RELATIONSHIPS BETWEEN MICROFLORA ON SENESCENT/NECROTIC TISSUE AND IN LIVING PHYLLOSPHERES

Succession on Ageing Aerial Plant Surfaces

One aspect of this relationship is a temporal one - the microflora of the living surface gives way to the microflora of the senescent and necrotic surface as time proceeds (Hudson and Webster, 1958; Hudson, 1968). On senescing tissue, yeast forms are the first to increase in numbers, followed by an increase in mycelial fungi (Dickinson, 1967). Certain microbes are particularly well-adapted to be in the vanguard of these earliest colonisers of necrotic tissue. Some pathogenic microbes are present at living-tissue surfaces as latent infections, unable to completely penetrate the plant surface defences, but well-placed to increase their biomass as soon as the plant surface integrity begins to wane, or as chemical signals (e.g. ethylene from fruit surfaces [Kolattukudy *et al.*, 1995]) indicate that the plant is undergoing senescence. Others in this group of early colonisers of senescent and necrotic plant parts are saprophytic or weakly parasitic fungi such as species of *Epicoccum, Cladosporium* and *Alternaria*. It is not clear that change in microbial communities on senescing tissue is always a succession, in which community composition is contingent upon a prior community; stochastic processes, including immigration, may be major factors of community change in some cases. But to the extent that we can manipulate the presence of organisms on living tissue (Wilson *et al.*, 1995), we may in some cases be able to affect their presence or dominance on newly-senescent and, later, necrotic surfaces (Sutton and Peng, 1993; Zhou and Reeleder, 1991). This sort of manipulation could be useful in biocontrol, as discussed below.

Necrotic Tissue as a Source of Beneficial Microbial Inoculum for Living Surfaces

In addition to a possible successional relationship, by which microflora of a living phyllosphere affects the microbial community that develops subsequently on senescent tissue, it is also possible for the microflora of senescent/necrotic tissue to have an effect on that of living tissue. The necrotic tissue can act as a source of inoculum for the microbial community of living plant surfaces.

One possibility that has not been much explored is the use of necrotic tissue as a source of beneficial microflora for communities on nearby living-plant surfaces. It has been noted (Kinkel, 1991; Andrews, 1992) that some phyllosphere communities are immigration-limited. Microflora of senescent and necrotic aerial plant surfaces, if present near the living plant surfaces, could be a continuing source for this immigration. Thus it may be possible to manage the microbial communities of necrotic tissues as a source of antagonistic (biocontrol) microflora for nearby, immigration-limited surfaces on living plants.

Necrotic Tissue as a Source of Phytopathogenic Fungi

Another manner in which the microflora of senescent and necrotic tissue affects the living phyllosphere microflora is as a source of inoculum of phytopathogenic organisms. There are several plant diseases in which colonisation of necrotic tissue plays an essential role in the disease cycle. In some cases, the pathogenic fungus uses energy derived from necrotic tissue to increase its inoculum potential sufficiently to attack contiguous living tissue. A prime example for foliar diseases is *Sclerotinia sclerotiorum* on several crops. On beans, the pathogen colonises senescent or dead blossoms, then grows from this tissue to living leaf or stem tissue on which the blossom has come to rest after being shed (Abawi and Grogan, 1975). The same pathogen attacks lettuce by first colonising senescent or dead lower leaves, then growing into healthy plant tissue (Sitepu and Wallace, 1984). *Botrytis cinerea* causes grey mold of strawberry fruit; it attacks the fruit after colonising the dead floral parts contiguous to it (Powelson, 1960; Sutton, 1990).

In other situations, the colonised necrotic tissue is not contiguous with the infection court, but is nonetheless critical for disease development. The pathogen can increase its population on senescent/necrotic tissue and then be disseminated as spores from this inoculum source to living plant surfaces. For some such diseases, the necrotic tissue is the site of saprophytic survival (*e.g.* overwinter) and the source of primary inoculum to begin the epidemic. *Venturia inaequalis* survives on dead apple leaves, and is dispersed as ascospores to begin epidemics on this host (Cullen and Andrews, 1984). Dead wheat straw on or above the soil surface is the survival habitat for *Pyrenophora tritici-repentis*, which produces ascocarps there and releases ascospores as primary inoculum for the tan spot disease of wheat (Adee and Pfender, 1989).

For yet other diseases, senescent and necrotic tissue in the crop may be a substratum for production of secondary inoculum that contributes to epidemic development during the season. Species of *Botrytis* are apt examples: *B. elliptica* on lilies (Köhl *et al.*, 1995) and *B. squamosa* on onions (Sutton, 1990) each can sporulate on necrotic host tissue in the crop canopy, producing secondary inoculum for epidemic development. Conidia of *B. cinerea* are produced primarily on dead, overwintered strawberry leaves in the crop (Sutton, 1990); they are dispersed to necrotic flower parts which they infect prior to attacking the fruit.

NECROTIC AND SENESCENT TISSUE AS A SITE FOR BIOLOGICAL CONTROL

For all such cases, in which phytopathogenic inoculum develops on necrotic aerial plant tissue, manipulation of the microbial community on these tissues could be an effective approach to biocontrol of the diseases. For this approach to be effective, however, the target disease must be one in which suppression of pathogen activity on senescent and necrotic tissues will result in a decrease in epidemic development. For *Botrytis* leaf spot of onion, a reduction in secondary inoculum production on necrotic tissue within the crop effectively reduces disease development (Köhl *et al.*, 1995). For tan spot of wheat, epidemic development is significantly lowered by a reduction in primary inoculum produced on necrotic tissue, even though there are subsequent secondary disease cycles from lesions on living leaves (Adee and Pfender, 1989). For the latter disease, however, a large reduction may be needed to obtain acceptable control: in field experiments with a very high infestation level, an 80% reduction in primary inoculum on straw residue by a biocontrol agent was insufficient to give a useful level of tan spot control (Pfender *et al.*, 1993).

Each of the diseases mentioned above has been the subject of research to develop biocontrol strategies targeted at pathogen colonisation of senescent or necrotic tissue. Antagonism against *S. sclerotiorum* on bean blossoms has been observed for a number of organisms, notably *Epicoccum purpurascens* (Boland and Inglis, 1988; Zhou and Reeleder, 1989), *Erwinia herbicola* (Yuen *et al.*, 1994), and species of *Fusarium* and *Trichoderma* (Boland and Inglis, 1988). Reeleder and co-workers determined that *Epicoccum purpurascens* in particular is an effective coloniser of blossoms (Zhou and Reeleder, 1991); in field and greenhouse trials, application of this antagonist significantly suppressed the ability of the pathogen to infect living tissue from colonised flowers (Zhou and Reeleder, 1989). *E. purpurascens*, as well as *Trichoderma viride* and *Alternaria alternata*, were each effective also against *S. sclerotiorum* on necrotic leaves of lettuce (Mercier and Reeleder, 1986).

Biocontrol of *Botrytis* spp. in the aerial necrotic tissue of several crops has been extensively studied as well. The importance of necrotic flower tissue as a source of infection for strawberry fruits was demonstrated decades ago (Powelson, 1960), and recent research has shown that *Gliocladium roseum*, *Trichoderma viride*, and *Penicillium* spp. are effective in reducing infection of flower parts and subsequent fruit infection (Peng and Sutton, 1991). These antagonists can also reduce sporulation of the pathogen on infected, necrotic leaves, especially if the antagonist is applied to leaves while they are still green and living (Sutton and Peng, 1993). This same pathogen is the subject of biocontrol research for diseases of lily and onion. Several antagonists, including the highly competitive coloniser *Ulocladium atrum*, can suppress sporulation of *Botrytis* spp. on necrotic leaves (Köhl *et al.*, 1995), which are the source of secondary inoculum to sustain epidemics in the crop (Köhl *et al.*, 1995).

Antagonism to pathogen sporulation in necrotic lesions on living leaves has been reported also. Saprophytic yeasts can suppress conidial production by *Botrytis cinerea* on leaf lesions of beans and tomatoes (Elad *et al.*, 1994). *Trichoderma harzianum* reduced sporulation capacity of *Cochliobolus sativus* on excised leaf lesions from infected wheat (Biles and Hill, 1988).

Two ascomycetous pathogens that produce primary inoculum on necrotic crop residue, *Venturia inaequalis* and *Pyrenophora tritici-repentis*, are susceptible to suppressive effects of basidiomycete antagonists during their survival and reproduction stages. The apple scab pathogen can be suppressed by *Athelia bombacina* (Heye and Andrews, 1983), and the wheat tan spot pathogen by *Limonomyces roseipellis* (Pfender, 1988; Pfender *et al.*, 1993).

ENVIRONMENTAL FACTORS IN BIOLOGICAL CONTROL ON SENESCENT AND NECROTIC AERIAL TISSUES

In order to manipulate more successfully microbial interactions to the detriment of phytopathogenic fungi in these aerial necrotic and senescent plant tissues, a better understanding of the ecology of the pathogens with respect to their environment and to other microbes in this habitat would be useful. Here we return to the considerations developed earlier in this chapter. What are the constraints on activity and the possibilities for interaction in this habitat?

Temperature extremes must have an impact on survival of microbes on dry, necrotic plant tissue, but this issue has been very little studied. Temperatures are affected by moisture status of the tissue, but these conditions have not commonly been measured or investigated for their effects on microbes and their interactions.

Moisture availability is undoubtedly the most important limiting factor for microbial activity on necrotic aerial plant surfaces. Without the boundary layer of humidified air that is present around living leaves, microbes on senescent and necrotic aerial surfaces are exposed to extreme desiccation interspersed with periods in which moisture is available and, commonly, saturating. The success of a microbe, whether pathogen or antagonist, in this habitat depends on its surviving desiccation, and also on the speed of its response to water when it becomes available. Microbes differ in their ability to grow under various degrees of moisture stress, and the outcome of antagonistic interactions between fungi is thus affected by water potential (Magan and Lacey, 1984; Pfender *et al.*, 1991). In most climates, aerial necrotic plant tissue is typically very dry most of the time, has periods of high water potential due to rain or dew events, and relatively brief time periods when it is in a transitional stage of intermediate water potential whilst it dries or becomes wet (Zhang and Pfender, 1992). Therefore, a microbe maximises its active time if it responds quickly when moisture becomes available, and if its range of activity includes lower water potentials that persist briefly as the tissue dries. These capabilities - survival of desiccation, quick response to moisture occurrence, and maximisation of water potential range - are integrated in the response of organisms to interrupted wetness periods. The suppressive effect of several antagonists to *Pyrenophora tritici-repentis* on wheat straw is affected by wetness period duration in intermittently wet tissue, and none of several antagonists tested was effective against the pathogen when these periods were less than 12 hours (Zhang and Pfender, 1993). Potential biocontrol agents for *Botrytis* spp. on dead onion leaves differ in their response to intermittent wetting: *Gliocladium roseum* is an efficient antagonist under continuously wet conditions but not under interrupted wetness periods, whereas antagonistic *Ulocladium* spp. suppressed the pathogen even when the wetness period on the onion leaves was interrupted (Köhl *et al.*, 1995).

MECHANISMS OF MICROBIAL INTERACTION THAT INHIBIT FUNGI ON SENESCENT AND NECROTIC AERIAL PLANT SURFACES

The microbial community develops within this context of temperature and moisture fluctuations, as microbes interact with each other and compete in various ways for the nutrient resource. To understand this development, and to be able to manipulate it to the detriment of phytopathogenic fungi in biocontrol, it would be useful to understand the mechanisms of interaction of antagonists with these pathogenic fungi. The general types of mechanisms are similar to mechanisms that occur on other plant surfaces, including the living phyllosphere. But because the senescent/necrotic surface differs from the living phyllosphere in a number of ways, one might expect that the relative importance of these mechanisms could be different here.

Although any given interaction may involve more than one mechanism, in the following discussion mechanisms are categorised (Wicklow, 1992) as exploitative competition for the resource, interference competition by antibiosis, and interference competition by direct consumption of competitors, i.e. mycoparasitism. A more detailed discussion of the definition of competition and models to assess its importance among micro-organisms on aerial plant surfaces is presented in the chapter by Kinkel *et al.* (this volume).

Nutrient Competition

Competition for nutrients has been demonstrated for a number of living-phyllosphere systems, and these data have been presented in prior volumes of this series. Especially noteworthy has been the elucidation of interactions between yeasts and the conidia of phytopathogenic fungi, where nutrient competition has been clearly implicated through the use of radio-labelled nutrient sources (Blakeman and Brodie, 1977). Such competition probably occurs naturally on most plant surfaces due to indigenous microbial populations, but may be subject to artificial enhancement in situations where nutrients accumulate too rapidly for indigenous populations to keep pace with the resource (Fokkema, 1993). The increase in nutrient availability as aerial plant tissue dies may be such a situation, especially because the loss of plant response permits non-pathogenic mycelial forms to penetrate into the plant tissue. Radio-label studies could perhaps demonstrate the relative competitive abilities of pathogen and antagonist in their partitioning of resources from senescent and necrotic plant tissue. Another approach that could be used to infer the nature of resource partitioning among competing fungi is the replacement series technique. In this method, borrowed from plant ecology (Silvertown, 1982), a resource is exposed to a population comprising two competitors. In different treatments, the initial populations of the two competitors are in various proportions, but the total initial population is the same in all treatments. By observing the biomass (population size, spore production) that each competitor is capable of producing in the presence of varying populations of the other competitor (but with a similar overall population density), interpretations about competition for resource components can be made. Specifically, one can observe whether competition within the species is more or less inhibitory to population growth than competition between the species; if different organisms are using different components of the resource they will compete less with each other than with conspecifics, for example. There are some difficulties in using this method for filamentous fungi (primarily in quantifying population size of a clonal organism), but it can be informative if interpreted carefully. It should be noted that this method does not necessarily identify antagonism as being due to nutrient competition - any inhibitory effect will be measurable - but it can provide information about use of a shared resource, to support other observations about the mechanism of interaction. This method was used to study competition between two phytopathogenic fungi in living and senescent tissue (Adee, *et al.*, 1990), and Wilson and Lindow (1995) have applied it to a study of saprophytic bacterial epiphytes. But apparently it has not been attempted in the study of saprotrophic fungal competition such as could occur in and on aerial senescent and necrotic plant tissues.

Destruction and Consumption of Fungi by Competitors

In interference competition, one organism actively interferes with the competitor's access to the resource, making it unavailable to that competitor regardless of whether or not the resource is depleted by the antagonist (Wicklow, 1992). An extreme form of this type of competition is attack and consumption of the competitor. When this occurs between fungi, it is generally called mycoparasitism, and that term is used here for convenience. Considering the nutritional aspects of the senescent/necrotic tissue habitat, mycoparasitism is a useful adaptation. As the nutrient resource is degraded and the metabolically refractory components predominate, the mycelia that have flourished on the relatively abundant nutrients of the previously senescent tissue now come to represent a nutrient resource in themselves. Filamentous organisms with adaptations to tap this resource will be able to extend through the matrix to exploit this food source. Because fungal walls contain chitin, chitinase production is likely to be one of the relevant adaptations.

Chitinase as a correlate of antagonism has been investigated in several biocontrol interactions (*e.g.* Cherif and Benhamou, 1990; Di Pietro *et al.*, 1993; Lorito *et al.*, 1993; Chernin *et al.*, 1995). Most of these studies have involved species of *Trichoderma* or *Gliocladium* as the antagonists, and have focused on their interactions with soil-borne pathogens. However, it is likely that these chitinase systems are active in antagonism to fungi on aerial senescent and necrotic tissue as well. For example, chitinases are inhibitory to spore germination and germ-tube elongation in *Botrytis cinerea* (Lorito *et al.*, 1994), which is a biocontrol target for some applications of *Trichoderma* to aerial plant surfaces (Biles and Hill, 1988; Peng and Sutton, 1991; Tronsmo, 1992). The chitinolytic systems of these fungi are complex, consisting of a diversity of enzymes with complementary modes of action that permit attack against a range of fungi (Haran *et al.*, 1995). Chitinases are also produced by some bacteria, and the possibility of inserting genes for microbial chitinases into other microbes to enhance the efficacy of antagonism to pathogens is being investigated (Shapira *et al.*, 1989; Lorito *et al.*, 1993).

The relationship of mycoparasitism and chitinase-producing ability to the ecological processes in necrotic tissue can be illustrated by our study of microbial ecology in conservation-tillage wheat straw infested with the pathogenic fungus *Pyrenophora tritici-repentis*. A community analysis of this necrotic tissue showed that fungi could be grouped in associations correlated with microenvironmental conditions (Pfender and Wootke, 1988). One association of interest was found initially on straw that was positioned on the soil surface and, later, on straw within the mulch layer, midway between the soil surface and the uppermost, exposed residue. In this association, the population of *P. tritici-repentis* appeared to be declining. Chitinolytic fungi, very common in buried straw, were somewhat less common here. But one chitinolytic fungus with a rapid growth rate was noted; among all the chitinolytic fungi recovered, it was the most similar in its pattern of occurrence to pioneer colonisers and plant parasites such as *Cladosporium* spp., *Epicoccum* sp., and *P. tritici-repentis*. This fungus, *Limonomyces roseipellis*, was found to be antagonistic to *P. tritici-repentis* on straw tissue (Pfender, 1988), and its chitinolytic capabilities are apparently involved in a mycoparasitic trophic habit, as demonstrated by microscopic observation (Pfender *et al.*, 1991). Our interpretation of these experiments is that *L. roseipellis* is an example of a fungus adapted for acquisition of nitrogen and other nutrients previously sequestered by *P. tritici-repentis* when they were relatively more available in the senescent tissue. If such chitinolytic organisms can be cultivated in the senescent/necrotic tissue habitat, so that they are present earlier in the succession than they would naturally occur, their mycoparasitic habit would find an abundant resource in the hyphae that have ramified the recently-defenceless plant tissue, and they could be effective biocontrol agents. Because the physical environment of this habitat is characterized by fluctuating moisture availability (see prior discussion), the most promising mycoparasites would be those that can tolerate moisture-stress and/or can reactivate relatively quickly when moisture becomes available after a drought-induced quiescence.

Antibiosis

Another manifestation of interference competition among microbes on plant surfaces is antibiosis. Observations of microbial antibiosis against other microbes under pure-culture conditions have a long history in microbiology, and the role of antibiotics as one component of phytopathogen inhibition has been demonstrated under culture conditions for fungal and bacterial antagonists (Kempf and Wolf, 1989; Wilhite *et al.*, 1994). The relevance of these observations to events on aerial plant surfaces has been the subject of controversy, however. Although there are cases in which antibiotic production has been clearly associated with inhibition of a pathogen by a bacterial culture applied to a leaf surface (*e.g.* Kempf and Wolf, 1989), it is clear that antibiotic-producing capabilities of an organism are not always expressed while growing in the phyllosphere - a common observation is that the culture filtrates of the antagonist are inhibitory to the pathogen on the leaf, but the washed cells of the antagonist are not (Chakraborty *et al.*, 1994). This failure may be due to poor growth of the applied antagonist, or to lack of antibiotic production under the nutritional and physical conditions of the leaf surface. Antibiotic production can require ample nutrients (Atlas and

Bartha, 1993 [p. 57]),and it may be that intact living plant surfaces provide insufficient nutrients to support antibiotic production (Blakeman, 1991).

Perhaps on senescent and necrotic aerial plant tissue, however, the nutritional situation is sufficiently different from that on living leaves that antibiotic production is more likely to occur. Inhibition of *Sclerotinia sclerotiorum* on bean blossoms appears to involve antibiotic production, at least as one component of the interaction. Reeleder and co-workers found no evidence for mycoparasitism of the pathogen by the antagonist *Epicoccum purpurascens*, and their observations suggested that an antibiotic is involved in the suppression in this pathosystem (Zhou *et al.*, 1991) and in inhibition of *S. sclerotiorum* by *E. purpurascens* on lettuce (Mercier and Reeleder, 1986). Also, observations of inhibition by strains of *Erwinia herbicola* on bean blossoms suggest antibiosis, perhaps via herbicolin; the temperature-dependent nature of the antagonism, as observed in the field (Yuen *et al.*, 1994), is correlated with differences in temperature-dependent production of the antibiotic by various strains of the bacteria (G. Yuen, personal communication). Definitive experiments to show antibiosis as the major mechanism for biocontrol of fungi on necrotic/senescent aerial plant tissue have not yet been presented. One approach to this definitive demonstration is the use of specific antibiotic-deficient mutants of antagonists to test correlation of antibiotic production capacity with antagonism (Kempf and Wolf, 1989). We (Pfender *et al.*, 1993) performed experiments of this type with *Pyrenophora*-infested necrotic wheat straw challenged by application of the well-studied *Pseudomonas fluorescens* strain Pf-5 (Howell and Stipanovic, 1980). Tn5 mutants of Pf-5, deficient in one or more of the several antibiotics it produces, were tested for ability to suppress the fungal pathogen in agar culture and on infested straw treated with washed bacterial cells. Antagonism was correlated with pyrrolnitrin production, both being absent in mutant JL3985, and both simultaneously restored when JL3985 was complemented with wild-type DNA of the Tn5-disrupted segment (Pfender *et al.*, 1993). Under conditions of the test, the bacteria persisted and evidently produced antibiotic *in situ*. Purified pyrrolnitrin also inhibited *Pyrenophora* on straw pieces. The objective of the research was to demonstrate involvement of an antibiotic in antagonism, in order to make possible the eventual transfer of antibiotic-production genes to other bacterial strains better suited than Pf-5 to survival and activity on necrotic tissue in the field. Pf-5 did not survive well on naturally-infested wheat straw incubated under conditions of moisture stress. This research did not present evidence of antibiosis under natural conditions, because the straw was incubated under conditions of constant moisture, and the nutritional status of the straw was artificial because it was sterilised before being infested with the pathogenic fungus. However, it indicates the possibility that antibiotic production in situ on necrotic tissue can be utilised in biocontrol. A more informative approach to assessing the role of antibiosis in antagonism would be the use of reporter gene fusions to indicate whether identified genes for antibiotic production are transcribed *in situ*.

Antibiosis and Stress

Further work with Pf-5 and its mutant JL3985 has revealed an aspect of bacterial antibiosis that again relates to the microbial environment on senescent and necrotic aerial plant tissue. Sarniguet *et al* (1996), working with the pyrrolnitrin-deficient mutant, determined that the DNA segment involved in the deficiency is not a structural gene, but a regulatory sequence homologous to the *rpoS* gene in *E. coli*. This gene codes for a sigma factor of RNA polymerase, and is induced under stress conditions (Hengge-Aronis, 1993). In Pf-5, a mutation at this locus confers atypical sensitivity of stationary-phase cells to hydrogen peroxide or high salt concentration, i.e. a loss of normal stress-response traits (Sarniguet *et al.*, 1996). That is, production of pyrrolnitrin, an antifungal antibiotic, is regulated coordinately with several other traits that confer resistance to stress, and are induced when the bacterial cell senses a stress of some kind. (The existence of coordinately-regulated gene clusters that respond to environmental conditions has been suggested to be a general characteristic of genetic organisation in microbes (Beattie and Lindow, 1994).) For *rpoS* in *E. coli*, one trigger for environmental stress is starvation (Hengge-Aronis, 1993); for Pf-5, it is not certain whether this is the triggering condition, or even one of the triggering conditions, for the response. Furthermore, regulation of antibiotic production in this organism is complex; several antibiotics of Pf-5, including pyrrolnitrin, are coordinately regulated by a different, two-component system with a sensor component homologous to LemA (Corbell and Loper, 1995), of the well-studied LemA/GacA regulatory system (Hrabak and Willis, 1992). Also, production of another Pf-5 antibiotic, 2,4-diacetylphloroglucinol, responds to carbon source (Nowak-Thompson and Gould, 1994). The interrelationships of such nutrient-stimulated antibiotic production, the two-component overall regulation and the stress-induced regulation of pyrrolnitrin is presently unknown. In any case, there is thus now evidence that for at least some antibiotics, stress conditions that occur on plant surfaces may be a key to expression of traits important in biological control. As more is learned about the stresses that trigger this coordinated response, it may be possible to better manipulate the microbial environment to obtain expression of antibiotic-producing capacity at the intended site of biocontrol interactions.

CONCLUSIONS

Senescent and necrotic aerial plant surfaces are a unique aerial plant-surface environment for microbial interactions, one that exists as a continuum from senescent to necrotic. This continuum of environments is the result of a transition, through time, of the integrity of the plant surface - as the plant's vigour weakens, the nature of the nutritional resource and physical microenvironmental changes. The resulting changes in availability of nutrients and their accessibility for microbes is a major determinant of the microbial interactions in this habitat. These interactions, which are crucial for microbial community development, may be subject to manipulation for achieving biocontrol of fungi. Biocontrol strategies for senescent and necrotic tissues can include: reduction in inoculum potential of pathogenic fungi on necrotic tissue contiguous to the infection court, reduction of primary or secondary inoculum produced on senescent/necrotic tissue, or possibly cultivation of beneficial microbes on necrotic tissues for dispersal to immigration-limited living phyllospheres where biocontrol is to occur. Through a better understanding of the physical, nutritional and biological nature of the microbial environment on senescent/necrotic aerial plant tissues, and the ways that microbes sense and respond to it, we may be able to manage the persistence and activity of microbial populations there to our benefit.

REFERENCES

Abawi, G. S. and Grogan, R. G. 1975, Source of primary inoculum and effects of temperature and moisture on infection of beans by *Whetzelinia sclerotiorum*. *Phytopathology* 65:300-309.

Adee, E. A. and Pfender, W. F. 1989, The effect of primary inoculum level of *Pyrenophora tritici-repentis* on tan spot epidemic development in wheat. *Phytopathology* 79:873-877.

Adee, S. R., Pfender, W. F. and Hartnett, D. C. 1990, Competition between *Pyrenophora tritici-repentis* and *Septoria nodorum* in the wheat leaf as measured with de Wit replacement series. *Phytopathology* 80:1177-1182.

Andrews, J. H. 1992, Biological control in the phyllosphere. *Annu. Rev. Phytopathol.* 30:603-635.

Atlas, R. M. and Bartha, R. 1993, *Microbial Ecology: Fundamentals and Applications.* USA: Benjamin-Cummings Publishing Co.

Baddeley, M.S. 1971, Biochemical aspects of senescence, pp. 415-429 In: Preece, T.F. and Dickinson, G.H., (eds.) *Ecology of Leaf Surface Micro-organisms*. London: Academic Press.

Beattie, G.A. and Lindow, S.E. 1994, Epiphytic fitness of phytopathogenic bacteria: Physiological adaptations for growth and survival, pp. 1-27 In: Dangl, J.L., (ed.) *Current Topics in Microbiology and Immunology*, vol. 129. Heidelberg: Springer-Verlag.

Biles, C. L.and Hill, J. P. 1988, Effect of *Trichoderma harzianum* on sporulation of *Cochliobolus sativus* on excised wheat seedling leaves. *Phytopathology* 78:656-659.

Blakeman, J. P. 1991, Foliar bacterial pathogens: epiphytic growth and interactions on leaves. *J. Appl. Bacteriol. Symposium Supplement* 70:49S-59S.

Blakeman, J. P. and Brodie, I. D. S. 1977, Competition for nutrients between epiphytic microorganisms and germination of spores of plant pathogens on beetroot leaves. *Physiol. Plant Pathol.* 10:29-42.

Boland, G. J. and Inglis, G. D. 1989, Antagonism of white mold (*Sclerotinia sclerotiorum*) of bean by fungi from bean and rapeseed flowers. *Can. J. Bot.* 67:1775-1781.

Burrage, S.W. 1976, Aerial microclimate around plant surfaces, pp. 173-184 In: Dickinson, C.H. and Preece, T.F., (eds.) *Microbiology of Aerial Plant Surfaces*. London: Academic Press.

Chakraborty, B. N., Chakraborty, U. and Basu, K. 1994, Antagonism of *Erwinia herbicola* towards *Leptosphaeria maculans* causing blackleg disease of *Brassica napus*. *Letters in Appl. Microbiol.* 18:74-76.

Cherif, M. and Benhamou, N. 1990, Cytochemical aspects of chitin breakdown during the parasitic action of a *Trichoderma* sp. on *Fusarium oxysporum* f. sp. *radicis-lycopersici*. *Phytopathology* 80:1406-1414.

Chernin, L., Ismailov, Z., Haran, S. and Chet, I. 1995, Chitinolytic *Enterobacter agglomerans* antagonistic to fungal plant pathogens. *Appl. Environ. Microbiol.* 61(5):1720-1726.

Corbell, N. and Loper, J. E. 1995, A global regulator of secondary metabolite production in *Pseudomonas fluorescens* strain Pf-5. *J. Bacteriol.* 177:6230-6236.

Cullen, D. and Andrews, J.H. 1984, Epiphytic microbes as biological control agents, pp. 381-399 In: Kosuge, T. and Nestler, E.W., (eds.) *Plant-microbe Interactions*. New York: MacMillan Publishing Company.

Di Pietro, A., Lorito, M., Hayes, C. K., Broadway, R. M. and Harman, G. E. 1993, Endochitinase from *Gliocladium virens*: isolation, characterization, and synergistic antifungal activity in combination with gliotoxin. *Phytopathology* 83:308-313.

Dickinson, C. H. 1967, Fungal colonization of *Pisum* leaves. *Can. J. Bot.* 45:915

Dik, A. J., Fokkema, N. J. and van Pelt, J. A. 1992, Influence of climatic and nutritional factors on yeast population dynamics in the phyllosphere of wheat. *Microb. Ecol.* 23:41-52.

Elad, Y., Köhl, J. and Fokkema, N. J. 1994, Control of infection and sporulation of *Botrytis cinerea* on bean and tomato by saprophytic yeasts. *Phytopathology* 84:1193-1200.

Fernandes, J. M. C., Sutton, J. C. and James, T. D. W. 1991, A sensor for monitoring moisture of wheat residues: Application in ascospore maturation of *Pyrenophora tritici-repentis*. *Plant Dis.* 75:1101-1105.

Fiala, V., Glad, C., Martin, M., Jolivet, E. and Derridj, S. 1990, Occurrence of soluble carbohydrates on the phylloplane of maize (*Zea mays* L.): variations in relation to leaf heterogeneity and position on the plant. *New Phytol.* 115:609-615.

Fokkema, N. J. 1993, Opportunities and problems of control of foliar pathogens with micro-organisms. *Pestic. Sci.* 37:411-416.

Haran, S., Schickler, H., Oppenheim, A. and Chet, I. 1995, New components of the chitinolytic system of *Trichoderma harzianum*. *Mycol. Res.* 99(4):441-446.

Hengge-Aronis, R. 1993, Survival of hunger and stress: The role of *rpoS* in early stationary phase gene regulation in *E. coli*. *Cell* 72:165-168.

Heye, C. C. and Andrews, J. H. 1983, Antagonism of *Athelia bombacina* and *Chaetomium globosum* to the apple scab pathogen, *Venturia inaequalis*. *Phytopathology* 73:650-654.

Howell, C. R. and Stipanovic, R. D. 1980, Suppression of *Pythium ultimum*-induced damping-off of cotton seedlings by *Pseudomonas fluorescens* and its antibiotic, pyoluteorin. *Phytopathology* 70:712-715.

Hrabak, E. M. and Willis, D. K. 1992, The *lemA* required for pathogenicity of *Pseudomonas syringae* pv. *syringae* on bean is a member of a family of two-component regulators. *J. Bacteriol.* 174:3011-3020.

Hudson, H. J. 1968, The ecology of fungi on plant remains above the soil. *New Phytol.* 67:837-874.

Hudson, H. J., Webster, J. 1958, Succession of fungi on decaying stems of *Agropyron repens*. *Trans. Brit. Mycol. Soc.* 41:165-177.

James, T. D. W., Sutton, J. C. and Rowell, P. M. 1984, Monitoring wetness of dead onion leaves in relation to *Botrytis* leaf blight. *Proc. British Crop Protection Conference* 2:627-632.

Kempf, H. J. and Wolf, G. 1989, *Erwinia herbicola* as a biocontrol agent of *Fusarium culmorum* and *Puccinia recondita* f. sp. *tritici* on wheat. *Phytopathology* 79:990-994.

Kinkel, L. 1991, Fungal community dynamics, pp. 253-270 In: Andrews, J.H. and Hirano, S.S., (eds.) *Microbial Ecology of Leaves.* New York: Springer-Verlag.

Köhl, J. and Fokkema, N.J. 1994, Fungal interactions on living and necrotic leaves, pp. 321-334 In: Blakeman, J.P. and Williamson, B., (eds.) *Ecology of Plant Pathogens.*, Oxon, UK: CAB International.

Köhl, J., Molhoek, W. M. L., van der Plas, C. H. and Fokkema, N. J. 1995, Suppression of sporulation of *Botrytis* spp. as a valid biocontrol strategy. *Eur. J. Plant Pathol.* 101:251-259.

Köhl, J., Molhoek, W. M. L., van der Plas, C. H. and Fokkema, N. J. 1995, Effect of *Ulocladium atrum* and other antagonists on sporulation of *Botrytis cinerea* on dead lily leaves exposed to field conditions. *Phytopathology* 85:393-401.

Köhl, J., van der Plas, C. H., Molhoek, W. M. L. and Fokkema, N. J. 1995, Effect of interrupted leaf wetness periods on suppression of sporulation of *Botrytis allii* and *B. cinerea* by antagonists on dead onion leaves. *Eur. J. Plant Pathol.* 101:627-637.

Kolattukudy, P. E., Rogers, L. M., Li, D., Hwang, C., and Flaishman, M. A. 1995, Surface signalling in pathogenesis. *Proc. Nat. Acad. Sci. USA* 92:4080-4087.

Lorito, M., Di Pietro, A., Hayes, C. K., Woo, S. L. and Harman, G. E. 1993, Antifungal, synergistic interaction between chitinolytic enzymes from *Trichoderma harzianum* and *Enterobacter cloacae*. *Phytopathology* 83:721-728.

Lorito, M., Hayes, C. K., Di Pietro, A., Woo, S. L. and Harman, G. E. 1994, Purification, characterization, and synergistic activity of a glucan 1,3-ß-glucosidase and an *N*-Acetyl-ß-Glucosaminidase from *Trichoderma harzianum*. *Phytopathology* 84:398-405.

Magan, N. and Lacey, J. 1984, Effect of water activity, temperature and substrate on interactions between field and storage fungi. *Trans. Brit. Mycol. Soc.* 82:83-93.

Mercier, J. and Reeleder, R. D. 1987, Interactions between *Sclerotinia sclerotiorum* and other fungi on the phylloplane of lettuce. *Can. J. Bot.* 65:1633-1637.

Morgan, J. V. and Tukey, H. B. 1964, Characterization of leachate from plant foliage. *Plant Physiol.* 39:590-593.

Nowak-Thompson, B. and Gould, S. J. 1994, Production of 2,4-diacetylphloroglucinol by the biocontrol agent *Pseudomonas fluorescens* Pf-5. *Can. J. Microbiol.* 40:1064-1066.

Peng, G., Sutton, J. C. 1991, Evaluation of microorganisms for biocontrol of *Botrytis cinerea* in strawberry. *Can J. Plant Pathol.* 13:247-257.

Pfender, W. F. 1988, Suppression of ascocarp formation in *Pyrenophora tritici-repentis* by *Limonomyces roseipellis*, a basidiomycete from reduced-tillage wheat straw. *Phytopathology* 78:1254-1258.

Pfender, W. F., King, L. G. and Rabe, J. R. 1991, Use of dual-stain fluorescence microscopy to observe antagonism of *Pyrenophora tritici-repentis* by *Limonomyces roseipellis* in wheat straw. *Phytopathology* 81:109-112.

Pfender, W. F., Kraus, J. and Loper, J. E. 1993, A genomic region from *Pseudomonas fluorescens* Pf-5 required for pyrrolnitrin production and inhibition of *Pyrenophora tritici-repentis* in wheat straw. *Phytopathology* 83:1223-1228.

Pfender, W. F., Sharma, U. and Zhang, W. 1991, Effect of water potential on microbial antagonism to *Pyrenophora tritici-repentis* in wheat residue. *Mycol. Res.* 95:308-314.

Pfender, W. F. and Wootke, S. L. 1988, Microbial communities of *Pyrenophora*-infested wheat straw as examined by multivariate analysis. *Microb. Ecol.* 15:95-113.

Pfender, W. F., Zhang, W. and Nus, A. 1993, Biological control to reduce inoculum of the tan spot pathogen *Pyrenophora tritici-repentis* in surface-borne residues of wheat fields. *Phytopathology* 83:371-375.

Powelson, R. L. 1960, Initiation of strawberry fruit rot caused by *Botrytis cinerea*. *Phytopathology* 50:491-494.

Sarniguet, A., Kraus, J., Henkels, M.D., Muelchen, A.M. and Loper, J.E. 1995, An *rpoS* homolog affects antibiotic production and biological control activity of *Pseudomonas fluorescens* Pf-5. *Proc. Nat. Acad. Sci. USA* 92:12255-12259.

Shapira, R., Ordentlich, A., Chet, I. and Oppenheim, A. B. 1989, Control of plant diseases by chitinase expressed from cloned DNA in *Escherichia coli*. *Phytopathology* 79:1246-1249.

Silvertown, J.W. 1982, Interactions in mixtures of species, pp. 147-165 In: Silvertown, J.W. *Introduction to Plant Population Ecology*. New York: Longman.

Sitepu, D. and Wallace, H. R. 1984, Biological control of *Sclerotinia sclerotiorum* in lettuce by *Fusarium lateritium*. *Aust . J. Exp. Agric. Anim. Husb.* 24:272-276.

Sutton, J. C. 1990, Epidemiology and management of botrytis leaf blight of onion and gray mold of strawberry: a comparative analysis. *Can. J. Plant Pathol.* 12:100-110.

Sutton, J. C. and Peng, G. 1993, Biocontrol of *Botrytis cinerea* in strawberry leaves. *Phytopathology* 83:615-621.

Tronsmo, A. 1992, Leaf and blossom epiphytes and endophytes as biological control agents, pp. 43-54 In: Tjamos, E.S. (ed.) *Biological Control of Plant Diseases.* New York: Plenum Press.

Wicklow, D.T. 1992, Interference competition, pp. 265-274 In: Carroll, G.C. and Wicklow, D.T (eds.) *The Fungal Community: Its Organization and Role in the Ecosystem.* 2nd ed. New York: Marcel Dekker, vol. 15.

Wilhite, S. E., Lumsden, R. D. and Straney, D. C. 1994, Mutational analysis of gliotoxin production by the biocontrol fungus *Gliocladium virens* in relation to suppression of Pythium damping-off. *Phytopathology* 84:816-821.

Wilson, M., Savka, M. A., Hwang, I., Farrand, K. and Lindow, S. E. 1995, Altered epiphytic colonization of mannityl opine-producing transgenic tobacco plants by a mannityl opine-catabolizing strain of *Pseudomonas syringae. Appl. Environ. Microbiol.* 61:2151-2158.

Yuen, G. Y., Craig, M. L., Kerr, E. D. and Steadman, J. R. 1994, Influences of antagonist population levels, blossom development stage, and canopy temperature on the inhibition of *Sclerotinia sclerotiorum* on dry edible bean by *Erwinia herbicola. Phytopathology* 84:495-501.

Zhang, W. and Pfender, W. F. 1992, Effect of residue management on wetness duration and ascocarp production by *Pyrenophora tritici-repentis* in wheat residue. *Phytopathology* 82:1434-1439.

Zhang, W. and Pfender, W. F. 1993, Effect of wetting-period duration on ascocarp suppression by selected antagonistic fungi in wheat straw infested with *Pyrenophora tritici-repentis. Phytopathology* 83:1288-1293.

Zhou, T. and Reeleder, R. D. 1989, Application of *Epicoccum purpurascens* spores to control white mold of snap bean. *Plant Dis.* 73:639-642.

Zhou, T. and Reeleder, R. D. 1991, Colonization of bean flowers by *Epicoccum purpurascens. Phytopathology* 81:774-778.

Zhou, T., Reeleder, R. D. and Sparace, S. A. 1991, Interactions between *Sclerotinia sclerotiorum* and *Epicoccum purpurascens. Can. J. Bot.* 69:2503-2510.

9

ANTAGONISM AND SYNERGISM IN THE PLANT SURFACE COLONISATION STRATEGIES OF FUNGI

Alan D. M. Rayner

School of Biology and Biochemistry
University of Bath
Claverton Down
Bath BA2 7AY
United Kingdom

LIVING ON THE BRINK: TO CAPITALISE ON OPPORTUNITY OR TO CO-OPERATE IN ADVERSITY?

As energy-using systems, all life forms must possess boundaries. However, these boundaries can neither be fully open nor fully closed to energy transfer if they are not to become infinitely dispersed or forever static. Instead, they define the reactive interfaces between the insides and outsides - i.e. the "dynamic contexts" - of living systems. The properties of these interfaces are fundamental to the origins of phenotypic patterns and the manner in which neighbouring life forms interact with one another. Moreover, these properties depend on the interaction of ingredients coming from inside, outside and within the boundaries themselves - not least the water which envelopes the genes of all life forms.

The paradigm that I have just been describing may be termed "contextual dynamicism". It can be expressed mathematically in the form of the equation:

$$P = O(g,e,c)$$

where P represents phenotype, O is a nonlinear function and g, e and c are variables respectively representing the content and expression of genetic information and external conditions. O is defined by the "organisational context" of a living system - i.e. the physico-chemical arena which exchanges energy with the external environment.

This paradigm puts genes in context not as *instigators*, but as *moderators* of phenotype, and the means by which particular sets of boundary conditions can be repeated in successive generations. Contextual dynamicism therefore contrasts with the currently predominant paradigm of "genetic determinism", which assumes that there is a direct relationship between phenotypic and genetic diversity, as expressed by the equation:

Aerial Plant Surface Microbiology, edited by Morris et al.
Plenum Press, New York, 1996

$$P = G(c)$$

where G is a linear genetic function determined by the DNA content of the specific organism under consideration. The inadequacies of genetic determinism as a means of understanding the phenotypic diversity and interactions of life forms arise from its most fundamental assumption - that natural selection operates on discrete units of genetic information contained in discrete individuals. For this to be true, the boundaries of genes and individuals must be absolute, a condition which, as already stated, cannot apply to any dynamic, energy-using system.

The aim of this paper is therefore to discuss how fungi establish on the surface and gain access to the interior of plants, in terms of contextual dynamicism. I will use contextual arguments to show *how* particular phenotypic attributes emerge as an implicit consequence of conditions at fungal and plant boundaries. On the other hand, I will use adaptational arguments to explain which of a realm of possible phenotypic attributes ("phenotypic potential") actually occur, *why* these persist and spread, and how they may be refined to suit the requirements of colonisation strategies in specific niches. I will also highlight the generality of contextual considerations by contrasting the phenotypic and interactive properties of fungi that are appropriate to colonisation of durable (corky or woody) plant surfaces with those properties that are apt on the non-durable surfaces more usually studied by phyllosphere biologists.

In these terms, fungi arriving at the boundary between a plant's internal and external environments encounter conditions that mix the threat of exposure to hostile biotic and abiotic influences with the promise of a nutrient supply. The latter can consist of materials present at the boundary itself, either deposited there from the outside or originating within living plant cells. Such materials furnish the plant surface as a theatre for the establishment of fungal colonies. On the other hand, the very nature of this surface makes it a dynamic, heterogeneous and responsive barrier that impedes access to nutrient sources within the relatively sheltered interior of the plant.

Any successful fungal colonist of plant surfaces therefore has to locate nutrients and convert them into a viable reproductive or migratory form without - or, at least, before - succumbing to competition with neighbours, unfavourable abiotic conditions or an adverse host response. The ability to do this depends fundamentally on the degree to which a fungus is capable, as a dynamic system, of obtaining and retaining resources and distributing them to colonisation foci - i.e. its "inoculum potential" (Garrett, 1970) or "energetic state". Once fashionable, the concept of inoculum potential has all but been superseded in the quest to isolate and characterise genetic factors determining resistance and virulence in particular host-pathogen combinations. However, the operation of these genetic determinants cannot be understood in isolation from contextual boundaries of the participating organisms.

Critical here are processes that enable neighbouring colonies (or components of colonies) to associate (integrate) and so pool resources, or dissociate (differentiate or disintegrate) and so gather, partition, redistribute and compete for resources. These processes involve the generation, degeneration, opening and sealing of contextual boundaries. For reasons about to be explored, dissociation is generally initiated when conditions in at least some part of the system are favourable for energy-gathering, whereas association is initiated when conditions are adverse.

Correspondingly, fungal colonisation of plant surfaces can be viewed as the outcome of a dynamic interplay between association and dissociation in two kinds of energetic fields. One of these fields, the fungus, succeeds by following and creating avenues of least resistance in the other field, the plant surface.

DEFINING ORGANISATIONAL CONTEXT - FUNGAL COLONIES AS HETEROGENEOUS, HYDRODYNAMIC FIELDS

Fungal colonies, whether consisting of unicells (*e.g.* yeasts) or hyphae (mycelia) have widely been regarded as if, ideally, they are reducible to purely additive assemblages of discrete units that grow in direct proportion to the amount that they assimilate. Such colonies can be assumed to exhibit fully predictable, readily calculable dynamics under any particular set of growth conditions (e.g. Trinci, 1978; Prosser, 1991, 1993, 1994a,b).

However, a different perspective becomes possible if it is appreciated that fungal mycelia are *intrinsically* heterogeneous, both in view of the environmental settings in which they operate naturally and because of the way they are organised as dynamic systems. Constancy of conditions and resource supply is not typical of most fungal habitats. Rather, there is variability, discontinuity and consequent unpredictability, though not a complete absence of order, both in space and in time.

The heterogeneity of fungal colonies is evident in the marked shifts in organisational pattern that can occur during the life span of an individual system (e.g. see Rayner, 1991a, 1994; Rayner, Griffith and Ainsworth, 1994a; Rayner, Griffith and Wildman 1994b; Rayner, Ramsdale and Watkins, 1995). When a spore germinates by taking up water and nutrients, it often expands isotropically at first and then "breaks symmetry" to allow the emergence of one or more apically extending, protoplasm-filled hyphal tubes. Alternatively, a determinate developmental pattern may be maintained for greater or lesser periods, resulting in the formation of "giant cells" and yeasts. Once polarity has been established, the hyphal tubes may be fully coenocytic or they may become internally partitioned by centripetal ingrowths or septa. Sooner or later, the tubes branch, either in a Christmas-tree-like or a delta-like pattern. The branches may diverge or they may converge and fuse (anastomose), so converting the initially radiate system at least partially into a network. Whilst some parts of the system may be intimately associated with the nutrient source, others may become sealed off or emerge beyond the immediate sites of assimilation. The branches may remain diffusely associated or they may aggregate to form protective, reproductive or migratory structures. The latter are especially produced by fungi growing in soil or humid tree canopies; they consist of cable-like arrays, mycelial cords and rhizomorphs, and can often extend an order of magnitude faster than individual hyphae. Whilst some parts of the boundary of an established mycelial system may continue to expand, others may stop growing and degenerative processes may set in.

The ecological versatility provided by such a changeable, indefinitely expandable (i.e. indeterminate) dynamic structure is revealed by the multifaceted nature of fungal colonisation processes in natural habitats - altogether, there is far more to mycelial life than growth! A good example is provided by the forest pathogen, *Heterobasidion annosum* (Stenlid and Rayner, 1989). Individual mycelial genotypes (genets) of this fungus arrive and establish at exposed woody surfaces by means of basidiospores. They spread along and transfer between the surfaces of living roots as a superficial, "ectotrophic" mycelium. They colonise heartwood and sapwood. They mate and compete with neighbours. They produce sexual basidiocarps and asexual conidiophores. They can survive for years in decaying roots, trunks and stumps, encased within pseudosclerotial boundaries. All of these activities are likely to occur, and in some cases are essential, within the life span of any one successful genet, but their timing is unpredictable. Such "predictable unpredictability" - in which the occurrence of particular circumstances is certain, but their spatial and temporal location is uncertain - can only be negotiated by a structure that possesses a malleable boundary and so is able to shift its organisational pattern spontaneously or in response to local circumstances. By contrast, a biotrophic rust fungus arriving at a leaf surface produces distinctive infection structures in a predictable sequence prescribed by the organism's developmental programme (Wanner *et al.*, 1985).

The versatility of indeterminate mycelial systems is also brought out by their patterns of development in "matrix" culture designs of the kind illustrated in Fig. 1. These designs consist of sets of chambers that are isolated from one another with respect to diffusion through the growth medium, but interconnected by passageways that allow particular portions of a mycelium to grow between and across separate domains. They therefore combine the discreteness that enables distinctive phenotypic states to be produced in accord with local circumstances, with the continuity which is fundamental to the operation of mycelia as integrated systems. The resultant enhancement both of the diversity and ordering of phenotypic patterns arises because of (1) *microenvironmental selection* - due to exposure to distinct local regimes; (2) *physical sieving* - due to the chance selection and amplification of those phenotypic forms that pass through the gaps between one chamber and the next; (3) *physical focusing* - due to the ability of those parts of the mycelial system that emerge through gaps, and so continue to extend, to act as nodes through which growth resources are gathered and redistributed; (4) *physical reinforcement* - due to the re-iterative use of mycelia connecting between chambers as distributive channels.

It may be possible to understand mycelial versatility in terms of purely contextual considerations based on nonlinear theory (Rayner *et al.*, 1994a,b, 1996). A start can be made by thinking about a colony in its most fundamental form as a self-fuelling or autocatalytic system that uses the energy it has gained from the environment to produce biomass that assimilates more energy. The resulting accretion of biomass is accompanied by the accumulation of water, causing the system to behave as a hydrodynamic field. The boundary of any such system cannot be entirely sealed if there is to be further assimilation, cannot be entirely open if a potential difference is to be sustained, and cannot be entirely static if further energy sources are to be located. There is therefore a counteractive (i.e. nonlinear) dynamic interplay between the expandability of the system and its resistances to boundary deformation, to passage of molecules from or to its environment and to displacement of its contents (Rayner, 1996). Together, the resistances to deformation and molecular passage comprise what may be defined as the degree of "insulation" of the system. Ways in which variations in these resistances could affect four fundamental processes governing the primary functions of mycelia are illustrated in Fig. 2.

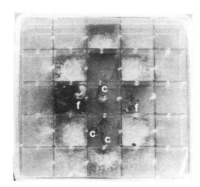

Figure 1. A 10 x 10 cm matrix in which the fungus *Coprinus radians* has been grown across an array of chambers containing low and high concentrations of nutrients. Note the formation of connective mycelial cords (c) and fruit bodies (f) in low nutrient chambers and assimilative mycelium in high nutrient chambers (courtesy of Louise Owen).

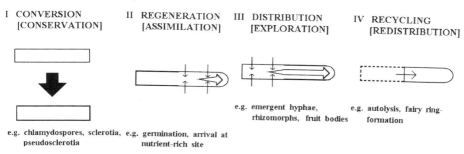

Figure 2. Four fundamental processes in elongated hydrodynamic systems, as determined by boundary deformability, permeability and internal partitioning. Rigid boundaries are shown as straight lines, deformable boundaries as curves, impermeable boundaries by thicker lines, degenerating boundaries as broken lines and protoplasmic disjunction by an internal dividing line. Simple arrows indicate input across permeable boundaries into metabolically active protoplasm, tapering arrows represent throughput due to displacement. (Devised during discussions with Philip Drazin and David Griffel; from Rayner *et al.*, 1995)

In these terms, the dissociative process of branching can be explained as the consequence of input exceeding throughput (conductive) capacity (Rayner *et al.*, 1994a,b, 1995; Rayner, 1996). The latter may be defined as the ability to displace contents smoothly to existing sites of deformation or discharge on the system boundary, and is equivalent to the equilibrium or carrying capacity of a population modelled by the logistic equation (Rayner, 1996). The frequency and pattern of branching will therefore depend on the degree of insulation of the lateral boundaries of hyphae and hence on whether they are in a regenerative (assimilative) or distributive (explorative) state. Assimilative hyphae may be expected to branch like river tributaries, because the sites of input to and outgrowth from the system coincide. By contrast, explorative hyphae will branch like distributaries because the sites of input are remote from the sites of emergence (Rayner, 1996; Rayner *et al.*, 1985). Any internal impedance to the displacement of hyphal contents (e.g. septa) or any external impedance to extension of hyphal apices will have the effect of reducing throughput capacity and increasing branching.

By replacing an in-series set of resistances with an in-parallel one, the integrative process of anastomosis greatly increases the conductivity of a mycelium, allowing it to exhibit coherent, macroscopic variations in biomass distribution not possible in a purely dendritic system. As shown in Fig. 3, many of these properties can be reproduced using a simple, nonlinear, reaction-diffusion model with four fundamental components: (i) a *diffusible substrate* which is the energy source of the system; (ii) *replenishment* of this substrate at a constant specific rate; (iii) an autocatalytic *activator* which facilitates the conversion of the substrate into energy, drives the proliferation of biomass and decays at a constant specific rate; (iv) *diffusion of the activator* at a rate which is inversely related to the system's resistance to throughput.

It should be noted, however, that the model shown could not sustain either an evenly growing margin without a decline in the capacity for substrate utilization in the interior, or the extension and expansion of activator peaks. The latter would correspond with the emergence of hyphal aggregates such as mycelial cords and fruit bodies, due to enhanced delivery to local sites on the boundary of an integrated mycelial system. These inabilities may be related to the fact that as a first step, the model treated the mycelium as a purely assimilative structure, unable to self-insulate, and so incapable of conservation or distribution as depicted in Fig. 2.

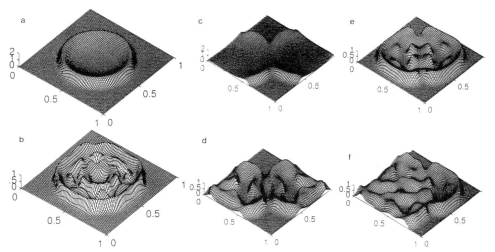

Figure 3. Use of a reaction-diffusion model to predict energy assimilation patterns in growing and interacting mycelial networks of varied resistance to throughput and replenishment of substrate. Numbers on the vertical and horizontal axes respectively represent units of activator concentration (and hence, biomass-generating capacity) and spatial intervals. (a) Fairy-ring-like travelling-wave solution produced when replenishment is low or absent. (b) Irregularly lobed pattern produced in a relatively high resistance system when replenishment is sufficient to offset decay. In low resistance systems, the heterogeneity is reduced and expansion ceases (but can be resumed as a travelling wave if replenishment is prevented). (c) Mutual extinction of the interface between colliding travelling waves, as is observed in natural fairy rings. (d) Formation of a demarcation zone between established, replenished systems - as commonly exhibited by self-inhibiting cultures. (e) Coalescence of immature, replenished systems - as commonly seen in self-pairings between mycelia. (f) Formation of ridges protruding from an established into an immature replenished system - simulating the commonly observed penetration of mycelial cords from one colony into another (from Davidson et al., 1996).

INFLUENCE OF GENETIC AND ENVIRONMENTAL PARAMETERS ON PHENOTYPIC PATTERNS

Given that fungal colonies are organised as hydrodynamic fields, it follows that they can generate varied phenotypic patterns by means of a complex interplay between genetic and environmental factors and the physical properties of boundary components. Of key concern here are mechanisms that result in the opening, sealing and relocation of colony boundaries depicted in Fig. 2.

For any energetically self-fuelling system to sustain itself, its boundaries need to be open to resource capture when external supplies are plentiful, but sealed when there is shortage. It is therefore of interest that it has long been recognized that depletion of nutrients by mycelial systems commonly results in a switchover from what have been termed primary metabolic pathways to secondary metabolic pathways (e.g. Bushell, 1989). Unlike primary metabolism, secondary metabolism is not essential for growth - a fact that has led to a long, still largely unresolved debate about its functional significance. However, many hydrophobic aromatic and terpenoid products of secondary metabolism could play an important role in reducing the permeability of hyphal walls, particularly when subject to polymerisation and depolymerisation in the presence of phenol-oxidising enzymes (e.g. Rayner *et al.*, 1994b). So too could the recently discovered hydrophobin proteins that have been shown to coat the walls of aerial hyphae and sporophores (e.g. Wessels, 1994). If so, then the switch between primary and secondary metabolism - or inductive and transductive metabolism as they might more appropriately be termed - could provide the basis for a feedback mechanism capable of varying hyphal insulation to accord with local availability of assimilable resources (Rayner *et al.*, 1994b).

It is equally important to recognise that the sealing of system boundaries serves not only to prevent dissipation of resources in energy poor environments, but can also provide protection against all kinds of potentially harmful agents. Boundary sealing may therefore enable fungi to survive in and invade what would otherwise be restrictive environments subject to desiccation or irradiation, or containing inhibitory chemicals and/or competitors.

In terrestrial habitats perhaps the most primary, and most commonly unsuspected, danger comes from oxygen (Halliwell and Gutteridge, 1989). The affinity of oxygen for electrons can result in the production of reactive oxygen species (ROS), including singlet oxygen, peroxide ions and superoxide and hydroxyl radicals capable of disrupting the chemical integrity of protoplasm and causing it to degenerate. As shown in Fig. 4, the resulting oxidative stress may profoundly influence boundary conditions affecting pattern generation, via its effects on metabolic processes.

Normally, the efficiency of the electron transport pathway in mitochondria succeeds in fully reducing oxygen to water, whilst minimising ROS formation and maximising production of the energy-rich molecule, ATP. Moreover, the production of enzymes such as superoxide dismutase and antioxidant molecules (many secondary metabolites can act as such) serves to minimise damage, and may indeed be induced by a limited amount of oxidative stress. However, if ROS are for any reason allowed to build up to concentrations where these intracellular mechanisms become overloaded, then the cell will be set on course to death.

In addition to mechanisms operating within protoplasm, another means by which some protection from oxidative stress can be afforded is by coating the cell exterior with substances capable of absorbing oxygen. Many of the aromatic, terpenoid and polypeptide compounds already described as possible insulators are capable of doing just this, particularly when engaged in polymerisation and tanning processes involving the action of phenol-oxidising enzymes and generation of free radicals. For reasons which will be made clearer

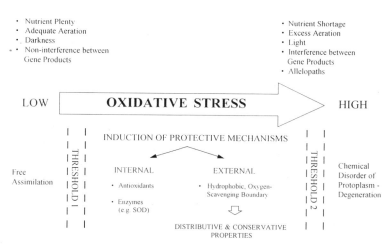

Figure 4. The toxic influence of oxygen on boundary conditions. Factors associated with low and high oxidative stress are listed above the arrow. Distinctive responses to different degrees of oxidative stress above and below two thresholds are listed below the arrow.

shortly, this mechanism may be especially important in the heterokaryotic mycelial systems of basidiomycetes.

The sensitivity of fungi to oxidative stress and involvement of free radical-generating processes in boundary insulation have important implications for the way these organisms respond to their biotic and abiotic environment.

To begin with, whilst the ability to produce particular kinds of molecules involved in pattern generation will depend on specific genetic information, the involvement of free radical reactions provides scope for an autocatalytic boundary chemistry to develop. Once set in motion, this chemistry will become largely independent of genetic control, whilst being very sensitive to local conditions affecting redox potential, pH etc. Control of phenotype generation may then largely be "hyperepigenetic", dependent on non-genetic feedback at the contextual boundaries - i.e. a kind of "learning" or "experiential" process (Rayner *et al.*, 1995). Indeed, the resilience of pre-established phenotypic patterns to changes in genetic information content that has recently been demonstrated in allopatric matings of the basidiomycete, *Heterobasidion annosum*, lends support to the existence of such a process (Ramsdale and Rayner, 1996).

Secondly, any external factor capable of increasing the degree of oxidative stress within fungal protoplasm will induce self-protective and degenerative processes. Exposure to light, for example, is well-known to have powerful effects on fungal morphogenesis and metabolism, at least some of which could be induced by the generation of singlet oxygen via haem and flavin-containing proteins in mitochondria (Halliwell and Gutteridge, 1989). Similarly, release of molecules that disrupt or reinforce hyphal insulation and/or cause mitochondrial dysfunction could play an important role in interfungal and fungus-plant interactions (Rayner et al 1994b; see below).

Thirdly, any intracellular interference between gene products that leads to enhanced susceptibility to oxidative stress may limit the capacity of adjacent mycelial systems to become functionally integrated. Such interference could largely be responsible for the somatic incompatibility reactions that follow anastomosis between different mycelial genets in many ascomycetes and basidiomycetes (Rayner, 1991a,b; see below). In basidiomycetes, the ability to override somatic rejection is essential if sexual outcrossing is to occur by means of heterokaryon formation. This ability may well be due to the production of insulated hyphal systems that resist oxygen overload. The greater propensity of heterokaryons to possess wall-bound phenoloxidase activity and produce distributive and conservative mycelial phases (e.g. rhizomorphs and pseudosclerotia) would readily be explained by this ability.

RELATING PHENOTYPIC PATTERNS TO CONDITIONS AND COLONISATION STRATEGIES AT PLANT SURFACES

The role of association and dissociation in determining patterns of establishment of plant-inhabiting fungal populations and communities may be expected to vary with the durability of host tissues. The greater the durability, the more important the incidence of neighbouring organisms, exposure to adverse external environmental factors and limitation of assimilable resources are likely to be. Such restrictive circumstances require fungal colonists to be combative - possessing antagonistic mechanisms that enable competitors to be overcome or resisted - and/or tolerant of abiotic stresses (e.g. Andrews and Rouse, 1982; Cooke and Rayner, 1984; Cooke and Whipps, 1987; Pugh, 1980). In other words, such colonists need to sustain a relatively non-dissipative (entropy-minimising), far-from-equilibrium energetic state or inoculum potential if they are to succeed. By contrast, less restrictive conditions in non-durable tissues require colonists to proliferate rapidly as relatively dissipative (entropy-maximising) systems, so that they can exploit readily accessible resources before these resources run out.

A rationale for linking metabolic processes affecting boundary-opening and boundary-sealing to classical theories of thermodynamics and the extremes of r- and K-selection in restrictive and non-restrictive environments is shown in Fig. 5.

Under r-selective conditions, when supplies of assimilable resources are temporarily plentiful, the emphasis is on boundary-opening and consequent regenerative processes that result in rapid proliferation of unicells or hyphal branches, associated with high input and respiration rates. Such processes are familiar in many non-durable plant-surface communities, e.g. on young leaves and ripening fruits and in honeydew and damaged tissues. However, they are only sustainable as long as there is continual enrichment of the communities with additional resource supplies. In the absence of such enrichment, it becomes necessary to seal or redistribute boundaries; in the most r-selected forms this most commonly involves production of dispersal spores. If this is not done, the result is genetic suicide, perhaps not only due to starvation, but abetted by oxidative stress. In fact experiments with *Heterobasidion annosum* have demonstrated a reciprocal relationship between the formation of viable colonies possessing wall-bound laccase (a phenol oxidase) activity and the density of conidial suspensions spread onto malt agar (Ramsdale and Rayner, unpublished). Apparently, the overcrowded colonies were unable to self-insulate - a situation reminiscent of overcrowded higher plant seedlings unable to "harden off".

Colonisation processes under r-selective conditions are therefore both highly dissociative and highly dissipative, and the resultant lack of coherence within and between colonies prevents the emergence of ordered patterns of spatial and temporal distribution of individual genotypes and species. Moreover, the lack of structure within species populations is compounded by the propensity of the organisms to reproduce clonally rather than sexually - and thereby with least energy requirement, but reducing genetic diversity.

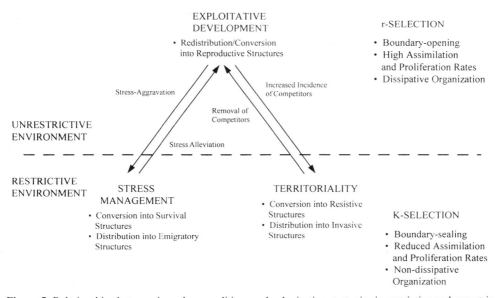

Figure 5. Relationships between boundary conditions and colonisation strategies in restrictive and unrestrictive environments.

When conditions change from *r*- to K-selective, as can often occur within the lifetime of an indeterminate fungal genet, the emphasis changes to boundary-sealing and relocating processes serving conservational, explorative and recycling functions. These functions are aided by anastomosis and - provided that sufficient resources have already been gathered into the system - result in much more coherently ordered, overtly territorial, patterns of colony development.

Territoriality is evident in the formation of antagonistic interactions between adjacent colonies, whether these be different species or different genets of the same species. It is fundamental to the development and change of spatial structure in natural populations and communities (e.g. Rayner and Todd, 1979).

Territoriality between species may sometimes be mediated by diffusible inhibitors, without the need for actual physical contact between colonies. This form of antagonism requires, however, that the responsible chemicals are not dissipated too rapidly, and its relevance in natural as opposed to artificial environments continues to be debated. It may be more usual amongst relatively uninsulated hyphal systems (e.g. those of many ascomycetous fungi) which tend to release rather than sequester secondary metabolites (cf. Rayner and Webber, 1984).

By contrast, many examples of interspecific antagonism only develop when hyphae are very close or in actual contact; this may most often apply to relatively insulated systems, notably those of many basidiomycetous fungi. In such circumstances, interactions can occur both at the scale of individual hyphae, where they result in hyphal interference (due to protoplasmic degeneration) and parasitism (leading to encoiling and penetration), and on macroscopic scales at the level of whole mycelia. In the latter case, the formation of emergent mycelial phases that resist or effect invasion at interaction interfaces, accompanied by redistribution of resources from other parts of the colony, are crucial to the outcome of individual confrontations (Griffith *et al.*, 1994 a,b,c). These explorative and conservational phases can be organised into mycelial barrages, sheets, fans, crusts, cords and rhizomorphs, and clearly depend on the energetic state of individual systems.

Similar kinds of process to those which dictate the outcome of interactions between fungal species can apply to fungus-plant interactions. What are often regarded as specialised infection structures in the form of appressoria, infection cushions and rhizomorphs all provide examples of relatively well-insulated mycelial phases capable of exerting considerable invasive force at an interface. Moreover, the ability to induce a degenerative protoplasmic response in the plant host can both hinder and aid establishment of infection, depending on whether it results in the sealing or opening of host boundaries.

Territoriality within species is expressed in the form of nonself-rejection (somatic incompatibility) reactions that result in the formation of demarcation zones between adjacent colonies, and also by various intracellular and extracellular mechanisms that lead to invasion and take-over of genetic territory (Rayner, 1991a,b). In ascomycetes, nonself-rejection is typically expressed between genetically different homokaryons. By contrast, in sexually outcrossing populations of basidiomycetes, it is expressed between heterokaryotic genets containing mating-type compatible combinations of nuclei or between homokaryons that are not competent to mate. Either way, rejection limits the potential for synergism between genets in these fungi (Rayner, 1991b). Paradoxically, the associative process involved in sexual recombination therefore provides the basis for antagonistic dissociation of individual genets, albeit that heterokaryon formation in basidiomycetes generates better-insulated colonies capable of sustaining a high energetic state. Arrival of large numbers of recombinatorially produced ascospores or basidiospores at a surface may therefore impede rather than enhance colonisation, as has been demonstrated by artificial inoculation of freshly cut beech logs (Coates and Rayner, 1985). On the other hand, arrival of genetically identical assemblages of propagules or hyphae can provide scope for integration and synergism. The ability of individuals and populations to reproduce either clonally or recombinatorially, depending on circumstances, may be critical here (Ainsworth, 1987; Stenlid and Rayner, 1989; Rayner, 1991b).

As has already been implied, the expression of territorial mechanisms between fungi depends on conditions having been favourable for assimilation at least at some stage during colonisation, in order that resources can be mobilised to interaction interfaces. Nonetheless, if adverse conditions are sustained, there may yet be scope for slow or sparse proliferation using what limited resources can be accessed without opening fungal systems to dissipation. This is the means by which stress-tolerant forms, emphasising conservation and (if migratory) recycling, eke out an existence, whilst interfering little with one another or with the functioning of plant hosts.

The ability to develop as stress-tolerant forms can, however, result in a significant territorial advantage if adverse conditions subsequently become alleviated, allowing actively exploitative (i.e. regenerative) states to be produced. This is an important consideration in understanding the colonisation strategies of the possibly very large number of plant-inhabiting fungi that include a latent or endophytic phase within their life cycle (Carroll, 1988; Boddy and Griffith, 1989). Such fungi can gradually build up inoculum until they reach a "critical mass", whence they can themselves overcome the microenvironmental limitations on colonisation. They do this by decomposing allelopaths or refractory compounds and by bringing about dysfunction of host tissues, cells or cell-components - i.e. by direct pathogenesis. Alternatively, they may benefit from predispositional effects brought about by extrinsic stress factors (e.g. drought), pathogenic or decomposer organisms, or intrinsic dysfunctional processes (e.g. senescence, competition between branches). Predisposition by other organisms provides the basis for a synergistic mechanism that does not require direct integration of colonies or activities.

To summarise, phenotypic properties appropriate to the colonisation of durable plant surfaces differ from those favoured on non-durable surfaces. Fungal colonists obviate adverse conditions due to competition, microenvironmental limitation or host response at durable surfaces in two ways. Either they infiltrate in cryptic form or they enhance their inoculum potential by means of synergistic interactions within or between colonies. Alternatively, they depend on damage caused by other agencies to provide a route into plant tissues. These different strategies result in distinctive territorial patterns in the distribution of individual genets, and have important implications regarding host plant survival and sensitivity to environmental change. These points will now be illustrated by referring to the specific case of colonisation of tree bark.

CASE STUDY: BARK AS A BARRIER TO AND THEATRE FOR ARRIVAL AND ESTABLISHMENT OF PATHOGENIC AND BENEFICIAL FUNGAL POPULATIONS

In keeping with themes developed earlier in this paper, tree bark is a durable, hydrophobic, highly oxidised, insulating layer that isolates the water- and assimilate-conducting conduits of xylem and phloem from a desiccating, oxygen-rich external environment. Any damage to or dysfunction within this layer inevitably exposes the underlying tissues to leakage, desiccation and aeration, thereby rendering them vulnerable to exploitative micro-organisms (Mullick, 1977; Rayner and Boddy, 1988). A healthy tree interior goes hand-in-glove with a healthy bark exterior. It is therefore critical to understand the dynamic and potentially complex relationships that can occur between bark function, environmental circumstances and patterns of fungal colonisation (cf. Bier, 1964). However, such understanding is limited by the relative lack of research interest that has been shown in bark as a habitat for fungi (Speer, 1981; Cotter and Blanchard, 1982).

Colonisation of bark can either be purely superficial or penetrate within the tissue itself. In the latter case, the inoculum source can either be external or within the wood, and the fungi involved vary markedly in the degree to which they can truly be regarded as pathogenic or capable of conferring some kind of benefit on the host.

Superficial colonisers of tree bark include both heterotrophs and autotrophs (e.g. lichens, algae, bryophytes) and can in time give rise to quite complex communities - especially if the bark becomes fissured - that illustrate, in microcosm, many of the classical concepts of plant succession. Carbon sources for heterotrophic colonists include the living or non-living tissues or remains of autotrophs - for example the basidiomycetes *Galerina hypnorum* and *Athelia epiphylla* ss. *lato* live respectively amongst bryophytes and algae and lichens. Other sources may be deposited from the atmosphere or arise from within the tree; for example *Mytilidion resinae* inhabits the resin of *Araucaria angustifolia* and *Melophia ophiospora* the cork of *Quercus suber* (Speer, 1981). Scale insects can be infected by members of the Septobasidiaceae, of which the most well known is *Septobasidium burtii*, which uses *Aspiodotus osborni* as a living bridge by which to tap into resources contained in medullary rays of oaks and fruit trees (Cooke, 1977).

Actively pathogenic fungi generally cause canker diseases and can be either necrotrophic (e.g. species of *Nectria, Cryphonectria, Hypoxylon, Diaporthe* etc.) or biotrophic (e.g. *Cronartium* and *Peridermium* species). In some cases they produce organised mycelial systems capable of exerting considerable invasive force. Certain heart rot fungi, such as *Irpex mollis* (Berry and Beaton, 1972) and *Phellinus chrysoloma* (Shigo, 1979) produce a thick pad of mycelium which locally prizes the bark from the tree and may rupture it completely. These fungi kill bark from inside-out. Conversely, the production of organised mycelial systems by fungi invading from outside-in is most generally associated with the ectotrophic habit of root-infecting fungi such as *Heterobasidion annosum* and *Armillaria* species (Garrett, 1970; see above). However, examples can also be found amongst aerial colonists, including various species of *Hymenochaete* and *Phellinus* that transfer from infected to healthy branches by means of extremely strong, melanised, mycelial bridges (Graves, 1914; Ainsworth and Rayner, 1990; Stenlid and Holmer, 1991). The ability to produce these bridging mycelia is correlated with an instability observable in agar culture between colony forms producing abundant, white aerial mycelium, and forms producing appressed, brown-pigmented mycelium. These colony forms have distinctive phenoloxidase activities and where they lie adjacent, a melanised pseudosclerotial plate develops, equivalent to that which covers the bridging mycelia formed in nature (Sharland, Burton and Rayner, 1986).

Not all parasitic fungi that inhabit bark necessarily cause significant disease, however. For example, *Ascodichaena rugosa*, which produces laterally elongated black stromata on smooth-barked trees, was long regarded as an innocuous saprotroph (Ellis and Everhart, 1892). However, more recent observations suggest that it may also have some potential as a biotrophic parasite (Speer, 1981).

A great many fungi can be isolated, ostensibly as endophytes, from surface-sterilised, healthy bark - especially from the outermost cork layers, but sometimes also adjacent to the vascular cambium. In hardwoods, these fungi commonly include species of *Cryptosporiopsis, Libertella, Phomopsis, Cylindrocarpon* and various members of the Xylariaceae. Recent studies (Hedges, Whitehouse and Rayner, unpublished) have indicated that the incidence of these fungi in sycamore (*Acer pseudoplatanus*) bark is significantly affected by the distribution of corticolous lichens. For example, the isolation frequency of most fungi under the lichens *Graphis scripta* and *Graphis elegans* was considerably reduced compared with that under *Enterographa crassa*.

Under certain circumstances, the presence of endophytic fungi in tree bark can have important consequences for the host. For example, under conditions where a tree, or part of a tree, becomes stressed, leading to a water deficit in its tissues, the release of microenvironmental limitation on the development of endophytes may enable certain of these fungi to switch into a more actively assimilative mode. A deleterious consequence may then be the formation of cankers (cf. Bier, 1964), although the degree to which the fungi should in such cases be regarded as primary pathogens is debatable. An example is diamond-bark disease of sycamore, associated with a variety of putatively endophytic fungi, including *Dichomera saubinetii*, which developed in the UK following the drought summer of 1976 (Bevercombe and Rayner, 1980). On the other hand, a beneficial consequence of the activation of endophytic fungi may be the limitation of disease agents. For example, the proliferation of *Phomopsis* in diseased elms inhibits the breeding of *Scolytus* beetles that spread Dutch elm disease (Webber, 1981).

Not all endophytic fungi that become active as a consequence of water deficit necessarily originate in bark. There is increasing evidence that many fungi are "specialised opportunists" that can develop cryptically within sapwood prior to becoming actively assimilative when this tissue becomes dysfunctional (Rayner and Boddy, 1988). These fungi include a variety of host-selective basidiomycetes (e.g. *Piptoporus betulinus* on *Betula*; *Stereum gausapatum* and *Peniophora quercina* on *Quercus*) and ascomycetes (e.g. *Biscogniauxia nummularia*, *Hypoxylon fragiforme* and *Eutypa spinosa* on *Fagus*; *Daldinia concentrica* on *Fraxinus*) that appear rapidly (often within a growing season) on recently dead parts of trunks or branches. Depending on circumstances, and to some extent on viewpoint (see below) they may be regarded as damaging to the tree or as beneficial in the hastening of recycling of resources that would otherwise remain sequestered in redundant tissue (Rayner and Boddy, 1988). Their distribution within the wood is exactly correlated with the position of dead cambium, and in partly living trunks or branches they occupy wedge-shaped columns underlying elongated strips of dead bark ("strip cankers") - which they also invade and sporulate upon. Commonly they are present as one or a few individual genets that can extend along the entire length of a strip canker (>10 m in *Eutypa spinosa*; Hendry, 1993). However, some species, such as *Hypoxylon fragiforme* produce innumerable individual genets (Chapela and Boddy, 1988). Interestingly, *Daldinia concentrica*, which forms extensive individual genets on *Fraxinus*, the tree upon which it is most commonly found fruiting in the southern UK (Boddy, Bardsley and Gibbon, 1985), can be isolated as numerous genets from healthy bark of sycamore (Whiteside, Hedges and Rayner, unpublished). As with other wood-inhabiting fungi (Rayner and Todd, 1979), the territorial limits of mutually exclusive individual genets of specialised opportunists are often detectable as interaction zone lines within the wood. However, there is evidence that such mutual exclusivity only becomes apparent if and when the wood becomes sufficiently aerated. Where a high moisture content is sustained, mixed populations of different species often occur and normally incompatible homokaryons of ascomycetes can produce heterokaryons (Hendry, 1993).

As with the activation of fungal endophytes originating in bark, the cause-effect relationship between the exploitative development of specialised opportunists in xylem and death of vascular cambium may be complex and delicately balanced. At one extreme of the spectrum, the presence of the fungus at some critical mass within the xylem may itself be sufficient to initiate dysfunction. Here the distinction between a specialised opportunist and a wound pathogen or vascular wilt fungus may become hard to define (Rayner and Boddy, 1988). At the other extreme, the activity of the fungus depends purely on dysfunction of the host brought about by other agencies. In between these extremes, enhanced activity of the fungus may contribute to host dysfunction. This provides the basis for a dynamic interplay, involving both positive and negative feedback, between boundary-opening and boundary-sealing processes, probably depending most fundamentally on responses to oxygen of both plant and fungus.

The existence of a dynamic interplay between host and fungus is evident in the oscillatory progress made by some fungal colonists both in bark and in wood. In bark, such progress causes "target cankers", and the formation of successive "necrophylactic periderms" as the result of dedifferentiation to produce an impermeable layer of cells (Mullick, 1977). In wood, it results in the formation of successive "reaction" and "barrier" zones (Pearce, 1991; Rayner and Boddy, 1988). The latter consist of cell layers that become impregnated with hydrophobic substances, including phenolic compounds and suberin. Reaction zones are typically dark in colour, delimit dysfunctional (and thereby colonised) sapwood from functional sapwood already present at the time of dysfunction, and are formed by dying parenchyma cells (Shain, 1979), putatively as a response to oxidative stress. Barrier zones are abnormal annual rings that separate wood present at the time of dysfunction from wood formed by subsequent cambial activity.

The formation of necrophylactic periderms, reaction and barrier zones can all be viewed as the product of non-specific boundary-sealing mechanisms that restrict the spread of dysfunction and so maintain conditions that are inimical to assimilative colonisation by fungi. However, where fungi are able to breach these barriers, including by means of non-assimilative (insulated) hyphal systems with reduced susceptibility to allelopaths, then they may contribute to the further spread of dysfunction until this is curtailed by formation of a new sealant zone.

As has been implied, any loss of function in bark, however brought about, is inevitably accompanied by the spread of dysfunction in sapwood - including when trees are wounded by natural or artificial agencies. In these terms, wounds serve both to expose sapwood to air-borne spores and, depending on the rate of spread of dysfunction and placement of sealant zones, to create conditions favourable for assimilative fungal development. In ecological terms, wounds are therefore an example of enrichment disturbance, initially favouring colonisation by *r*-selected organisms or life cycle stages, prior to favouring more territorial forms capable of degrading refractory substrates (Cooke and Rayner, 1984).

As might therefore be expected, many studies, especially in angiosperms, have revealed that micro-organisms capable of causing discoloration but not decay commonly predominate prior to decay-causing fungi in wounded sapwood.

However, taken together with observations of the confinement of decay and discoloration within reaction and barrier zones, this finding has widely been interpreted as evidence for a synergistic interaction enabling fungi to overcome a tree's defences (Shigo, 1984). Correspondingly, the non-decay fungi have been envisaged to degrade allelopaths produced within physico-chemical barriers produced by the tree as a means of "compartmentalising" infection.

Whether this interpretation is necessary, or even logical, is a moot point (Boddy and Rayner, 1983; Rayner and Boddy, 1988; Rayner, 1993). However, it does draw attention to the complex interplays possible, and difficulties in discriminating between cause and effect, when the associative and dissociative hydrodynamic fields of plants and fungi collide.

CONCLUSIONS

Contextual considerations, based on mechanisms of boundary opening and sealing, provide a perspective on the interactive processes that lead to fungal colonisation of plant surfaces that complements views based purely on genetic fitness. As a consequence, the critical importance of the energetic states of host and invasive organisms, previously embodied in the concept of inoculum potential, is given renewed emphasis. The study of factors affecting boundary conditions, such as oxidative stress, and their relation to genetic mechanisms, may therefore be a productive avenue for future research in plant surface microbiology.

REFERENCES

Ainsworth, A.M., 1987, Occurrence and interactions of outcrossing and non-outcrossing populations of *Stereum, Phanerochaete* and *Coniophora.*, pp. 285-299 In: Rayner, A.D.M., Brasier, C.M. and Moor, D. (eds.) *Evolutionary Biology of the Fungi*, Cambridge University Press.

Ainsworth, A.M. and Rayner, A.D.M., 1990, Aerial mycelial transfer by *Hymenochaete corrugata* between stems of hazel and other trees. *Mycol. Res.* 94:263-266.

Andrews, J.H. and Rouse, D.I., 1982, Plant pathogens and the theory of r- and K-selection. *Am. Nat.* 120:283-296.

Berry, F.H. and Beaton, J.A., 1972, Decay in oak in the central hardwood region. *USDA For. Res. Paper*, NE-242.

Bevercombe, G.P., and Rayner, A.D.M., 1980, Diamond-bark diseases of sycamore in Britain. *New Phytol.* 86:379-392.

Bier, J.E. 1964, The relation of some bark factors to canker susceptibility. *Phytopathology* 54:272-275.

Boddy, L., Bardsley, D.W. and Gibbon, O.M., 1985, Fungal communities in attached ash branches. *New Phytol.* 107:143-154.

Boddy, L. and Griffith, G.S., 1989, Role of endophytes and latent invasion in the development of decay communities in sapwood of angiospermous trees. *Sydowia* 41:41-73.

Boddy, L. and Rayner, A.D.M., 1983, Origins of decay in deciduous trees: the role of moisture content and a re-appraisal of the expanded concept of tree decay. *New Phytol.* 94:623-641.

Bushell, M.E., 1989, The process physiology of secondary metabolite production. *Symp. Soc. Gen. Microbiol.* 44:95-120.

Carroll, G.C., 1988, Fungal endophytes in stems and leaves: from latent pathogen to mutualistic symbiont. *Ecology* 69: 2-9.

Chapela, I.H. and Boddy, L., 1988, Fungal colonization of attached beech branches. II. Spatial and temporal organization of communities arising from latent invaders in bark and functional sapwood, under different moisture regimes. *New Phytol.* 110:47-57.

Coates, D. and Rayner, A.D.M., 1985, Fungal population and community development in beech logs. I. Establishment via the aerial cut surface. *New Phytol.* 101:153-171.

Cooke, R.C., 1977, *The Biology of Symbiotic Fungi.* John Wiley: Chichester.

Cooke, R.C. and Rayner, A.D.M., 1984, *Ecology of Saprotrophic Fungi.* London: Longman.

Cooke, R.C. and Whipps, J.M., 1987, Saprotrophy, stress and symbiosis, pp. 137-148 In: Rayner, A.D.M., Brasier, C.M. and Moor, D. (eds.) *Evolutionary Biology of the Fungi*, Cambridge University Press.

Cotter, H. Van T., and Blanchard, R.O., 1982, The fungal flora of bark of *Fagus grandifolia. Mycologia* 74:836-843.

Davidson, F. A., Sleeman, B. D., Rayner, A. D. M., Crawford, J. W. and Ritz, K. 1996, Context-dependent microscopic patterns in growing and interacting mycelial networks. *Proc. R. Soc. Lond. B.* 263: 373-380.

Ellis, J.B. and Everhart, B.M., 1892, *The North American Pyrenomycetes.* Ellis and Everhart: Newfield, New Jersey.

Garrett, S.D., 1970, *Pathogenic Root-Infecting Fungi.* Cambridge University Press.

Graves, A.H., 1914, Parasitism in *Hymenochaete agglutinans. Mycologia* 6:279-284.

Griffith, G.S., Rayner, A.D.M. and Wildman, H.G., 1994a, Interspecific interactions and mycelial morphogenesis of *Hypholoma fasciculare* (Agaricaceae). *Nova Hedw.* 59:47-75.

Griffith, G.S., Rayner, A.D.M. and Wildman, H.G., 1994b, Interspecific interactions, mycelial morphogenesis and extracellular metabolite production in *Phlebia radiata* (Aphyllophorales). *Nova Hedw.* 59:331-344.

Griffith, G.S., Rayner, A.D.M. and Wildman, H.G., 1994c, Extracellular metabolites and mycelial morphogenesis of *Hypholoma fasciculare* and *Phlebia radiata* (Hymenomycetes). *Nova Hedw.* 59:311-329.

Halliwell, B. and Gutteridge, J.M.C., 1989, *Free radicals in Biology and Medicine*, 2nd ed. Clarendon Press: Oxford.

Hendry, S.J. 1993, Strip-cankering in relation to the ecology of Xylariaceae and Diatrypaceae in beech (*Fagus sylvatica* L.). *PhD thesis*, University of Wales.

Mullick, D.B., 1977, The non-specific nature of defense in bark and wood during wounding, insect and pathogen attack, pp. 395-441 In: Loewus, F.A. and Runeckles, V.C. (eds.) *The Structure, Biosynthesis and Degradation of Wood*, Plenum Press: New York.

Pearce, R.B. 1991, Reaction zone relics and the dynamics of fungal spread in the xylem of woody angiosperms. *Physiol. Mol. Plant Pathol.* 39:41-5.

Prosser, J.I., 1991, Mathematical modelling of vegetative growth of filamentous fungi, pp. 591-623 In: Arora, D.H., Rai, B., Mukerji, K.G. and Knudsen, G.R. (eds.) *Handbook of Applied Biology, vol. 1*, New York: Marcel Dekker.

Prosser, J.I. 1993, Growth kinetics of mycelial colonies and aggregates of ascomycetes. *Mycol. Res.* 97:513-528.

Prosser, J.I. 1994a, Kinetics of filamentous growth and branching, pp. 301-318 In: Gow, N.A.R. and Gadd, G.M. (eds.) *The Growing Fungus*, Chapman & Hall: London.

Prosser, J.I., 1994b, Mathematical modelling of fungal growth, pp. 319-335 In: Gow, N.A.R. and Gadd, G.M. (eds.) *The Growing Fungus*, Chapman and Hall: London.

Pugh, G.J.F. 1980, Strategies in fungal ecology. *Trans. Br. Mycol. Soc.* 75:1-14.

Ramsdale, M. and Rayner, A.D.M. 1996, Imbalanced nuclear ratios, post-germination mortality and phenotype-genotype relationships in allopatrically derived heterokaryons of *Heterobasidion annosum*. *New Phytol.* 133: 303-319.

Rayner, A.D.M. 1991a, The challenge of the individualistic mycelium. *Mycologia*, 83:48-71.

Rayner, A.D.M. 1991b, The phytopathological significance of mycelial individualism. *Annu. Rev. Phytopathol.* 29:305-323.

Rayner, A.D.M. 1993, New avenues for understanding processes of tree decay. *Aboricultural J.* 17:171-189.

Rayner, A.D.M. 1994, Pattern-generating processes in fungal communities, pp. 247-258 In: Ritz, K., Dighton, J. and Giller, K.E. (eds.) *Beyond the Biomass*, Wiley-Sayce: Chichester.

Rayner, A.D.M. 1996, Has chaos theory a place in environmental mycology?, pp. 317-341 In: Frankland, J.C., Magan, N. and Gadd, G.M. (eds.) *Fungi and Environmental Change*, Cambridge University Press.

Rayner, A.D.M. and Boddy, L. 1988, *Fungal Decomposition of Wood*, John Wiley: Chichester.

Rayner, A.D.M., Griffith, G.S. and Ainsworth, A.M. 1994a, Mycelial interconnectedness, pp. 21-40 In: Gow, N.A.R. and Gadd, G.M. (eds.) *The Growing Fungus*, Chapman and Hall: London.

Rayner, A.D.M., Griffith, G.S. and Wildman, H.G. 1994b, Differential insulation and the generation of mycelial patterns, pp. 293-312 In: Ingram, D.S. (ed.) *Shape and Form in Plants and Fungi*, London: Academic Press.

Rayner, A.D.M., Ramsdale, M. and Watkins, Z.R. 1995. Origins and significance of genetic and epigenetic instability in mycelial systems. *Can. J. Bot.* 73: S1241-S1258.

Rayner, A.D.M. and Todd, N.K. 1979, Population and community structure and dynamics of fungi in decaying wood. *Adv. Bot. Res.* 7:333-420.

Rayner, A.D.M. and Webber, J.F. 1984, Interspecific mycelial interactions - an overview, pp. 383-417 In: Rayner, A.D.M. and Jennings, D.H. (eds.) *The Ecology and Physiology of the Fungal Mycelium*, Cambridge University Press.

Shain, L. 1979, Dynamic responses of differentiated sapwood to injury and infection. *Phytopathology* 69:1143-1147.

Sharland, P.R., Burton, J.L. and Rayner, A.D.M. 1986, Mycelial dimorphism, interactions and pseudosclerotial plate formation in *Hymenochaete corrugata*. *Trans. Br. Mycol. Soc.* 86:158-163.

Shigo, A.L. 1979, Tree decay - an expanded concept. *USDA For. Ser. Info. Bull.* no. 419.

Shigo, A.L. 1984, Compartmentalization: a conceptual framework for understanding how trees grow and defend themselves. *Annu. Rev. Phytopathol.* 22:189-214.

Speer, E.O. 1981, Recherches sur la biologie des champignons corticoles des ligneux. *D. Phil. Thesis*, Université Louis Pasteur de Strasbourg.

Stenlid, J. and Holmer, L. 1991, Infection strategy of *Hymenochaete tabacina*. *Eur. J. For. Pathol.* 21:313-318.

Stenlid, J. and Rayner, A.D.M. 1989, Tansley Review No. 19. Environmental and endogenous controls of developmental pathways: variation and its significance in the forest pathogen, *Heterobasidion annosum*. *New Phytol.* 113: 245-258.

Trinci, A.P.J. 1978, The duplication cycle and vegetative development in moulds, pp. 132-163 In: Smith, J.E. and Berry, D.R. (eds.) *The Filamentous Fungi, vol. 3*, London: Arnold.

Wanner, R., Förster, H., Mendgen, K. and Staples, R.C. 1985, Synthesis of differentiation-specific proteins in germlings of the wheat stem rust fungus after heat shock. *Exp. Mycol.* 9:279-283.

Webber, J.F. 1981, A natural biological control of Dutch elm disease. *Nature* 292:449-451.

Wessels, J.G.H. 1994, Developmental regulation of fungal cell wall formation. *Annu. Rev. Phytopathol.* 32:413-437.

10

ROLE OF IMMIGRATION AND OTHER PROCESSES IN DETERMINING EPIPHYTIC BACTERIAL POPULATIONS

Implications for Disease Management

Steven E. Lindow

Department of Environmental Sciences,
    Policy, and Management
University of California
151 Hilgard Hall
Berkeley, California 94720-3110

INTRODUCTION

Conceptually, there are 4 major processes that can influence the size and composition of bacterial populations on leaves. Bacterial population sizes on a leaf can increase by both multiplication on that leaf as well as by immigration of bacteria from other leaves. In contrast, both the death of bacterial cells and their migration from a leaf can contribute to decreases in population size. It appears that most workers have assumed that growth and death of bacteria are the predominant processes that determine population sizes. These processes are obviously much more easy to study than immigration and emigration since they can be done in isolation in the laboratory or greenhouse. For example, the study of the multiplication of bacteria on plants can be easily studied by inoculating plants that are isolated from other plants in incubation chambers; increases in population size are directly attributable to multiplication of the bacteria. In contrast, studies of the importance of immigration of bacteria to the population sizes of bacteria on a leaf require a source of immigrant bacteria and that conditions facilitating immigration be maintained. Since the conditions which favour the immigration of bacteria to plants are not yet well understood, such studies must be done under field conditions. They therefore face the complications of variable environmental conditions which make the experiments difficult to reproduce. For this reason, most information relevant to understanding processes that occur on plants are inferences made from simple field or laboratory observations. Most observations have not been sufficiently detailed to partition the many factors that can influence epiphytic bacterial populations. To date, there have been few studies designed to understand the processes that contribute to epiphytic bacterial populations. While a few excellent and detailed studies have addressed the processes such as immigration that lead to the development of fungal populations on

Aerial Plant Surface Microbiology, edited by Morris et al.
Plenum Press, New York, 1996

plants (Kinkel *et al.*, 1989; Kinkel, 1991), this review will focus exclusively on the processes that contribute to the development of epiphytic bacterial populations. Reference will primarily be made to the processes influencing the colonisation of leaves, but it should be recognized that similar concepts apply to the colonisation of other above-ground plant parts such as flowers. Strategies to manage bacterial population sizes on plants to control disease and other deleterious effects such as frost damage conferred by plant-associated bacteria obviously must address the predominant factors that affect bacterial abundance on plants to be successful.

GROWTH OF BACTERIA ON LEAVES

Net increases of bacterial population sizes on plants in the field have usually been attributed to growth of the bacteria on leaves. There have been many studies in which the population size of bacteria have been measured two or more times on the same set of plants, enabling estimates of net changes in population size to be made (Lindow *et al.*, 1978; Lindow, 1982, 1983, 1985a, 1985b , 1987; Mew and Kennedy, 1982; Gross *et al.*, 1983; Legard and Schwartz, 1987; Timmer *et al.*, 1987; Lindow and Panopoulos, 1988; Malvick and Moore, 1988; Hirano and Upper, 1989; Andersen *et al.*, 1991; Hirano *et al.*, 1991; 1995; Beattie and Lindow, 1994b; Jacques *et al.*, 1995). Many of these reports describe net increases in culturable bacteria with time. Usually, authors attribute this net increase in population size to growth. This attribution is usually based on the following observations: 1) Bacterial multiplication can be shown to occur under various conditions in the greenhouse on isolated plants where immigration does not occur. 2) The rates of increase of bacterial population size on plants in the greenhouse are often higher than that observed in the field. 3) Thus growth, but at a rate less than that under more optimum conditions, is presumed to occur in the field. Obviously, this logic allows us to address whether multiplication is *sufficient* to account for temporal changes in bacterial population size on plants. Unfortunately, it does not account for the potential contribution of alternative processes such as immigration.

There are several limitations to the study of multiplication of bacterial cells on plants. A major obstacle to a study of this process is that it can not be studied continuously on the same leaf. That is, leaves are destroyed when they are detached and washed or macerated to estimate bacterial populations. Similarly, viewing of bacteria by light or scanning electron microscopy (SEM) requires leaves to be fixed in a way that kills the bacteria. Thus it is not possible to return to the same leaf or portion of a leaf to compare bacterial populations at a later time. A second major obstacle in estimating bacterial multiplication is that population sizes are almost always quantified by plate counts of cells removed from leaves. Obviously only bacteria that are culturable will be quantified. Changes in the number of viable cells with time may be less than the change in culturable cells under some conditions in aquatic habitats (Rosak *et al.*, 1984; Rosak and Calwell, 1987; Wilson and Lindow, 1992). This distinction will be particularly important if bacteria enter a viable but non-culturable (VBNC) state. Underestimation of bacterial population sizes will also result if bacteria are cultured on media that are not conducive for growth, and hence for which the plating efficiency is low relative to a medium more conducive to growth. Many bacteria enter the VBNC state when exposed to stressful conditions, particularly when starved for nutrients (Rosak *et al.*, 1984; Rosak and Calwell, 1987). The occurrence of the VBNC state in epiphytic bacteria has only been examined in the case of a *Pseudomonas syringae* strain (Wilson and Lindow, 1992). All viable cells of this strain were found to be culturable under most conditions on leaf surfaces (Wilson and Lindow, 1992). However, when cells had been on moist leaves for many days, conditions which allowed the population to increase to a high and stable size for several days, as many as 60% of the viable cells were no longer culturable (Wilson and

Lindow, 1992). The VBNC state in this epiphyte was much less pronounced than bacteria in other habitats, such as in oligotrophic waters or soil; often fewer than 1% of the viable bacteria are culturable in such habitats (Rosak et al., 1984; Rosak and Calwell, 1987). Thus the VBNC state apparently occurs among epiphytic bacteria on plants but may be less pronounced than in other habitats. Any occurrence of the VBNC state, however, complicates estimates of growth rates of bacteria made from plate counts of cells removed from plants, especially if the proportion of the population that is VBNC is variable (Wilson and Lindow, 1992).

While net changes in the numbers of bacteria are often associated with their growth on plants, sampling must be conducted at frequencies approaching the estimated generation time of the bacteria in order that accurate estimates of growth rates be obtained. For example, for purposes of expediency, measurements of bacterial populations are often made at intervals of several days (Lindow et al., 1978; Lindow, 1982, 1983, 1985a, 1985b, 1987; Mew and Kennedy, 1982; Gross et al., 1983; Legard and Schwartz, 1987; Timmer et al., 1987; Lindow and Panopoulos, 1988; Malvick and Moore, 1988; Andersen et al., 1991; Beattie and Lindow, 1994b; Jacques et al., 1995). While changes in net population size over this scale allow estimates of *average* growth rates to be made, the apparent rapid changes in bacterial numbers that can occur at shorter time scales (Hirano and Upper, 1989; Hirano et al., 1995) would not allow estimates of the *maximum* growth rates to be made from measurements taken only over long time periods.

Despite the many limitations to estimating rates of bacterial multiplication on plants, some clear demonstrations of bacterial growth on plants have appeared. The most compelling evidence of bacterial multiplication on leaves has come from visualisation of bacterial cells on plants with SEM at sequential time points (Leben, 1969; Bashan et al., 1981; Mew et al., 1984; DeCleene, 1989; Mansvelt and Hattingh, 1987, 1989; Wilson et al., 1989). Such studies often reveal increasing spatial densities of bacterial cells with increasing time after inoculation of plants. These data corroborate the observation of increasing numbers of culturable cells with time (Wilson et al., 1989). Unfortunately, most of these studies have been conducted under controlled conditions and suffer from a lack of quantification. Because the numbers of bacterial cells among leaves are variable (Hirano et al., 1982) and even vary over small distances on the same leaf (Kinkel et al., 1995), quantifying bacterial populations visually requires counting cell numbers on too many fields of view. Compelling evidence for rapid multiplication of bacteria on plants under field conditions has come from frequent sampling of leaves to estimate population sizes of *P. syringae* and other bacteria (Hirano and Upper, 1989; Hirano et al., 1995). By measuring bacterial population sizes at a frequency of 2 hours Hirano and Upper (1989) were able to describe *P. syringae* populations that increased with an apparent generation time of 3 to 5 hours. Unfortunately, even though elegant, these estimates of *P. syringae* populations suffer from the uncertainty that the culturability of the cells remained constant during the sampling period and that there were not important immigration or death events. That is, the apparent increase in numbers of *P. syringae* cells may have been due to changes in the environment of the cells that caused their culturability to increase. Clearly, more knowledge of the culturability of bacterial cells in the phylloplane of plants growing in the field will be needed before the contribution of cell growth to changes in bacterial population size with time can be appreciated.

DEATH

Decreases in net population size that have been described on leaves (Lindow et al., 1978; Lindow, 1982, 1983, 1985a, 1985b, 1987; Mew and Kennedy, 1982; Gross et al., 1983; Legard and Schwartz, 1987; Timmer et al., 1987; Lindow and Panopoulos, 1988; Malvick and Moore, 1988; Andersen *et al.*, 1991; Beattie and Lindow, 1994b) have often been attributed to death of the epiphytes. For example, the population size of epiphytic bacteria frequently decreases following periods of dry weather (Henis and Bashan, 1986). The sensitivity of many bacteria to decreases in water availability (Griffin, 1981) suggests that epiphytes succumb to this environmental stress on leaves. Many of the same issues that were important for assessing the growth of bacteria on leaves also influence the interpretation of decreases of net bacterial population size. For example, physical stresses on the leaf may reduce the culturability of epiphytes without reducing their viability. Cells that would enter the VBNC state thus could not be said to be "dead" but would be indistinguishable from dead cells by standard plate-count methods of estimating bacterial populations. The culturability of bacteria on leaves exposed to harsh environmental conditions has only been addressed in one study. The culturability of *P. syringae* cells did not decrease upon their introduction to dry bean leaves (Wilson and Lindow, 1992). While the total culturable population size of the *P. syringae* strain decreased by several orders of magnitude, the number of viable cells also decreased the same amount (Wilson and Lindow, 1992). The conditions, if any, under which epiphytes may become unculturable on leaves needs to be better understood to assess death as a process that influences epiphytic bacterial population sizes.

EMIGRATION

Cells of epiphytic bacteria have been demonstrated to be removed from plants in at least two different ways. The most attention has been placed on rain as a vector by which cells are removed from leaves. Rain can wash off a substantial proportion of the cells from a leaf (up to 50%) (Haas and Rotem, 1976; Lindemann and Upper, 1985; Constantinidou *et al.*, 1990; Upper and Hirano, 1991). While a large fraction of epiphytic bacterial cells can be washed from leaves during heavy rain, this loss is small relative to the magnitude of variation in population size normally seen on leaves at a given time. For example, the population size of a given bacterial species such as *P. syringae* can vary by as much as 1 million-fold on a collection of leaves from a single plant population (Hirano and Upper, 1989; Hirano *et al.*, 1995). While rain can wash some of the cells from leaves, a large variation in bacterial population sizes (although a little smaller overall) will remain among these leaves (Hirano *et al.*, 1995). In addition, the reductions in bacterial population that occur during some rain events (Ercolani *et al.*, 1974) is sometimes small compared to the magnitude of increases that occur following rain (Hirano *et al.*, 1995). The relatively small reductions in bacterial populations during rain may be due in part to the relative efficiency with which bacterial cells attach to plants. While epiphytes are relatively easy to dislodge from leaves by washing or sonication (O'Brien and Lindow, 1989) suggesting that they are not irreversibly attached to plants surfaces, cells in aqueous suspensions often attach to leaves within a few seconds (Romantschuk, 1992; Romantschuk *et al.*, this volume). While many epiphytes may be dislodged during rain, many may reattach to other leaves before they are lost from the plant canopy. Some of the bacteria that are dislodged from plants during rain become incorporated into small droplets of water that can be dispersed away from the plant on which they were generated (Venette and Kennedy, 1975; Venette, 1982). Some of these aerosol droplets can be recaptured by adjacent plants by impaction and sedimentation, thereby minimising the loss of bacteria from the phylloplane during rain (Lindemann *et al.*, 1982; Venette, 1982; Lindemann and Upper, 1985; Upper and Hirano, 1991).

It has recently been appreciated that epiphytic bacteria leave the surface of plants in substantial numbers in dry aerosol particles. A net upward flux of bacteria has been noted above plant canopies, especially during mid-day when plants are dry and winds are at their maximum velocity (Lindemann *et al.*, 1982; Lindemann and Upper, 1985; Upper and Hirano,

1991; Lighthart and Shaffer, 1995). In fact, the upward flux of bacteria from dry plants is much greater than that from either wet plants or from bare soil (Lindemann et al., 1982; Lindemann and Upper, 1985; Upper and Hirano, 1991). A net daily introduction to the atmosphere of about 5×10^6 particles bearing viable bacteria per square meter of canopy of plants having large epiphytic bacterial population sizes (ca. 10^6 cells/cm^2 of leaf) has been observed (Lindemann et al., 1982). If one assumes a leaf area index of 5.0 for the bean canopies under which such fluxes were estimated, and assume that the entire soil surface was covered by plants, then the total epiphytic bacterial population in the canopy is about 5×10^{10} cells/m^2. Thus the loss of about 5×10^6 cells/m^2 represents an emigration of only about 0.0001% of the total epiphytic population each day (assuming that aerosol particles harboured one viable cell). Even if the number of viable cells per aerosol particle were as high as 10 (based on the average size of the particles that contained viable bacteria (Lindemann et al., 1982; Lindemann and Upper, 1985; Upper and Hirano, 1991)) the fraction of the epiphytic bacteria that would be lost as aerosol particles would still be very small. Only if a very high fraction of the emigrant cells lost viability or entered a VBNC state soon after removal from a plant canopy (and hence would not have been counted) could the number of total emigrant cells approach even a few percent of the total bacterial population. Clearly a better understanding of the significance of bacterial emigration could be obtained by the use of methods of estimating bacterial numbers that did not rely on culturing. Thus while an impressive number of bacteria can be released from a plant canopy having a high epiphytic bacterial population during dry conditions, emigration away from a plant appears to contribute little to the reduction of epiphytic bacterial populations on a plant. Other modes of removal of bacteria from plant surfaces, such as by insect vectoring, or physical removal by abrasion of adjacent leaves are probably inconsequential, but have not been quantified.

IMMIGRATION

Immigration of bacteria to a leaf is coupled strongly to their emigration from another leaf since most epiphytes have leaves as their primary habitat. Thus, for bacteria to immigrate to a leaf they must first emigrate from another plant having epiphytic bacterial populations. While, as discussed above, emigration of bacteria from a plant has been shown to be a common phenomenon, it is also probably small in magnitude compared to the population sizes that are often observed on plants.

Mechanisms of Immigration

Immigration to a leaf may occur via several modes of transportation. 1) Many bacteria can be transported to a leaf via rain splash. While quantitatively important in releasing bacteria from plants, rain deposits a large percentage of the bacteria released from plants onto the soil (Ercolani et al., 1974; Lindemann et al., 1982; Lindemann and Upper, 1985; Constantinidou et al., 1990; Upper and Hirano, 1991). Nonetheless, substantial lateral movement of bacteria can occur during rain (Venette, 1982). For example, deposition of a P. syringae strain increased over 5,000 fold during rain compared to deposition during dry conditions (Lindemann and Upper, 1985). The retention of immigrant bacteria in rain water during a rainstorm (as opposed to the number washed from leaves) has not been estimated, however. In general, it is believed that there is a net decrease of bacteria on leaves during a rain event. More bacteria are removed from a plant with an established epiphytic microflora than are deposited from adjacent plants (Upper and Hirano, 1991; Hirano et al., 1995). 2) A number of phytopathogenic bacteria can be transferred from infected plants to healthy plants by insect vectors. While there are some bacterial pathogens that are disseminated due to intimate associations of the pathogen with the insect vector (Harrison *et al.*, 1980; Venette, 1982), most often bacteria are transmitted via insects which are contaminated during their foraging or nectar collecting activities (Harrison *et al.*, 1980). This later phenomenon is most well studied in the case of the vectoring of *Erwinia amylovora* from cankers or infected flowers of pear and apple trees to newly opened flowers where infection can occur (Harrison *et al.*, 1980). While such transmission has been demonstrated, the number of cells that are transferred is unknown, but probably very small. 3) Plant pathogenic bacteria might be disseminated with infected leaves that become airborne. Since bacteria can survive for long periods of time on dead and/or dry infected leaves (Leben, 1981; Henis and Bashan, 1986) it is likely that as leaf fragments are dispersed in the wind, cells of phytopathogens could be transferred to healthy leaves (Venette, 1982). Again, the prevalence of such a phenomenon and the number of bacteria that might potentially be transferred to new leaves by this process is unknown. 4) Many bacterial plant pathogens produce exopolysaccharides. This slime can be quite substantial in the case of certain pathogens such as *E. amylovora*. These slimes can produce strands that are as much as 10 cm long in infected tissues and it has been speculated that these strands could fragment as they dry and disperse as small particles to other plants (Eden-Green and Billing, 1972). 5) Aerosols can be produced under both wet and dry conditions. As noted above, wind blowing over dry foliage is a source of substantial numbers of aerosol particles. The number of dry aerosol particles containing bacteria increased in proportion to wind speed and thus exhibited a diurnal periodicity, being highest at mid-day (Lindemann and Upper, 1985). The production of dry aerosol particles is proportional to the source strength of the plants over which the wind has blown (Lindemann *et al.*, 1982; Lindemann and Upper, 1985; Upper and Hirano, 1991). The deposition of dry aerosol particles has been rather well studied. On average, about 10^3 particles containing viable bacteria were found to be deposited in an area the size of a bean leaf (ca. 100 cm^2) each day (Lindemann and Upper, 1985; Upper and Hirano, 1991; Hirano *et al.*, 1995; Lindow, unpublished). Small water droplets containing bacteria cells can also be generated by raindrops hitting healthy foliage containing epiphytic bacterial populations or infected leaves. While many aerosol droplets are scrubbed from the air by other rain drops, some can disperse beyond the immediate site of release and potentially can be deposited onto other plants (Lindemann and Upper, 1985; Upper and Hirano, 1991; Hirano *et al.*, 1995). The release of such wet aerosol particles is apparently less frequent than the release of dry aerosol particles. The contribution of immigrant wet aerosol particles to the size of epiphytic populations has not been well studied, but is probably less than that of dry particles.

Significance of Immigration

Immigration is significant for populations of epiphytic bacteria in at least two different ways: 1) Immigrant cells are a source of inoculum of bacteria that subsequently multiply on leaves, and 2) Immigrant cells contribute directly to the population size of leaves on plants on which little growth is possible.

The role of immigrant cells as inoculum of plant pathogens that subsequently proliferate on plants has been examined most closely. A popular model of the microbiology of leaves is based on "island biogeography" theory (Andrews *et al.*, 1987). In this model, the population structure of living beings on an island is strongly influenced by the nature and size of immigrants to the islands that are isolated from each other and are initially void of colonists (Andrews *et al.*, 1987). Many leaves emerge from buds or seeds with few if any bacterial colonists (Lindow *et al.*, 1978; Lindow, 1982, 1983, 1985a, 1985b , 1987; Mew and Kennedy, 1982; Gross *et al.*, 1983; Lindow and Panopoulos, 1988; Hirano and Upper, 1989; Andersen *et al.*, 1991; Hirano *et al.*, 1991, 1995; Beattie and Lindow, 1994b; Jacques *et al.*, 1995). Such leaves usually have resources that can support the growth of bacteria inoculated onto them (Lindow *et al.*, 1978, 1983; Lindow, 1982, 1983, 1985a, 1985b, 1987; Lindow and Panopoulos, 1988; Andersen *et al.*, 1991; Beattie and Lindow, 1994b). Such leaves thus have similarities to isolated islands in their initial lack of colonists. The relatively low population size of newly emerged tissues is thus often apparently due to a lack of inoculum of appropriate epiphytic bacterial strains. Adjacent leaves often harbour quantitatively and qualitatively different resident microflora (Hirano *et al.*, 1982, 1995). Similarly, the diversity of bacteria on a given leaf may be either very high or very limited (Hirano *et al.*, 1995). Therefore the colonisation of leaves may be a rather stochastic process involving repeated (but limited) introduction of genotypes of bacteria capable of colonisation of leaves. Occurrence of a particular species (such as a plant pathogen) on a particular leaf will be dependent on its successful immigration (at least once) from another source. This has led to the interest in identifying "collateral hosts", plants with can support epiphytic growth of the pathogen but not lesion development, which can serve as a source of immigrant inoculum of plant pathogenic bacteria (Venette, 1982; Hirano and Upper, 1983, 1990). Such alternate hosts may be very important in the epidemiology of plant pathogens in enabling the pathogen to survive epiphytically for extended periods, even in the absence of susceptible hosts (Venette, 1982; Hirano and Upper, 1983, 1990; Henis and Bashan, 1986; Beattie and Lindow, 1995).

Conceptually, the impact of immigrant bacterial cells in determining the composition of populations on a leaf is influenced by the number of bacterial cells already present on the leaf. As the number of established bacterial cells increases, the contribution of an immigrant cell to the relative population size of the leaf will decrease (Figure 1). A similar decreasing effect of novel immigrant cells to the number of genotypes on a leaf would also be expected (Figure 2). If it is assumed that each strain on a leaf is similar in competitiveness (probably a poor assumption) and the population sizes of each strain each increase rapidly and at the same rate, then it is clear that the first immigrants to a leaf will (at least initially) be well represented in the phylloplane population. Obvious differences in competition, changes in physical environment, changes in host resources, and other perturbations will cause the proportion of different strains on a leaf to change with time; that is, succession will occur. Therefore, immigration to uncolonised leaves is very important in providing inoculum to exploit that habitat. Immigration to more colonised plants is less important quantitatively,

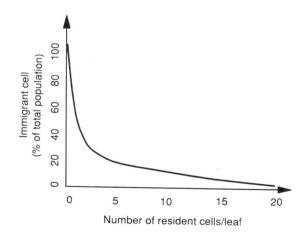

Figure 1. Contribution of a single bacterial cell to the proportion of the total population size of a leaf having the number of resident cells indicated on the abscissa.

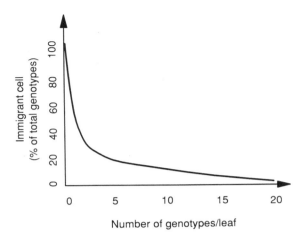

Figure 2. Influence of the addition of a single bacterial cell having a unique genotype on the number of genotypes present on a leaf already harbouring cells of the number of genotypes indicated on the abscissa.

but may be particularly important in contributing inoculum of strains which might exploit unfilled niches on leaf surfaces or of strains that are superior competitors or phytopathogenic.

While immigration has been recognized as contributing qualitatively to the composition of phylloplane bacterial populations, there has been little appreciation of its contribution to the size of these populations. Some communities, including those of some fungal epiphytes, respond directly to enhancements in recruitment or immigration; by definition, these communities are said to be immigration limited (Kinkel *et al.*, 1989; Hughes, 1990; Kinkel, 1991). For example, Kinkel *et al.* (1989) found that *Aureobasidium pullulans* population sizes on apple leaves in the field to which large numbers of this species had been inoculated were significantly larger than population sizes on leaves that had not been inoculated, even after substantial apparent growth of this species on the leaves. This direct response in population size to an increase in the numbers of immigrants indicated that *A. pullulans* is immigration-limited (Kinkel *et al.*, 1989). Similarly, Jacques *et al.* (1995) found that population sizes could be increased by up to 100-fold when the exposure of emerging leaves to air was increased, particularly later in the growing season when the abundant epiphytic bacterial populations that existed on adjacent plants could apparently serve as a source of immigrant inoculum. This is described in more detail in the chapter by Jacques (this volume). Unfortunately, few other studies have addressed this for immigration of fungal or bacterial species on leaves.

Immigration may be relatively more important in determining population size on some plant species than on others. A wide range of bacterial population sizes occurs on plants, even after leaves have been inoculated with epiphytic strains and allowed to incubate under moist conditions conducive for growth for long periods of time (Lindow *et al.*, 1978; Lindow, 1982, 1985; O'Brien and Lindow, 1989). While maximum population sizes for some plants such as bean and cucumber exceed 10^7 cells/cm^2, some plant species, such as citrus species and conifers, support less than about 10^3 cells/cm^2(Lindow *et al.*, 1978; Lindow, 1982, 1985; O'Brien and Lindow, 1989). The rate of deposition of bacteria in landscapes that include an abundance of plants having high epiphytic bacterial populations has been estimated to be about 10^3 cells/100 cm^2/day (Lindemann and Upper, 1985; Upper and Hirano, 1991; Hirano *et al.*, 1995; Lindow and Anderson, 1996). This rate of deposition would yield a cumulative immigrant population of about 10^4 cells per month on an average sized leaf (such as a bean

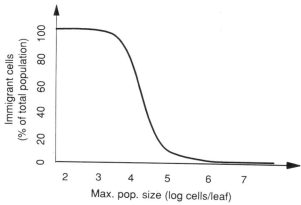

Figure 3. Contribution of immigrant cells to population sizes on plants having different carrying capacities. The relationship shown is for when as many as 10^4 cells immigrated to a leaf of various plant species that, at maturity, harbour the number of epiphytic bacteria indicated on the abscissa.

leaf). If we assume that most leaves in nature are subject to immigration of about 10^4 total cells per month then immigrant cells could account for a substantial fraction of the bacterial population observed on plants having a relatively low population size but only a very small proportion on plants like bean and cucumber having relatively high population sizes (Figure 3). For example, navel orange leaves in California seldom harbour more than about 10^5 cells (Lindow, 1982). Since these leaves are long-lived and have been shown to be exposed to an average of about 10^4 immigrant cells per month (Lindow and Anderson, 1996) most of the cells on their leaves may be attributable to immigration. Further support for such a conjecture is obtained from examining the growth of genetically-marked bacterial strains on citrus leaves after inoculation. Little growth of common epiphytic bacteria such as *P. syringae*, *Erwinia herbicola*, and *Pseudomonas fluorescens* was observed (Lindow and Anderson, 1996). Thus, many plant species having small epiphytic bacterial population sizes may not support the epiphytic growth of bacteria. Instead, such plant species may simply harbour "casual" occupants (immigrants) that did not arise by growth on the plants. As shown by Jacques *et al.* (1995), immigration may be important in determining the epiphytic population size on emerging leaves of plant species that harbour large populations on their mature leaves.

IMPLICATIONS OF THE ROLE OF IMMIGRATION AND OTHER PROCESSES IN DETERMINING BACTERIAL POPULATIONS ON LEAVES

The contribution of immigration relative to growth of bacteria to epiphytic population size has several practical and fundamental implications for the biology of plants and for management of deleterious bacteria. Bacteria on mature plants that have low population sizes are probably mostly immigrants that do not grow on the plant surface and thus may be largely quiescent. Such plants apparently do not provide sufficient nutrients or appropriate physical environments for microbial growth. For this reason, cells may be metabolically inactive. Because no multiplication is occurring on leaves (and also apparently little, if any, utilization of nutrients resources from the plant) the colonists on such plants may not interact as much as on other plants. For example, there is substantial information that bacteria on plants such as bean, which permit substantial bacterial growth, compete for the same limiting nutrients for growth (Wilson and Lindow 1994a, b). Some bacteria on plants may also interact with each other by producing mutually inhibitory antibiotics (Blakeman and Fokkema, 1982); the potential for production of such compounds would seem low on a plant that does not provide sufficient nutrients for cell multiplication. For this reason, biological control of epiphytic bacteria on plants that always have a low epiphytic bacterial population (harbouring primarily immigrant cells) would likely be poor; the potential for successful biological control should increase as the sustainable epiphytic bacterial population size (carrying capacity) increases.

The diversity of bacteria on plants may be expected to increase with decreasing carrying capacity of a plant species. For example, on a plant species such as navel orange, having a low carrying capacity and on which most of the epiphytes are apparently immigrants, a wide variety of bacterial strains is observed (Lindow, unpublished). Since immigrants to a leaf can be from both local as well as distant sources (Upper and Hirano, 1991), the immigrants would be expected to be composed of epiphytes that multiplied on a variety of plants growing under different conditions. In contrast, the young leaves of plant species such as bean which can permit the rapid and extensive growth of inoculated bacteria should harbour an epiphytic population composed primarily of the initial immigrants to the leaf, especially soon after development. The observed diversity of bacteria on these two plant species matches this expectation (Lindow, unpublished). Obviously, process such as competition for limited resources and differential susceptibility to environmental extremes could lead to successional changes in populations with time and blur this distinction. It is clear, however, that more detailed studies of both the rates of immigration and the rate of increase in bacterial populations on plants, similar to that of fungal populations (Kinkel et al., 1989), are needed to better understand the processes that lead to the diversity of bacteria seen on plants.

The epiphytic bacterial populations on many plant species may be expected to be immigration limited. Such immigration limitation has already been demonstrated for several crop plants such as almond, corn, bean, potato, navel orange, tomato, pear and broad-leaves endive (Lindow, 1982, 1983, 1985a, 1985b, 1987; Lindow et al., 1978, 1983; Lindow and Panopoulos, 1988; Andersen et al., 1991; Beattie and Lindow, 1994b; Jacques et al., 1995). Some of these plant species have relatively large carrying capacities, suggesting that immigration limitation will not be restricted to those plants in which little epiphytic growth occurs. Immigration limitation may be most prominent early in the development of an epiphytic microflora, before many of the resources that the plant provides are depleted. For example, newly opened flowers of trees such as pear apparently possess nutrient resources capable of supporting the development of large ($> 10^6$ cells/flower) populations. Frequently such populations are not achieved for many days or weeks unless flowers are inoculated with a suitable bacterial strain (Lindow, 1982, 1983; Andersen et al., 1991; Lindow, unpublished).

Knowing whether epiphytic bacterial populations are immigration or growth limited will have important implications on the strategy of management of these populations. For example, the population sizes of bacteria on plants that do not support much bacterial growth or the young leaves of flowers of many plants may be expected to be immigration-limited. The manipulation of immigrant inoculum may alter the epiphytic microflora on a plant. Growing of crop plants in areas distant from the presence of bacterial pathogens on crop or non-crop plants has been recognized as a way to limit its presence on the crop (Venette, 1982). In a classic example of this phenomenon, brown spot disease of bean has been shown to be associated with proximity to vetch, a non-host plant that harbours P. syringae (Ercolani et al., 1974). Likewise, epiphytic populations of P. syringae and brown spot disease were lower on beans planted in regions where other beans were absent than in a bean-growing region (Lindemann et al., 1984). Quantitative effects of nearby sources of immigrant bacteria on epiphytic populations on crop plants have also been observed. For instance, total culturable bacteria and ice nucleation active (Ice$^+$) bacteria were more numerous on navel orange trees grown near plants having high epiphytic bacterial populations (Lindow, 1996). Many cover crop plant species grown under pear trees in California had much higher epiphytic bacterial populations than pear, especially early in the spring when pear flowers and leaves were just emerging from buds (Lindow, unpublished). The population size of both total culturable bacteria and Ice$^+$ bacteria were higher on trees grown above such cover crop plant species than on trees grown above bare soil (Lindow, unpublished). Furthermore, the population sizes of bacteria on pear were highly correlated with the population sizes of bacteria on different plant species grown under the trees (Lindow, unpublished). Thus, an appreciation of the contribution of immigrant bacteria to epiphytic bacterial population sizes can be used to devise management strategies that minimise the development of large populations of deleterious bacteria on plants. The efficacy of bactericides to manage epiphytic bacterial populations on plants where immigration plays the predominant role in population size increases would be expected to be much less than on plants where growth was responsible for the majority of bacterial immigrants. Similarly, the management of sources of immigrant bacteria, such as by altering the number or proximity of other plants that could contribute immigrants bacteria, would be expected to be less effective in altering epiphytic bacterial populations on plants where rapid epiphytic growth occurs than on plants where it is not. Studies of processes determining populations sizes of epiphytes thus can contribute directly to improvements in control procedures.

REFERENCES

Andersen, G.L., Menkissoglu, O. and Lindow, S.E. 1991, Occurrence and properties of copper-tolerant strains of *Pseudomonas syringae* isolated from fruit trees in California, *Phytopathology* 81:648-656.

Andrews, J.H., Kinkel, L.L., Berbee, F.M. and Nordheim, E.V. 1987, Fungi, leaves and the theory of island biogeography, *Microb. Ecol.* 14:277-290.

Bashan, Y., Sharon, E., Okon, Y. and Henis, Y. 1981, Scanning electron and light microscopy of infection and symptom development in tomato leaves infected with *Pseudomonas tomato*, *Physiol. Plant Pathol.* 19:139-144.

Beattie, G.A. and Lindow, S.E. 1995, The secret life of foliar bacterial pathogens on leaves, *Annu. Rev. Phytopathol.* 33:145-172.

Beattie, G.A. and Lindow, S.E. 1994a, Survival, growth and localization of epiphytic fitness mutants of *Pseudomonas syringae* on leaves, *Appl. Environ. Microbiol.* 60:3790-3798.

Beattie, G.A. and Lindow, S.E. 1994b, Comparison of the behavior of epiphytic fitness mutants of *Pseudomonas syringae* under controlled and field conditions, *Appl. Environ. Microbiol.* 60:3799-3808.

Bedford, D.E., MacNeill, B.H., Bonn, W.G. and Dirks, V.A. 1988, Population dynamics of *Pseudomonas syringae* pv. *papulans* on Mutsu apple, *Can. J. Plant Pathol.* 10:23-29.

Blakeman, J.P. and Fokkema, N.J. 1982, Potential for biological control of plant diseases on the phylloplane, *Annu. Rev. Phytopathology* 20:167-192.

Constantinidou, H.A., Hirano, S.S., Baker, L.S. and Upper, C.D. 1990, Atmospheric dispersal of ice nucleation-active bacteria: the role of rain, *Phytopathology* 80:934-937.

Crosse, J.E. 1959, Bacterial canker of stone-fruits. IV. Investigation of a method for measuring the inoculum potential of cherry trees. *Ann. Appl. Biol.* 47:306-317.

DeCleene, M. 1989, Scanning electron microscopy of the establishment of compatible and incompatible *Xanthomonas campestris* pathovars on the leaf surface of Italian ryegrass and maize, *EPPO Bull.* 19:81-88.

Eden-Green, S.J. and Billing, E. 1972, Fire blight: Occurrence of bacterial strands on various hosts under glasshouse conditions, *Plant Pathol.* 21:121-123.

Ercolani, G.L., Hagedorn, D.J., Kelman, A. and Rand, R.E. 1974, Epiphytic survival of *Pseudomonas syringae* on hairy vetch in relation to epidemiology of bacterial brown spot of bean in Wisconsin, *Phytopathology* 64:1330-1339.

Fryda, S.J. and Otta, J.D. 1978, Epiphytic movement and survival of *Pseudomonas syringae* on spring wheat, *Phytopathology* 48:209-211.

Griffin, D.M. 1981, Water potential as a selective factor in the microbial ecology of soil, pp. 23-95. In: Parr, J.F., Gardner, M.R. and Elliott, L.F. (eds.), *Water Potential Relations in Soil Microbiology*, Soil Science Society of America, special publication no. 9.

Gross, D.C., Cody, Y.S., Proebsting, E.L., Radamaker, G.K. and Spotts, R.A. 1983, Distribution, population dynamics and characteristics of ice nucleation active bacteria in deciduous fruit tree orchards, *Appl. Environ. Microbiol.* 46:1370-1379.

Haas, J.H. and Rotem, J. 1976, *Pseudomonas lachrymans* inoculum on infected cucumber leaves subjected to dew- and rain-type wetting, *Phytopathology* 66:1219-1223.

Harrison, M.D., Brewer, J.W. and Merrill, L.D. 1980, Insect transmission of bacterial plant pathogens, pp. 201-292. In: Harris, K.F. and Maramorosch, K. (eds.). *Vectors of Plant Pathogens*. Academic Press, New York.

Henis, Y. and Bashan, Y. 1986, Epiphytic survival of bacterial leaf pathogens, pp. 252-268. In: Fokkema, N.J. and van den Heuvel; J. (eds.), *Microbiology of the Phyllosphere*. Cambridge University Press, New York.

Hirano, S.S., Rouse, D.I., Clayton, M.K. and Upper, C.D. 1995, *Pseudomonas syringae* pv. *syringae* and bacterial brown spot of bean: A study of epiphytic phytopathogenic bacteria and associated disease, *Plant Dis.* 79:1085-1093.

Hirano, S.S., Nordheim, E.V., Arny, D.C. and Upper, C.D. 1982, Lognormal distribution of epiphytic bacterial populations on leaf surfaces, *Appl. Environ. Microbiol.* 44:695-700.

Hirano, S.S. and Upper, C.D. 1991, Bacterial community dynamics, pp. 271-294. In: Andrews, J.H. and Hirano, S.S. (eds.), *Microbial Ecology of Leaves*. Springer-Verlag, New York.

Hirano, S.S. and Upper, C.D. 1990, Population biology and epidemiology of *Pseudomonas syringae*, *Annu. Rev. Phytopathology* 28:155-177.

Hirano, S.S. and Upper, C.D. 1989, Diel variation in population size and ice nucleation activity of *Pseudomonas syringae* on snap bean leaflets, *Appl. Environ. Microbiol.* 55:623-630.

Hirano, S.S. and Upper, C.D. 1983, Ecology and epidemiology of foliar bacterial plant pathogens, *Annu. Rev. Phytopathol.* 21:243-269.

Hughes, T.P. 1990, Recruitment limitation, mortality and population regulation in open systems: a case study, *Ecology* 71:12-20.

Jacques, M-A, Kinkel, L.L. and Morris, C.E. 1995, Population sizes, immigration and growth of epiphytic bacteria on leaves of different ages and positions of field-grown endive (*Cichorium endivia* var. *latifolia*), *Appl. Environ. Microbiol.* 61:899-906.

Jones, J.B., Pohrezny, K.L., Stall, R.E. and Jones, J.P. 1986, Survival of *Xanthomonas campestris* pv. *vesicatoria* in Florida on tomato crop residue, weeds, seeds and volunteer tomato plants, *Phytopathology* 76:430-434.

Kinkel, L. 1991, Fungal community dynamics. pp. 253-270. In: Andrews, J.H. and Hirano, S.S. (eds.), *Microbial Ecology of Leaves*. Springer-Verlag, New York.

Kinkel, L., Wilson, M. and Lindow, S.E. 1995, Effects of scale on estimates of epiphytic bacterial populations, *Microb. Ecol.* 29:283-297.

Kinkel, L.L. Andrews, J.H. and Nordheim, E.V. 1989, Fungal immigration dynamics and community development on apple leaves, *Microb. Ecol.* 18:45-58.

Leben, C. 1981, How plant-pathogenic bacteria survive, *Plant Dis.* 65:633-637.

Leben, C. 1969, Colonization of soybean buds by bacteria: observations with the scanning electron microscope, *Can. J. Microbiol.* 15:319-320.

Leben, C. 1965, Epiphytic microorganisms in relation to plant disease, *Annu. Rev. Phytopathology* 3:209-230.

Legard, D.E. and Schwartz, H.F. 1987, Sources and management of *Pseudomonas syringae* pv. *syringae* epiphytes on dry beans in Colorado, *Phytopathology* 77:1503-1509.

Lighthart, B. and Shaffer, B.T. 1995, Airborne bacteria in the atmospheric surface layer: Temporal distribution above a grass seed field, *Appl. Environ. Microbiol.* 61:1492-1496.

Lindemann, J. and Upper, C.D. 1985, Aerial dispersal of epiphytic bacteria over bean plants, *Appl. Environ. Microbiol.* 50:1229-1232.

Lindemann, J., Constantinidou, H.A., Barchet, W.R. and Upper, C.D. 1982, Plants as sources of airborne bacteria, including ice nucleation-active bacteria, *Appl. Environ. Microbiol.* 44:1059-1063.

Lindemann, J., Arny, D.C. and Upper, C.D. 1984, Epiphytic populations of *Pseudomonas syringae* pv. *syringae* on snap bean and nonhost plants and the incidence of bacterial brown spot disease in relation to cropping patterns, *Phytopathology* 74:1329-1333.

Lindow, S.E. 1987, Competitive exclusion of epiphytic bacteria by Ice⁻ mutants of *Pseudomonas syringae*, *Appl. Environ. Microbiol.* 53:2520-2527.

Lindow, S.E. 1983, Methods of preventing frost injury caused by epiphytic ice nucleation active bacteria, *Plant Dis.* 67:327-333.

Lindow, S.E. 1982, Population dynamics of epiphytic ice nucleation active bacteria on frost sensitive plants and frost control by means of antagonistic bacteria, pp. 395-416. In: Li, P.H. and Sakai, A. (eds.), *Plant Cold Hardiness*. Academic Press, New York.

Lindow, S.E. 1985a, Ecology of *Pseudomonas syringae* relevant to the field use of Ice⁻ deletion mutants constructed *in vitro* for plant frost control, pp. 23-35. In: Halvorson, H.O., Pramer, D. and Rogul, M. (eds.), *Engineered Organisms in the Environment: Scientific Issues*. ASM, Washington.

Lindow, S.E. 1985b, Integrated control and role of antibiosis in biological control of fireblight and frost injury, pp. 83-115. In: Windels, C. and Lindow, S.E. (eds.), *Biological Control on the Phylloplane*. American Phytopathological Society Press, Minneapolis.

Lindow, S.E. and Anderson, G. L. 1996, Influence of immigration on epiphitc bacterial populations on navel orange leaves. *Appl. Environ. Microbiol.* 62: 2978-2987.

Lindow, S.E., Arny, D.C. and Upper, C.D. 1983, Biological control of frost injury II: Establishment and effects of an antagonistic *Erwinia herbicola* isolate on corn in the field, *Phytopathology* 73:1102-1106.

Lindow, S.E., Arny, D.C., Barchet, W.R. and Upper, C.D. 1978, The role of bacterial ice nuclei in frost injury to sensitive plants, pp. 249-263 In: Li, P. (ed.), *Plant Cold Hardiness and Freezing Stress*, Academic Press, New York.

Lindow, S.E., Arny, D.C. and Upper, C.D. 1978, Distribution of ice nucleation active bacteria on plants in nature, *Appl. Environ. Microbiol.* 36:831-838.

Lindow, S.E. and Panopoulos, N.J. 1988, Field test of recombinant Ice- *Pseudomonas syringae* for biological frost control in potato, pp. 121-138. In: Sussman, M., Collins, C.H., Skinner, F.A. and Stewart-Tull, D.E. (eds.), *The Release of Genetically Engineered Micro-organisms*. Academic Press, London.

Malvick, D.K. and Moore, L.W. 1988, Survival and dispersal of a marked strain of *Pseudomonas syringae* in a maple nursery, *Plant Pathol.* 37:573-580.

Mansvelt, D.E. and Hattingh, M.J. 1989, Scanning electron microscopy of invasion of apple leaves and blossoms by *Pseudomonas syringae* pv. *syringae*, *Appl. Environ. Microbiol.* 55:533-538.

Mansvelt, E.L. and Hattingh, M.J. 1987, Scanning electron microscopy of colonization of pear leaves by *Pseudomonas syringae* pv. *syringae*, *Can. J. Bot.* 65:2517-2522.

Mariano, R.L.R. and McCarter, S.M. 1993, Epiphytic survival of *Pseudomonas viridiflava* on tomato and selected weed species, *Microb. Ecol.* 26:47-58.

McInnes, T.B., Gitaitis, R.D., McCarter, S.M., Jaworski, C.A. and Phatak, S.C. 1988, Airborne dispersal of bacteria in pepper transplant fields, *Plant Dis.* 72:575-579.

Mew, T.W. and Kennedy, B.W. 1982, Seasonal variation in populations of pathogenic Pseudomonads on soybean leaves, *Phytopathology* 72:103-105.

Mew, T.W., Mew, I.P.C. and Huang, J.S. 1984, Scanning electron microscopy of virulent and avirulent strains of *Xanthomonas campestris* pv. *oryzae* on rice leaves, *Phytopathology* 74:635-641.

O'Brien, R.D. and Lindow, S.E. 1989, Effect of plant species and environmental conditions on epiphytic population sizes of *Pseudomonas syringae* and other bacteria, *Phytopathology* 79:619-627.

Romantschuk, M. 1992., Attachment of plant pathogenic bacteria to plant surfaces, *Annu. Rev. Phytopathology* 30:225-244.

Roos, I.M.M. and Hattingh, M.J. 1983, Scanning electron microscopy of *Pseudomonas syringae* pv. *morsprunorum* on sweet cherry leaves, *Phytopathol. Z.* 180:18-25.

Rosak, D.B. and Colwell, R.R. 1987, Survival strategies of bacteria in the natural environment, *Microbiol. Rev.* 51:365-379.

Rosak, D.B., Grimes, D.J. and Colwell, R.R. 1984, Viable but nonrecoverable stage of *Salmonella enteridis* in aquatic systems, *Can. J. Microbiol.* 30:334-338.

Seidler, R.J., Walter, M.V., Hern, S., Fieland, V., Schmedding, D. and Lindow, S.E. 1994, Measuring the dispersal and reentrainment of recombinant *Pseudomonas syringae* at California test sites, *Microbial Releases* 2:209-216.

Timmer, L.W., Marois, J.J. and Achor, D. 1987, Growth and survival of xanthomonads under conditions nonconducive to disease development, *Phytopathology* 77:1341-1345.

Upper, C.D. and Hirano, S.S. 1991, Aerial dispersal of bacteria, pp. 75-94. In: Ginzburg, L.R. (ed.), *Assessing Ecological Risks of Biotechnology*. Butterworth-Heinemann, Stoneham, Massachusetts.

Vanneste, J.L., Yu, J. and Beer, S.V. 1992, Role of antibiotic production by *Erwinia herbicola* Eh252 in biological control of *Erwinia amylovora*, *J. Bacteriol.* 174:2785-2796.

Venette, J.R. 1982, How bacteria find their hosts, pp. 75-94. In: Mount, M.S. and Lacy, G.H. (eds.). *Phytopathogenic Prokaryotes*, vol. 2. Academic Press, New York.

Venette, J.R. and Kennedy, B.W. 1975, Naturally produced aerosols of *Pseudomonas glycinea*, *Phytopathology* 65:737-738.

Wilson, M., Epton, H.A.S. and Sigee, D.C. 1989, *Erwinia amylovora* infection of hawthorn blossom: II. The stigma. *J. Phytopathol.* 1127:15-28.

Wilson, M. and Lindow, S.E. 1992, Relationship of total and culturable cells in epiphytic populations of *Pseudomonas syringae*, *Appl. Environ. Microbiol.* 58:3908-3913.

Wilson, M. and Lindow, S.E. 1993, Interactions between the biological control agent *Pseudomonas fluorescens* A506 and *Erwinia amylovora* in pear blossoms, *Phytopathology* 83:117-123.

Wilson, M. and Lindow, S.E. 1994a, Ecological differentiation and coexistence between epiphytic Ice$^+$ *Pseudomonas syringae* strains and an Ice$^-$ biological control agent, *Appl. Environ. Microbiol.* 60:3128-3137.

Wilson, M. and Lindow, S.E. 1994b, Coexistence among epiphytic bacterial populations mediated through nutritional resource partitioning, *Appl. Environ. Microbiol.* 60:4468-4477.

11

INTEGRATED CONTROL OF *BOTRYTIS CINEREA* ON GREENHOUSE TOMATOES

Philippe C. Nicot[1] and Alain Baille[2]

INRA
[1] Station de Pathologie Végétale
[2] Station de Bioclimatologie
Centre de Recherches d'Avignon
France

INTRODUCTION

Tomato (*Lycopersicon esculentum*) is presently the most produced vegetable, with about 70 million tons per year, representing 17% of total vegetable production (Kraeutler, 1995). In the USA, nearly 80% of tomato production is for processing, while in the European Union, over 50% of the tomatoes are produced for the fresh market. Fresh market tomatoes are typically produced as protected crops, and those for processing are grown in the field (Kraeutler, 1995). A variety of structures may be used for greenhouse production of vegetables, from the most simple, unheated, plastic-covered tunnels, to the most sophisticated high-investment glasshouses where plants are grown year-round in hydroponic systems with computer-controlled fertilisation, climate, and atmosphere composition (Jarvis, 1992). In Europe, glasshouse production is more typical of the northern countries while tunnels are prevalent in the south. Both types of greenhouses are used in France for tomato production, with *ca* . 850 hectares of minimally or unheated tunnels and 1300 hectares of high-investment, heated glass- or plastic-covered houses (Kraeutler, 1995).

Whatever the type of structure used, grey mould, caused by *Botrytis cinerea* Pers.:Fr., is often considered a major problem in greenhouse production of ornamentals and vegetables, including tomato (Jarvis, 1992). In a study conducted by our laboratory in 15 greenhouses in southern France, incidence of grey mould in May and June 1991 ranged from 32 to 100%, and plant mortality as high as 46% was encountered (Table 1). In a larger survey conducted in the same region in late April 1993, attacks by *B. cinerea* were found in 58 of 73 tomato greenhouses (Terrentroy, 1994). Plants killed by the fungus were recorded in 31 of those greenhouses.

The concurrence of several factors concerning the pathogen, the host and the greenhouse environment render the control of this disease a very difficult task. *B. cinerea* is a ubiquitous phylloplane inhabitant both as a saprophyte and a pathogen. It has been isolated at different times of the year from the aerial surfaces of a variety of cultivated and wild plants

Aerial Plant Surface Microbiology, edited by Morris et al.
Plenum Press, New York, 1996

169

Table 1. Incidence of grey mould (*Botrytis cinerea*) in 15 commercial tomato greenhouses in southern France in May-June 1991

Greenhouse	Incidence (%) of plants with lesions on			% healthy plants	% dead plants
	Leaves	Fruits	Stems		
1	90.0	0.0	86.0	0.0	10.0
2	86.0	6.0	96.0	0.0	10.0
3	82.0	0.0	84.0	0.0	16.0
4	63.2	4.4	27.9	26.5	2.9
5	56.0	6.0	94.0	0.0	20.0
6	46.0	0.0	90.0	0.0	46.0
7	25.4	0.0	65.7	25.4	9.0
8	23.4	0.0	28.8	55.0	3.6
9	18.8	0.0	31.3	55.4	2.7
10	0.0	0.0	96.2	3.8	34.6
11	0.0	0.0	52.1	47.9	19.8
12	0.0	0.0	47.2	52.8	7.5
13	0.0	0.0	45.5	54.5	0.0
14	0.0	0.0	32.1	67.9	14.7
15	63.5	92.3	3.8	3.8	0.0

(Blakeman, 1980). As a pathogen, it has a range of over 200 host species which includes angiosperms, gymnosperms, pteridophytes and bryophytes (Jarvis, 1977, 1980a). This fungus is prolific and has a rapid asexual reproductive cycle. In conditions conducive to disease development, spore production can be detected two days after inoculation on tomato stem tissue, and reach several million spores per gram of fresh tissue in less than a week (Nicot *et al.*, 1996). This abundant inoculum is easily transported by air movements and may contribute to the development of explosive epidemics (Jarvis, 1980b).

Many of the cultural practices in greenhouse production of tomatoes promote disease development, particularly in situations of intensive production, where plants are often subjected to various stresses that may leave them more susceptible to disease. The most significant factors are probably those related to the microclimate of the greenhouse. The range of temperature that is favourable for the development of *B. cinerea* largely overlaps the range desired by growers for optimal production (Elad and Shtienberg, 1995). Condensation may occur on the plants, fostering the development of disease, as episodes of high relative humidity are frequent and difficult (and costly) to avoid. Another disease-enhancing cultural practice is the pruning of the plants to remove axillary buds and leaves. This practice is common in greenhouses where indeterminate tomato varieties are trained on vertical strings. This creates a drier microclimate in the lower part of the plants near the ground, but also provides numerous points of entry for the pathogen. Infections by *B. cinerea* on the pruning wounds often result in the development of stem cankers, which may eventually girdle the stem, resulting in plant death. The length of the growing season, which can reach 10-11 months in glasshouse production, is an aggravating factor, as it may allow symptoms to reach their maximum severity, and epidemics to develop fully.

Finally, the control of this disease in tomato greenhouses is further complicated by the absence of commercial varieties resistant to *B. cinerea* and by the very limited arsenal available for chemical control. This is typically a situation where one single control method may not be efficient and an integrated approach has to be taken. As several reviews have recently been published on the management of diseases, including grey mould, in greenhouse crops (Elad and Shtienberg, 1995; Jarvis, 1992), the present chapter will focus on recent research advances providing potential tools for an integrated approach of the control of *B.*

cinerea on tomatoes, with particular emphasis with our own knowledge of the French cropping system.

DISEASE MANAGEMENT THROUGH CLIMATE CONTROL IN THE GREENHOUSE

Microclimatic parameters have long been recognised as key factors in the development of diseases caused by fungal pathogens on aerial plant surfaces. The study of their effect has been used to develop risk prediction models and warning systems in order to help the grower devise better timing of control interventions, including disease escape (Campbell and Madden, 1990). In a greenhouse environment, where the grower has some ability to intervene on the regulation of climatic parameters, epidemiological models may also be used to limit the occurrence of conditions conducive to disease development.

Development of Epidemiological and Risk Prediction Models

The effects of micro-climatic parameters on the different steps of the development of *Botrytis* sp. on its host (namely, spore germination, germ tube elongation, penetration, tissue colonisation and symptom development, sporulation, spore dispersal, survival) have been studied in great detail for a variety of crops. Despite remaining gaps in our knowledge, and occasional lack of consistency in the results obtained in different conditions, these studies, reviewed by Jarvis (1980b, 1992) and Elad and Shtienberg (1995), have greatly increased our understanding on the impact of the environment, and particularly of the essential role of humidity and free water, on the epidemiology of grey mould. Much of the data concern field crops such as onions, strawberry and grapes. Epidemiological models of infection have also been developed for greenhouse ornamentals such as geranium (Hausbeck and Pennypacker, 1991; Sirjusingh and Sutton, 1996), gerbera (Salinas *et al.*, 1989; Kerssies, 1993) and rose (Kerssies *et al.*, 1995). In contrast, information on greenhouse vegetables is more limited. A model was recently developed to forecast outbreaks of grey mould epidemics in non-heated cucumber greenhouses (Yunis *et al.*, 1994). Similar efforts remain to be done for other vegetables crops, as available data are more fragmentary and concern a few specific steps in the development of the fungus, *e.g.* spore germination on tomato leaves (Nicot and Allex, 1991).

Several epidemiological models have been developed and integrated into functional forecast systems, usually coupled with weather forecast, for field crops including grapes and onions (Sutton *et al.*, 1986; Vincelli and Lorbeer, 1989; Broome *et al.*, 1995). Key elements in predicting disease are the occurrence of free water, as rain or dew, and high relative humidity. Although not totally independent from the outside weather, those factors as well as temperature would presumably be more easy to predict in greenhouses, especially those with climate control systems.

Modification of the Greenhouse Climate for Disease Control

Appropriate management of the climate is a key task for the greenhouse grower. The main decisions concern the proper timing for the activation of the heating system and the opening of vents in order to maintain temperature and relative humidity as close as possible to the desired climate schedule. This schedule is often complex and varies according to the physiological state of the crop and the outside weather. Much diversity can be found among greenhouse vegetable growers in the sophistication of climate control equipment and management (Jarvis, 1992). Opening of vents, sometimes coupled with a forced ventilation system, allows the renewal of air. It is used to evacuate the water vapour released by the transpiration of the crop and to avoid build-up of excess humidity inside the greenhouse. Dehumidification in the greenhouse can also be achieved with other devices, for example by passing the air over hygroscopic solutions (Jaffrin *et al.*, 1992; Chraibi *et al.*, 1995). However, such devices, mostly in the experimental stage, are usually not cost effective in the current economic context. Under current practices, ventilation and/or heating remain the principal means of avoiding excess humidity in the greenhouse. If the temperature of the outside air is lower than that desired inside the greenhouse, ventilation must be accompanied by heating. The result of these interventions may be limited by three main factors: (1) the outside climate (for example, ventilation may not suffice to dehumidify a greenhouse if the outside air is saturated as during fog or rain events); (2) the technical capacity of the greenhouse equipment (for example, a furnace with low heating capacity may not suffice to heat adequately a greenhouse during a particularly cold period); and (3) the cost of the energy necessary to maintain the desired levels of temperature and humidity.

In commercial situations, the primary concern for the greenhouse manager is to maintain a climate favourable for plant development and productivity. However, the need to keep relative humidity low and avoid condensation on the plants to limit the development of *B. cinerea* on greenhouse crops have long been recognised and widely communicated by extension services in many regions. Utilisation of climate management for disease control is increasingly regarded by tomato growers as one of the most efficient tools against *B. cinerea* (Terrentroy, 1994). In a survey of 73 greenhouses in southern France in April 1994, Terrentroy reported that symptoms of *B. cinerea* were less frequent in greenhouses (glasshouses and plastic houses) equiped with climate regulation facilities. In these structures, lesions on leaves and fruits were seldom if ever found, while they were prevalent in unheated or minimally heated tunnels, where excess humidity and condensation on the plants occur commonly (Table 2).

A concrete demonstration of the benefits of reducing relative humidity was provided by Winspear *et al.* (1970). These authors showed that the incidence of ghost spots caused by *B. cinerea* on tomato fruits could be reduced substantially in a greenhouse where dehumidification was activated whenever relative humidity became greater than 90%, while disease was almost totally inhibited in a regime of dehumidification set at 75%. An important limit in the large scale implementation of such a control measure is the high energy cost for dehumidification. One possible strategy would be to rationalise the timing of dehumidification based on the estimated risk of disease development, as provided by the epidemiological models presented earlier, and possibly refined to take into account a quantitative measure of inoculum present in the greenhouse. A key factor in many of these epidemiological models is the presence of free water for several hours on plant surfaces. Avoiding the occurrence of

Table 2. Relation between the type of greenhouse and the incidence of grey mould in a survey of 73 commercial tomato greenhouses of Bouches du Rhône district (southern France) in April 1994 (compiled from data of Terrentroy, 1994)

Type of greenhouse	Number of greenhouses with grey mould symptoms on		
	mostly stems	mostly leaves / fruits	no symptoms
Glasshouse	29	0	14
Unheated tunnel	1	11	2
Plastic house	13	4	1

such events may be less energy-costly to implement than permanently keeping the green-house atmosphere below a certain set point of low relative humidity.

Presently, sensors providing pertinent information for climate control and management, such as temperature, relative humidity, and radiation sensors, are available. In the last years, many studies have dealt with the development of wetness sensors able to measure the time period during which condensation occurs on plant surfaces (Barthakur, 1985; Mermier and Fauvel, 1996; review in Sutton *et al.*, 1988). Models to predict the occurrence and duration of dew are also available (see review by Huber and Gillespie, 1992). All these methodological advances in the knowledge and prediction of condensation episodes in the greenhouse will presumably be of great benefit for a more comprehensive management of the climate and the crop. Some attempts to integrate this knowledge in a decision support system for climate management are under study and validation by teams of the French "Institut National de la Recherche Agronomique" (Lagier *et al.*, 1995, 1996). Information concerning *B. cinerea* is taken into account in this system and this aspect may be expanded as more epidemiological data become available on tomato.

Control in Unheated Greenhouses

The plastic covered greenhouses most prevalent in southern Europe are usually not heated. During the winter months, ventilation of these tunnels is restricted to a minimum to avoid entry of cold outside air, particularly at night (Meneses and Monteiro, 1990). This causes build up of excess humidity which often results in abundant disease development on aerial plant parts.

Several regimes of natural ventilation have been tested to decrease humidity in unheated tomato greenhouses during winter and spring months in Portugal. These studies demonstrated that it was possible to achieve both a significant reduction in air humidity at night (Abreu and Meneses, 1994) and satisfactory tomato production (Abreu *et al.*, 1994) if continuous ventilation was combined with modulation in the degree of opening of the vents. The reduction in humidity was sufficient to decrease sharply the incidence of grey mould in the continuously ventilated greenhouse as compared to that closed at night according to traditional winter practice (Meneses and Monteiro, 1990; Meneses *et al.*, 1994). This translated into substantial differences in commercial yield in one experiment (Meneses and Monteiro, 1990), but not in a later experiment in which diseased tissue was regularly removed to avoid the development of full-fledged epidemics (Meneses *et al.*, 1994). This sanitation measure presumably resulted in much lower yield losses than would have occurred in a commercial situation. In this experiment, commercial yields were similar under both types of ventilation regimes, because the losses due to disease in the greenhouse with traditional ventilation were comparable to the yield reduction due to lower winter temperature in the continuously ventilated greenhouse.

Reduction of relative humidity in plastic-covered greenhouses may also be obtained by the use of photoselective cladding materials. Vakalounakis (1992) demonstrated that covering unheated tomato greenhouses with film treated to absorb long-wave infrared light (IRA film) increased air temperature as compared to control film, thus reducing relative humidity. The combined disease index of early blight (*Alternaria solani*), leaf mould (*Cladosporium fulvum*), and grey mould (*B. cinerea*) was reduced by nearly 50% under the IRA film as compared to the control. Disease control may have resulted directly from the reduction in relative humidity, but also perhaps indirectly from an increase in plant vigour. Plant height, stem diameter, inflorescence number, and fruit production were all greater under the IRA than under the control film. Furthermore, the first mature fruits were harvested several weeks earlier under the IRA than under the control film.

The combined use of such films and continuous modulated ventilation, possibly with more cold-hardy tomato varieties, offers great potential for the control of humidity-dependant diseases of greenhouse vegetables in the Mediterranean context. Furthermore, they do not imply great changes in cropping practices, which could facilitate their adoption by the growers, as well as their integration with other control methods.

CHEMICAL CONTROL

Control of fungal diseases, including those caused by *B. cinerea* , still relies largely on chemical control although it has come under increasing public scrutiny over the last decade, and its implementation has been complicated both by legal restrictions on the use of certain fungicides and by the multiplication of resistant strains of the pathogens (Gullino, 1992, Gullino and Kuijpers, 1994, Ragsdale and Sisler, 1994). In face of these difficulties, much research has been devoted to the development of strategies for a more rational use of fungicides (Gullino, 1992). In practical terms, the grower faces two key questions for the implementation of chemical control: what is the most appropriate time for treatments and what fungicides should be used? A third question, concerning doses and methods of application will not be addressed here.

Timing of Application

For many diseases of field crops, including some caused by *Botrytis* spp., preventive chemical control still relies on calendar-based schedules of fungicide applications. Knowledge on the epidemiology of the disease and the identification of periods conducive to disease development may lead to substantial reductions in the number of fungicide applications, as illustrated by a four-year study in a perennial strawberry planting (Wilcox and Seem, 1994). These authors demonstrated that the incidence of grey mould (*B. cinerea*) on the berries at harvest was highly correlated with the occurrence of environmental conditions conducive to infection during the bloom stage, and that two sprays applied at that time provided the same level of annual disease control as four or five treatments from bloom to harvest. Conversely, a weak correlation was observed between disease incidence and environmental factors recorded after bloom, and fungicide applications made after that stage provided limited disease control (Wilcox and Seem, 1994).

Reductions in the number of fungicide applications may also be obtained with the help of disease warning systems based on epidemiological models, that allow growers to begin spray programs according to the predicted risk of disease onset. Many such systems have been developed for a variety of host/disease combinations (Campbell and Madden, 1990). An example of a predictive system implicating *Botrytis* is "BOTCAST", used to time the initiation of fungicide programs for the protection of onions against leaf blight caused by *B. squamosa* (Sutton *et al.*, 1986).

More complex models have also been developed to provide guidance for the proper timing of both the initial and subsequent fungicide applications. In California, the implementation of such an advisory system for the control of downy mildew of lettuce reduced the total number of sprays in seven trials by 67%, relative to the calendar-based schedule, with no difference in disease intensity (Scherm *et al.* 1995). Substantial reductions in fungicide use -combined with adequate disease control- have also been obtained in field experiments in North America with "BLIGHT-ALERT", a system developed to predict periods of onion infection by *B. squamosa* (Vincelli and Lorbeer, 1989) and in vineyards in Chile with a model of grape infection by *B. cinerea* (Broome *et al.*, 1995).

Such forecast systems would be useful for the control of *Botrytis* in tomato green-houses when epidemiological models become available. The implementation of such strate-gies may present benefits in terms of production costs (savings in cost of materials as well as labour) and in terms of environmental safety (reduced exposure of growers while spraying and reduced residues on the produce and in the environment). It may also decrease the selection pressure for resistance to fungicides (Gullino and Garibaldi, 1982; Löcher *et al.*, 1987).

Choice of the Active Ingredient

Resistance of *Botrytis* spp. to several families of fungicides is well documented and the risks of cross-resistance for fungicides belonging to a same chemical family have been extensively studied (Barak and Edgington, 1984; Lorenz, 1988; Smith, 1988; review by Gullino, 1992). Studies have also been conducted to determine the relationship between field efficacy of specific fungicides and the frequency of resistant strains, and to elaborate strategies of alternation or combination of chemicals with different modes of action (Leroux and Clerjeau, 1985; Löcher *et al.*; 1987; Lorenz, 1988; Gullino, 1992). The aim of maintain-ing satisfactory levels of control in the presence of strains resistant to one of the fungicides used in the alternation, and of preventing an increase in the incidence of resistance, was sometimes achieved for certain combinations of multisite fungicides with dicarboximides or benzimidazoles (*e.g.* Gullino and Garibaldi, 1982; Löcher *et al.*, 1987; Northover, 1988; Gullino *et al.*, 1989). The implementation of such strategies in the context of greenhouse vegetable production is attractive, but complicated by the small number of chemicals available for control (Gullino and Kuijpers, 1994). In France, for example, only 6 fungicides were registered in 1995 for use against grey mould on tomatoes (Cluzeau, 1995), of which three belong to the dicarboximides.

The importance of the decision that the grower must take when choosing a fungicide for a treatment is illustrated by the results of a survey of the frequency of resistance to benzimidazoles (carbendazime) and dicarboximides (procymidone and vinclozolin) in 11 tomato greenhouses in southern France in 1991. Although resistance to either type of fungicides was found nearly everywhere, the frequency of resistance differed widely among specific greenhouses (Table 3). Combined resistance to both types of fungicides occurred, but it concerned a small proportion of the strains (<12%) in all but one of the greenhouses. Furthermore, the proportion of strains sensitive to one but not the other type of fungicides was high (>85%) in nearly half of the greenhouses, suggesting that efficient control might have been achieved by choosing the appropriate chemical. Conversely, it is likely that sprays with the wrong choice of fungicide (that is, fungicides for which resistance was present in a high proportion of strains) would have provided little if any control in most greenhouses. In addition to the waste of time and money, and the useless impact on the environment, such action also presents the risk of further increasing the prevalence of resistant strains by maintaining a selection pressure.

In such a situation, it would be beneficial for growers to base their choice of fungicide on data showing the frequency of resistance to available fungicides among strains present in their own greenhouse just prior to a scheduled treatment. To this end, one needs a method to test rapidly and reliably a large number of individual isolates, and a sampling scheme providing a sample representative of the populations present in the greenhouse. The tests and methods developed for large scale monitoring programs of resistance to a variety of fungicides may be very useful (Lorenz, 1988; Trivellas, 1988). Of particular interest is a rapid spore germination test (Leroux and Gredt, 1981; Leroux and Clerjeau, 1985) adapted by Gullino and Garibaldi (1986) for the detection of resistance in tomato greenhouses. Recently developed methods may further enhance the convenience, rapidity or reliability of

Table 3. Frequency of resistance of *Botrytis cinerea* to
benzimidazoles (carbendazime) and dicarboximides (procymidone and vinclozolin)
in 11 commercialtomato greenhouses in southern France in 1991

| | % strains resistant to | | | |
| | benzimidazoles | dicarboximides | | |
Greenhouse	only	only	both	neither
1	0.0	35.0	0.0	65.0
2	26.3	0.0	5.3	68.4
3	0.0	33.3	6.7	60.0
4	4.8	47.6	33.3	14.3
5	95.0	0.0	0.0	5.0
6	38.9	5.6	11.1	44.4
7	0.0	88.9	0.0	11.1
8	0.0	91.3	8.7	0.0
9	0.0	88.9	0.0	11.1
10	0.0	100.0	0.0	0.0
11	0.0	0.0	0.0	100.0

those tests, based on the use of pH indicators (Bardinelli *et al.*, 1989; Germeier et al, 1994), automatic data acquisition systems (Raposo *et al.*, 1995), or the polymerase chain reaction (Luck and Gillings, 1995). Other authors proposed to complement the germination tests with observations of sporulation 6-10 days after inoculation (Johnson *et al.*, 1994). Bioassays have also been employed successfully (Wang *et al.*, 1986; Moorman and Lease, 1992), and a sunflower seedling assay was used to develop a reliable and inexpensive kit allowing farmers to identify resistance to dicarboximides in 3-7 days directly from infected tissue, without an intermediate isolation step in the laboratory (Moorman *et al.*, 1994). Direct quantification of resistant strains in the air spora has also been achieved by exposing plates containing a medium selective to *B*. cinerea amended with the test fungicides (Elad and Shtienberg, 1995).

While rapidity and reliability are essential qualities of such tests, the cost of determining the frequency of resistance among strains present in a greenhouse is also likely to be a key factor for their utilisation in a situation of commercial production. One aspect of this cost may be the duration of validity of the results of a test. Does a grower need to run a test before every treatment he envisions, or is the structure of the *B. cinerea* population evolving slowly enough in a greenhouse that he can rely on data from previous tests?

Evolution of the Frequency of Resistant Strains during a Growing Season

As a result of numerous monitoring programs in different countries, much knowledge has accumulated on the dynamics of *B. cinerea* populations resistant to benzimidazoles and dicarboximides, mostly in field crops and particularly grapes (Leroux and Clerjeau, 1985; Gouot, 1988; Smith, 1988). Those field trials, as well as fitness studies in the laboratory, concurred to suggest that resistance to benzimidazoles was stable and persisted in field populations as the result of a high fitness of resistant strains (Leroux and Clerjeau, 1985; Smith, 1988). In contrast, the frequency of resistance to dicarboximides usually decreased rapidly in the populations after treatments were stopped, presumably due to low fitness of resistant strains or dilution with sensitive strains from adjacent plots (Gouot, 1988; Vali and Moorman, 1992). In studies where populations were monitored several times per year, selection for dicarboximide resistant strains of *B. cinerea* appeared to occur most rapidly during periods most favourable for disease development, while a decrease in the frequency of resistance was observed over the winter in many (but not all) cases (Leroux and Clerjeau, 1985; Löcher *et al.*, 1987; Northover, 1988; Pak *et al.*, 1990; Beever *et al.*, 1991; Johnson *et al.*, 1994).

Knowledge on seasonal changes in the frequency of resistance in *B. cinerea* populations in greenhouse crops is more fragmentary and the results appear more variable in different settings (Gouot, 1988). One may expect the dynamics of resistant populations in greenhouse crops to differ substantially from that in grape production and other seasonal field crops, as susceptible host tissue and conditions conducive to disease development remain present for long portions of the year (up to 11 months in many glasshouses), and the somewhat confined environment may limit the blend between populations inside and outside of the greenhouse.

To investigate the dynamics of resistant populations of *B. cinerea* at the scale of a growing season in a greenhouse environment, we studied the monthly changes in the frequency of resistance to several fungicides in four commercial tomato greenhouses in southern France. From January to August 1994, *ca.* 40 isolates were taken monthly from the air spora of each greenhouse, on a *Botrytis*-selective medium (Kerssies, 1990), and another set of *ca.* 24 samples was taken from sporulating lesions on diseased plants. Throughout this study, no resistance to the mixture carbendazime+diethofencarb was encountered. For dicarboximides (vinclozolin) and benzimidazoles (carbendazime), large changes (>25%) in the frequency of resistance were sometimes observed from one month to the next, particularly in the air spora (Figure 1). Some (but not all) of the increases in resistance frequency coincided with sprays of fungicides in the same family as that tested. Overall, the frequency of resistance to dicarboximides increased over the season in all four greenhouses, including two where no dicarboximides were used, while that for benzimidazoles decreased or appeared erratic even in greenhouses where carbendazime was applied alone or mixed with diethofencarb (Figure 2).

Interestingly, some of these results do not correspond to what would be expected, based on knowledge on the behaviour of resistant strains in the field: a decrease in the proportion of presumably less fit dicarboximide-resistant strains in absence of treatments, and a stability in benzimidazole-resistant strains, whether those fungicides were used or not (Gouot, 1988; Smith, 1988). Prevalence of vinchloziline-resistant strains was also observed by Moorman and Leese (1995) over a two-year period in a commercial greenhouse where no dicarboximides had ever been applied. One possible complicating factor for the interpretation of such results may be the presence of strains with multiple resistance to several types of fungicides. It is conceivable that the proportion of strains resistant to one type of fungicide not used in a greenhouse could increase as a result of selection pressure from the utilisation of another type of fungicide, if strains resistant to both types of chemicals are present. Further complication could originate from the common practice by tomato growers to apply fungicides as a paste over individual stem lesions as an attempt to prevent or delay stem girdling and early death of the plants. While this method is sometimes successful, it is not rare to observe the lesions resume their extension and the fungus sporulate through the fungicide (Table 4). Similar observations have been reported in Morocco on stem lesions pasted with a mixture of thiophanate-methyl (a benzimidazole) and copper+maneb+zineb (Besri and Diatta, 1985). All the isolates collected from those lesions were resistant to benomyl. Furthermore, these same authors (1992) demonstrated that pasting lesions resulted in higher increases in the frequency of resistance than sprays. Depending on the types of fungicides used by the growers and the incidence of such phenomenon, the contribution of this source of inoculum to the resistant population of a greenhouse may not be negligible.

As *B. cinerea* is a polyphagous pathogen present on numerous plants, another hypothetical complicating factor may be an intrusion of inoculum from outside the greenhouse, possibly from neighbouring crops where spray schedules and selection pressures may

A resistance to vinclozolin

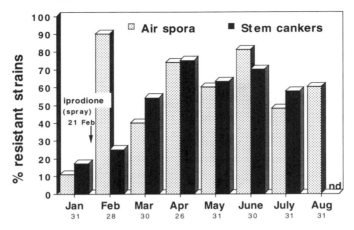

B resistance to carbendazime

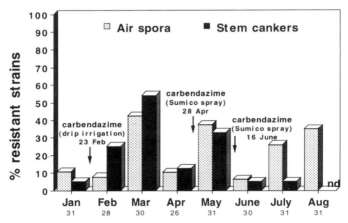

Figure 1. Evolution of the frequency of resistance of *Botrytis cinerea* populations to carbendazime (A) and vinclozolin (B) in a commercial tomato greenhouse in Chateaurenard (France) during the 1994 growing season.

be different. This phenomenon, and the possible emigration of spores from the greenhouse, could increase or dilute the proportion of resistant strains in the population inside. Although greenhouses have somewhat confined environments, exchange of air with the outside occurs continuously as a function of wind speed and the degree of aperture of the vents, even in a well closed modern glasshouse (Groen, 1988; Fernandez and Bailey, 1992). On the average in southern France, the air of a heated plastic-covered greenhouse may be renewed 3-4 times per hour during calm winter nights when all the vents are closed, and 50-60 times per hour during breezy summer days when the vents are wide open (Boulard and Baille, 1995; Boulard and Draoui, 1995). These air movements may facilitate the dispersion of inoculum within a greenhouse, but also exchanges with outside populations. Some of the results on the frequency of resistance to dicarboximides and bezimidazoles in our 1994 study lend support to the hypothesis that there may have been occasional (but significant) influxes of *B. cinerea* spores into the greenhouses that were monitored. Despite an overall concordance between

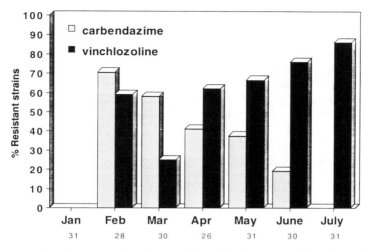

Figure 2. Evolution of the frequency of resistance of *Botrytis cinerea* populations from stem lesions to carbendazime and vinclozolin in a commercial tomato greenhouse in Berre (France) during the 1994 growing season. No dicarboximides were applied in 1993 and 1994. Carbendazime was applied 5 times in 1993 and 3 times between January and June 1994.

the frequencies of resistance among the samples from the air spora and those from sporulating lesions on diseased plants, there were also occasional differences (some large) between the two populations, particularly in February and in summer and late spring, as illustrated by Figure 1. This suggests that although the strains present in the air spora probably originated for the most part from the sporulating lesions inside the greenhouse, there may also have been some episodes of significant immigration of inoculum from the outside, with frequencies of resistance much different from the inside population.

    To increase our understanding of the population dynamics of resistant strains in greenhouses, it appears essential to find ways to determine the relative importance of internal sources of inoculum and of immigration/emigration as determinants of the phenotypic composition of the air spora inside the greenhouse. Approaches to study these phenomena

Table 4. Efficacicy of treating *B. cinerea* stem lesions with fungicide pastes in a survey of 10 commercial tomato greenhouses in 1991 in southern France

Greenhouse	% stem lesions treated in greenhouse	% stem lesions with fungal sporulation	
		untreated lesions	treated lesions
1	62.3	53.8	11.6
2	45.9	75.0	23.5
3	75.9	100.0	13.6
4	40.0	58.3	6.3
5	58.8	64.3	25.0
6	55.0	85.2	27.3
7	45.7	48.0	52.4
8	59.6	47.4	7.1
9	51.7	58.6	19.4
10	57.4	100.0	19.4

for other epiphytic micro-organisms are described by Lindow (this volume). Information on medium and long-range transportation of *Botrytis* inoculum and its effect on spore infectivity would also be useful. If exchange of inoculum is common among neighbouring greenhouses, strategies for the management of resistance need to be considered at a scale larger than that of an individual greenhouse, possibly that of a small region, as in the case of field crops (Trivellas, 1988). In the mean time, it seems advisable that the frequency of resistance to fungicides be determined in an individual greenhouse shortly before a scheduled treatment, in order to provide adequate information to the grower.

BIOLOGICAL CONTROL

Over the years, a wealth of antagonistic micro-organisms have been selected for their efficacy in protecting a variety of crops. Although few have resulted in commercially available products thus far, recent technical advances and the changing socio-economic context for plant protection, as described for post-harvest diseases by Droby *et al.* (this volume), have prompted a new surge of interest in the development of biological control methods against a wide array of plant pathogens.

Recent Advances in Biological Control

In the last few years, great progress has been made in our understanding of the mechanisms of interactions between antagonists, pathogens, and their hosts (see reviews by Droby *et al.*, by Pfender, and by Rayner, this volume). Such work may foster the discovery of new microbial compounds useful for plant protection (see for example Janisiewicz *et al.*, 1991; Hajlaou *et al.*, 1994; Droby *et al.*, 1995; Leifert *et al.*, 1995; Milner *et al.*, 1995; Reglinski *et al.*, 1995). It may also provide useful criteria for the selection of more efficient strains and for quality control in the industrial production of biocontrol agents. Increasingly, research also encompasses methods to enhance the efficacy and the survival of antagonists on plant surfaces, as well as improvement of mass production and formulation (Dubos, 1992; Klijnstra *et al.*, 1993; Thonard *et al.*, 1993).

Much work has been devoted to the biological control of diseases caused by *Botrytis*. Reviews and a historical perspective of this work have recently been presented (*e.g.* Dubos, 1992; Marois, 1992; Elad and Shtienberg, 1995; Gullino, 1995, Elad *et al.*, 1996). On greenhouse tomatoes, successful biological control of grey mould was reported in Israel with a strain of *Trichoderma harzianum* now available in several countries as a commercial product (Elad and Zimand, 1992). Lack of efficacy was occasionally observed in these trials, and was related to climatic conditions particularly conducive to disease development (Elad and Shtienberg, 1995). Other strains of *Trichoderma* have also been tested (Gullino *et al.*, 1991, Malathrakis and Klironomou, 1992; Migheli *et al.*, 1994; Decognet *et al.*, 1995; Dik *et al.*, 1995; Elad *et al.*, 1995). Other micro-organisms, including various bacteria, yeasts and fungi, have been selected for their ability to inhibit infection of tomato leaves or fruits by *B. cinerea* (Malathrakis and Klironomou, 1992; Nicot *et al.*, 1993a, 1993b; Elad *et al.*, 1994a, 1994b; Dik *et al.*, 1995; Elad *et al.*, 1996). Using a screening procedure targeted specifically at the protection of pruning wounds on tomatoes, Nicot and co-workers selected several antagonists of *B. cinerea* that were highly efficient in reducing the incidence and expansion of stem lesions in growth chambers and greenhouse trials (Nicot *et al.*, 1993a, 1994; Decognet *et al.*, 1995). Some of these strains were also efficient on other crops including cucumber (Nicot *et al.*, unpublished; Koning and Köhl, 1994) and strawberries (Guinebretiere *et al.*, 1993). All these studies concerned a "pre-emptive" biocontrol strategy, mostly aimed at preventing infection and secondarily, at slowing the development of symptoms.

An other biocontrol strategy, developed against *Botrytis* diseases on field crops such as strawberries (Sutton and Peng, 1993a, 1993b) and onions (Köhl *et al.*, 1992, 1995b), may be particularly useful in the context of greenhouse production. This approach aims at reducing the large amounts of spores produced by the pathogen on necrotic tissue, thereby affecting the development of polycyclic epidemics (Köhl *et al.*, 1995b). Several microbial strains have been selected for their ability to compete with *B. cinerea* in the colonisation of necrotic tissue, effectively inhibiting sporulation of the pathogen, and to withstand dry conditions (Elad *et al.*, 1994a, 1994b; Köhl *et al.*, 1995a). Work on other host/pathogen models, and a review of the rapidly increasing body of knowledge on factors affecting colonisation and microbial interactions on senescing and necrotic plant tissues are presented by Pfender in this volume. The advantages of this type of strategy (see Köhl *et al.*, 1995b) make it a very promising approach for the control of *Botrytis* diseases in the greenhouse. Furthermore, the possibility of using pollinating insects to deliver such biocontrol agents, as demonstrated by Peng and coworkers (1992), may be of great value to limit the colonisation of senescing flower parts, a very common phenomenon in greenhouse crops. Colonised flower parts may serve as a food base for the fungus to attack the growing fruit, and they may also act as an efficient inoculum source when they fall on other plant organs. The fact that commercially produced bumble bee hives are increasingly used for the pollination of greenhouse vegetables such as tomatoes offers interesting perspectives.

Integration of Biological Control with Other Control Methods

One of the most significant factors in promoting the use of biological control in a commercial situation may be the recent progress in combining biological methods with other control methods. The selection of antagonists resistant to several fungicides made it possible to evaluate the potential of combining chemical and biological methods to control *B. cinerea* on vegetable crops. The alternation or combination of fungicides with antagonists such as *Trichoderma harzianum* provided better control than the antagonist applied alone (Gullino *et al.*, 1991; Elad and Zimand, 1992; Malathrakis and Klironomou, 1992; Elad *et al.*, 1993, 1995). In many cases, it also provided better control than the schemes with fungicides alone.

An important step further was taken by Elad and co-workers, when they combined these methods with observations on the effect of climatic conditions on the relative efficacy of biological and chemical control methods. Work is in progress to evaluate the performance of a decision support system, "BOTMAN", which was developed to advise the growers to use either a fungicide or a biocontrol agent, or to abstain from any application, based on climatic records of the preceding days in the greenhouse and weather forecast for the 4 following days (Elad and Shtienberg, 1995). As the body of knowledge increases rapidly on the precise effect of environmental parameters on the efficacy of biological control in various host-pathogen-antagonist systems (*e.g.* Elad and Kirshner, 1993; Elad *et al.*, 1994a; Kôhl *et al.*, 1995a; Hannusch and Boland, 1996; review in Elad *et. al.*, 1996), this promising approach could conceivably be adapted to most greenhouse crops and possibly combined with sophisticated epidemiological models such as those mentioned earlier in this chapter.

OTHER CULTURAL PRACTICES

A variety of other cultural measures may be used by greenhouse managers to limit the development of *Botrytis* epidemics. For example, several studies have shown that pruning wounds on tomato plants are less likely to become infected by *B. cinerea* if leaves are cut close to the stem than if a fragment of petiole is left on the stem (Martin *et al.*, 1994; Dik and Buitelaar, 1995). Other measures such as sanitation and factors affecting the sensitivity of the crops (*e.g.* fertilisation, application of plant hormones) have been reviewed in detail by Jarvis (1992), Elad and Shtienberg (1995) and by Elad and Evensen (1995). This presentation will focus on the potential use of photoselective cladding materials to reduce the production of secondary inoculum by *B. cinerea* in the greenhouse.

Photoselective films may contribute to the control of several tomato diseases by modifying the climate inside greenhouses, as mentioned earlier in this chapter. The light spectrum may also be altered to inhibit sporulation of *B. cinerea*. Films absorbing near ultraviolet light (nUV) have been shown to reduce spore production *in vitro* by several fungal pathogens of vegetable crops (Honda *et al.*, 1977; Sasaki *et al.*, 1985; Reuveni *et al.*, 1989; Vakalounakis, 1991). A similar effect was also observed *in vivo*, but the efficacy of nUV-absorbing films in inhibiting sporulation appeared to vary with the host species and the type of plant organ (Nicot, 1992; Reuveni and Raviv, 1992; Nicot *et al.*, 1996). In greenhouse experiments with cucumbers and tomatoes in Japan (Honda *et al.*, 1977) and in Israel (Reuveni *et al.*, 1989; Reuveni and Raviv, 1992), use of nUV-absorbing films resulted in reduced incidence of grey mould as compared to control films.

In the 1970's, much work (reviewed by Epton and Richmond, 1980) was devoted to elucidating the specific effect of various portions of the light spectrum on the sporulation of *B. cinerea*. It was demonstrated that nUV and blue light had a stimulatory and an inhibitory effect, respectively (Tan, 1974). Accordingly, the highest efficiency of nUV-absorbing films in inhibiting spore production was associated, both *in vitro* and on tomato plants, with films having the highest blue/ultraviolet ratio of the transmitted light (Reuveni *et al.*, 1989; Reuveni and Raviv, 1992). One may hypothesise that the efficacy of these films could be further improved with additives absorbing far-red light, as this portion of the spectrum was shown by Tan (1975) to re-promote sporulation previously inhibited by blue light.

While this potential control method offers much promise for greenhouse vegetable growers, several points merit attention:

1. Little is known on the possible effect of nUV-absorbing films on the development and vigour of the crop. In an experiment where both disease incidence and yield were monitored, the reduction in incidence of grey mould under the nUV-absorbing films did not always translate into increases in yield (Honda *et al.*, 1977; Reuveni and Raviv, 1992). It was hypothesised that this may be related to the somewhat lower transmissivity of the nUV-absorbing films for the photosynthetically active radiation (PAR), as compared to the controls (Reuveni and Raviv, 1992). The production of improved films combining high PAR transmissivity with high blue/nUV ratios reported by these authors will provide the means to test this hypothesis.

2. Greenhouses must be ventilated daily to avoid or limit the build-up of excess humidity resulting from the transpiration of the crop. In most cases, this is achieved by manual or automated opening of variously shaped vents on the sides or on the roof of the greenhouse (Jarvis, 1992). This operation may interfere in at least two ways with the inhibition of spore production by the pathogen. Depending on the size and shape of the vents and their orientation relative to the sun, their opening may expose the plants inside the greenhouse to direct sun light during part of the day, and therefore reduce the efficacy of the nUV filtering films. Whether this may be of practical concern in certain situations remains to be elucidated. Secondly, ventilation is likely to foster the penetration into the greenhouse of exogenous airborne inoculum, as well as release to the outside environment some of the inoculum produced on diseased plants inside the greenhouse.

As discussed earlier in this chapter, we believe that significant influxes of *B. cinerea* spores from the outside may occasionally occur, although to our knowledge, immigration of fungal spores into a greenhouse has not been studied quantitatively. To further evaluate the possible impact of this phenomenon on the efficacy of this control method, experiments must be conducted to determine the relative importance of the outside environment and the crop itself as sources of the inoculum present in a greenhouse.

3. Finally, isolates of *B. cinerea* have been shown to differ in their sensitivity to the inhibitory effect of a nUV-absorbing film (Nicot *et al.*, 1996). In view of these results, one may wonder whether the use of nUV-absorbing films on commercial tomato greenhouses could lead to the selection of "resistant" strains which would eventually jeopardise the efficacy of this control method. However, the prospects presented by the data reported so far do not appear to warrant extreme concern. The least nUV-sensitive isolate reported by Nicot *et al.* (1996) produced approximately 10 times more spores than the most sensitive isolate under the nUV-absorbing film. But even for this isolate, there was still a *ca*. 230-fold reduction in spore production under the nUV absorbing film as compared to the control film. Although an epidemiological study relating spore load in the greenhouse and the development of grey mould on tomatoes would be needed to test this hypothesis, it is probable that a 230-fold reduction in spore production would be sufficient to exert significant control on the evolution of the epidemic. In a field study of onion leaf spot, Köhl and co-workers (1995b) demonstrated that a reduction of about 50 % in airborne inoculum of *Botrytis* had significant effects on the progress of the disease. More information is needed on the frequency of less sensitive strains in populations of *B. cinerea* and on factors, such as parasitic fitness, that might influence their selection.

These points clearly deserve attention to further evaluate the potential of using photoselective films for the control of grey mould, possibly in association with other (*e.g.* biological) means of reducing the production of secondary inoculum on diseased plants. As glass may vary substantially in its transmissivity of nUV light (Hite, 1973), application of photoselective compounds on the panels of glasshouses might also be of benefit.

FUTURE CHALLENGES AND OPPORTUNITIES FOR INTEGRATED CONTROL

In the complex task of controlling diseases in a tomato greenhouse, *Botrytis* probably offers the greatest challenge to the imagination and the skills of both the growers and the researchers. An accumulating body of evidence suggests that it may be unrealistic to expect to achieve successful control of grey mould with the help of a single method, such as for example, the application of fungicides (Gullino, 1992). Rather, it is likely that control will increasingly rely on a combination of several (possibly many) different types of interventions concurring to strengthen the resistance of the host plant or weaken the development of the pathogen. Recent research, including that on new fungicides (*e.g.* Leroux, 1994; Akagi *et al*, 1995; Milling and Richardson, 1995), provides useful tools for this purpose. Much of the work has been focused on preventing or inhibiting infection, and more recently on reducing spore production, while two other key steps in the disease cycle have received far less attention: symptom expression and dissemination of inoculum. One possible avenue of research concerns the selection of micro-organisms capable of colonising *Botrytis* stem lesions and inhibiting their gradual expansion before they girdle the stem. Beside saprophytic phyllosphere microbes, hypo-aggressive strains of *B. cinerea* might also be very useful as "curative antagonists", particularly if this trait is associated with virus-like particles that can be transmitted to aggressive strains. Virus-like particles have been found in several isolates of *B. cinerea* in New Zealand, and double-stranded RNA in 143 of 200 isolates examined (Howitt *et al.*, 1995). However, in contrast with other fungi (*e.g. Sclerotinia sclerotiorum*, see Boland *et al.*, 1993, 1995), this did not appear to be correlated to reduction in aggressiveness. In preliminary trials in our laboratory, a slow-growing, hypo-aggressive strain of *B. cinerea* was able to inhibit the development of lesions by an aggressive strain in pruning wounds (unpublished). Results were not consistent and may be complicated by mycelial incompatibility factors as described by Beever and Parkes (1993). Much remains to be done to explore this potential control method.

The great majority of work on control methods against grey mould has been focused mainly on the fungal part of the host-pathogen-environment interaction, and much less on the plant. While the effect of fertilisation and plant hormones on susceptibility to grey mould has been studied for several vegetable crops, much less is known, for example, on the effect of climatic parameters on the tomato plant's defence mechanisms to *B. cinerea* or on the healing of pruning wounds. Varieties resistant or less susceptible to *B. cinerea* have been reported for several plant species such as rose (Hammer and Evensen, 1994) and onion (Lin *et al.*, 1995), for example (see review by Elad and Evensen, 1995). Dominant monogenic resistance of tomato to *B. cinerea* remains to be discovered and it does not seem to exist in related wild species where resistance to many other tomato diseases was found (Laterrot, 1990). However, small differences in susceptibility to grey mould may be noticed among commercial tomato varieties in the greenhouse. These differences might be due to the establishment of different microclimates in the canopy in relation to plant architecture. They could also arise from a variety of physiological factors as reviewed by Elad and Evensen (1995). The complexity of breeding for (probably) polygenic resistance may explain why this potential remains largely un-exploited for tomatoes.

Truly integrated approaches combining chemical treatments with climate management strategies and other cultural methods are already in use in certain tomato greenhouses, and it is realistic to envision an extension and a refinement of their application in the coming years. A greater challenge will be to expand on these models to take into account other diseases (for example powdery mildew) as well as insect pests and to integrate economic considerations. Much work remains to be done to estimate thresholds of tolerable damage and yield loss, and to evaluate the cost-effectiveness of the different control measures.

REFERENCES

Abreu, P.E. and Meneses, J.F. 1993, Climatic characterization of two plastic covered greenhouses under different natural ventilation methods, with a cool season tomato crop, *Acta Horticulturae* 366:183-194.

Abreu, P.E., Monteiro, A.A. and Meneses, J.F. 1993, Response of untreated plastic covered greenhouse tomatoes during the cool season and under two different natural ventilation methods, *Acta Horticulturae* 366:195-200.

Akagi, T., Mitani, S., Komyoji, T.and Nagatani, K. 1995, Quantitative structure-activity relationships of fluazinam and related fungicidal n-phenylpyridinamines: preventive activity against *Botrytis cinerea*, *J. Pestic. Sci.* 20:279-290.

Barak, E. and Edgington, L.V. 1984, Cross-resistance of *Botrytis cinerea* to captan, thiram, chlorothalonil and related fungicides, *Can. J. Plant Pathol.* 6:318-320.

Bardinelli, T.R., Butterfield, E.J. and Jones, T.L. 1989, Diagnostic media for the detection of fungi (*Botrtyis cinerea*) resistant to vinclozolin and benomyl, *Phytopathology* 79:1212-1213.

Bartharuk, N.N. 1985, A comparative study of radiometric and electronic leaf wetness sensors. *Agric. For. Meteorol.* 36:83-90.

Beever, R.E., Pak, H.A., Laracy, E.P. 1991, An hypothesis to account for the behaviour of dicarboximide resistant strains of *Botrytis cinereae* in vineyards, *Plant Pathol.* 40:342-346.

Beever, R.E. and Parkes, S.L. 1993, Mating behavior and genetics of fungicide resistance of *Botrytis cinerea* in New Zealand. *N. Z. J. Crop Hort. Sci.* 21:303-310.

Besri, M. and Diatta, F. 1985, Résistance de *B. cinerea*, agent de la pourriture grise de la tomate, aux benzimidazoles, dicarboximides et sulfamides, *EPPO Bulletin* 15:379-385.

Besri, M. and Diatta, F. 1992, Effect of fungicide application techniques on the control of *Botrytis cinerea* and development of fungal resistance, pp. 248-251 In: Verhoeff, K., Malathrakis, N.E. and Williamson, B. (eds.) *Recent Advances in* Botrytis *Research,* Pudoc, Wageningen.

Blakeman, J.P. 1980, Behaviour of conidia on aerial plant surfaces, pp. 115-151 In: Coley-Smith, J.R., Verhoeff, K. and Jarvis, W.R. (eds.) *The Biology of Botrytis*, Academic Press, London.

Boland, G.J., Melzer, M.S. and Zhou, T. 1995, Hypovirulence in *Sclerotinia* species, p. 48 In: *Proc. 6th Int. Symp. Microbiology of Aerial Plant Pathogens*, Bandol, France.

Boland, G. J., Mould, M.J.R. and Robb, J. 1993, Ultrastructure of a hypovirulent isolate of *Sclerotinia sclerotium*, *Physiol. Mol. Plant Pathol.* 43:21-32.

Boulard, T. and Baille, A. 1995, Modelling of air exchange rate in a greenhouse equiped with continuous roof vents, *J. agric. Engng. Res.* 61:37-48.

Boulard, T. and Draoui, B., 1995, Natural ventilation of a greenhouse with continuous roof vents: measurements and data analysis, *J. agric. Engng. Res.* 61:37-48.

Broome, J.C., English, J.T., Marois, J.J., Latrore, B.A. and Aviles, J.C., 1995, Development of an infection model for *Botrytis* bunch rot of grapes based on wetness duration and temperature, *Phytopathology* 85:97-102.

Campbell, C.L. and Madden, L.V. 1990, *Introduction to Plant Disease Epidemiology*, John Wiley and Sons, New York.

Chraibi, A., Jaffrin, A., Makhlouf, S. and Bentounes, N. 1995, Déhumidificatin de l'air d'une serre par contact direct à courants croisés avec une solution hygroscopique organique, *J. Phys. III France* 5:1055-1074.

Cluzeau, S., 1995, *Index Phytosanitaire 1995*, ACTA, Paris, France.

Decognet, V., Minuto, G., Nicot, P.C. and Mezzalama, M. 1995, Biocontrol of *Botrytis cinerea* on tomato crops in Italy and France with *Pseudomonas fluorescens* and *Trichoderma harzianum*, p. 70 In: *Proc. 6th Int. Symp. Microbiology of Aerial Plant Pathogens*, Bandol, France.

Dik, A. J. and Buitelaar, K. 1995, Voorlopig niet breken maar snijden of knippen, *Groenten en Fruit / Glasgroenten* 5(3):14-15.

Dik, A.J., Köhl, J., Fokkema, N.J., Elad, Y. and Shtienberg, D. 1995, Biological control of *Botrytis cinerea* in cucumber and tomato, p. 71 In: *Proc. 6th Int. Symp. Microbiology of Aerial Plant Pathogens*, Bandol, France.

Droby, S., Chalupovicz, L., Chalutz, E., Wisniewski, M.E. and Wilson, C.L. 1995, Inhibitory activity of yeast cell wall materials against postharvest fungal pathogens, *Phytopathology* 85:1123.

Dubos, B. 1992, Biological control of *Botrytis*: State of the art, pp. 169-178 In: Verhoeff, K., Malathrakis, N.E. and Williamson, B. (eds.) *Recent Advances in* Botrytis *Research*, Pudoc, Wageningen.

Elad, Y. and Evensen, K. 1995, Physiological aspects of resistance to *Botrytis cinerea, Phytopathology* 85:637-643.

Elad Y., Gullino M.L., Shtienberg D. and Aloi C. 1995, Managing *Botrytis cinerea* on tomatoes in greenhouses in the Mediterranean, *Crop Protect.* 14:105-109.

Elad, Y. and Kirshner, B. 1993, Survival in the phylloplane of an introduced biocontrol agent (*Trichoderma harzianum*) and populations of the plant pathogen *Botrytis cinerea* as modified by abiotic conditions, *Phytoparasitica* 21:303-313.

Elad, Y., Köhl, J. and Fokkema, N.J. 1994a, Control of infection and sprulation of *Botrytis cinerea* on bean and tomato by saprophytic yeasts, *Phytopathology* 84:1193-1200.

Elad, Y., Köhl, J. and Fokkema, N.J. 1994b, Control of infection and sprulation of *Botrytis cinerea* on bean and tomato by saprophytic bacteria and fungi, *Eur. J. Plant Pathol.* 100:315-336.

Elad, Y., Malathrakis, N.E. and Dik, A.J. 1996, Biological control of *Botrytis*-incited diseases and powdery mildews in greenhouse crops, *Crop Protect.* (in press).

Elad, Y. and Shtienberg, D. 1995. *Botrytis cinerea* in greenhouse vegetables: chemical, cultural, physiological and biological controls and their integration, *Integrated Pest Management Reviews*, 1:15-29.

Elad, Y. and Volpin, H. 1993, Reduced developement of grey mould (*Botrytis cinerea*) in bean and tomato plants by calcium nutrition, *J. Phytopathol.* 139:146-156.

Elad, Y. and Zimand, G. 1992, Integration of biological and chemical control for grey mould, pp. 272-276 In: Verhoeff, K., Malathrakis, N.E. and Williamson, B. (eds.) *Recent Advances in* Botrytis *Research*, Pudoc, Wageningen.

Elad, Y., Zimand, G., Zaqs, Y., Zuriel, S. and Chet, I, 1993. Use of *Trichoderma harzianum* in combination or alternation with fungicides to control cucumber grey mould (*Botrytis cinerea*) under commercial greenhouse conditions, *Plant Pathol.*42:324-332.

Epton, H.A.S. and Richmond, D.V., 1980, Formation, structure and germination of conidia; pp. 41-83 In: Coley-Smith, J.R., Verhoeff, K. and Jarvis, W.R. (eds.) *The Biology of Botrytis*, Academic Press, London.

Fernandez, J.E. and Bailey, B.J. 1992, Measurement and prediction of greenhouse ventilation rates. *Agric. For. Meteorol.* 58:229-245.

Germeier C., Hedke K. and Vontiedemann, A. 1994, The use of pH-indicators in diagnostic media for acid-producing plant pathogens, *Z. PflKrankh. PflSchutz* 101:498-507.

Gouot, J.M. 1988, Characteristics and population dynamics of *Botrytis cinerea* and other pathogens resistant to dicarboximides, pp. 53-55 In: Delp, C.J. (ed.) *Fungicide Resistance in North America.*, American Phytopathological Society, St Paul, MN.

Groen, J. 1988, Temperatuurverschillen in de kas motor van luchtstroming. *Vakblad voor de Bloemisterij* 43:36-37.

Guinebretière, M.H., Voiblet, C., Nguyen-Thé, C. and Nicot, P.C. 1993, Inhibitory and physiological properties of 12 yeasts and bacteria antagonistic to *Botrytis cinerea* on strawberry fruits, *IOBC WRS Bull.* 16(11):123-126.

Gullino, M.L., 1992, Chemical control of *Botrytis* spp. pp. 217-222 In: Verhoeff, K., Malathrakis, N.E. and Williamson, B. (eds.) *Recent Advances in* Botrytis *Research*, Pudoc, Wageningen.

Gullino, M.L. 1995, Biological control of *Botrytis* spp.: overview and future approaches, *IOBC WPRS Bull.* 18(3):125-130.

Gullino, M.L., Aloi, C., Benzi, D. and Garibaldi, A. 1991, Biological and integrated control of grey mould in vegetable crops. *Petria* 1:149-150.

Gullino, M.L., Aloi, C. and Garibaldi, A. 1989, Influence of spray schedules on fungicide resistant populations of *Botrytis cinerea* Pers. on grapevine, *Neth. J. Plant Pathol.* 95(supplement 1):87-94.

Gullino, M.L. and Garibaldi, A. 1982, Use of mixture or alternation of fungicides with the aim of reducing the risk of appearance of strains of *Botrytis cinerea* resistant to dicarboximides. *EPPO Bulletin*, 12:151-156.

Gullino, M.L. and Garibaldi, A. 1986, Fungicide resistance monitoring as an aid to tomato grey mould management, *Proc. Brighton Crop Prot. Conf.* 2:499-505.

Gullino, M.L. and Kuijpers, L.A.M. 1994, Social and political implications of managing plant diseases with restricted fungicides in Europe, *Annu. Rev. Phytopathol.* 32:559-579.

Hajlaou, M.R., Traquair,J.A., Jarvis, W.R. and Bélanger, R.R. 1994, Antfungal activity of extracellular metabolites produced by *Sporothrix flocculosa*, *Biocontrol Sci. Technol.* 4:229-237.

Hammer, P.E. and Evenson,K.B. 1994, Differences between rose cultivars in susceptibility to infection by *Botrytis cinerea*, *Phytopathology* 84:1305-1312.

Hannusch, D.J. and Boland, G.J. 1996, Interactions of air temperature, relative humidity and biological control agents on grey mold of bean, *Eur. J. Plant Pathol.* 102:133-142.

Hausbeck, M.K. and Pennypacker, S.P. 1991, Influence of time intervals among wounding, inoculation, and incubation on stem blight of geranium caused by *Botrytis cinerea*; *Plant Dis.* 75:1168-1172.

Hite, R.E. 1973, The effect of irradiation on the growth and asexual reproduction of *Botrytis cinera, Plant Dis. Rep.* 57:131-135.

Honda, Y., Toki T. and Yunoki, T, 1977, Control of gray mold of greenhouse cucumber and tomato by inhibiting sporulation, *Plant Dis. Rep.* 61:1041-1044.

Howitt, R.L.J., Beever, R.E., Pearson, M.N. and Forster, R.L.S. 1995, Presence of double-stranded RNA and virus-like particles in *Botrytis cinerea*. *Mycol. Res.* 99:1472-1478.

Huber, L. and Gillespie, T.J. 1992, Modeling leaf wetness in relation to plant disease epidemiology, *Annu. Rev. Phytopathol.* 30:553-577.

Jaffrin, A., Charibi A. and Blondeau, P. 1992, Greenhouse humidity control with enthalpy recovery, pp. 441-456 In: *Proc. CEC Thermie European Seminar "Business Opportunities for Energy Technologie in the Field of Greenhouse Horticulture in Southern Europe"*, Heraklion, Greece.

Janisiewicz, W.J., Yourman, L., Roitman, J and Mahoney, N. 1991, Postharvest control of blue mold of apples and pears by dip treatment with pyrrolnitrin, a metabolite of *Pseudomonas cepacia*, *Plant Dis.* 75:490-494.

Jarvis, W.R 1977, Botryotinia *and* Botrytis *species: Taxonomy,Physiology and Pathology*, Monogr. 15, Research Branch, Canada Department of Agriculture, Ottawa.

Jarvis, W.R. 1980a. Taxonomy, pp. 1-18 In: Coley-Smith, J.R., Verhoeff, K. and Jarvis, W.R. (eds.) *The Biology of Botrytis*, Academic Press, London.

Jarvis, W.R. 1980b. Epidemiology, pp. 219-250 In: Coley-Smith, J.R., Verhoeff, K. and Jarvis, W.R. (eds.) *The Biology of Botrytis*, Academic Press, London.

Jarvis, W.R. 1989, Managing diseases in greenhouse crops. *Plant Dis.* 73:190-194.

Jarvis, W.R. 1992, *Managing Diseases in Greenhouse Crops*. American Phytopathological Society, St Paul, MN.

Johnson, K.B., Sawyer, T.L. and Powelson, M.L, 1994, Frequency of benzimidazole- and dicarboximide-resistant strains of *Botrytis cinerea* in western Oregon small fruit and snap bean plantings. *Plant Dis.* 78:572-577.

Kerssies, A. 1990, A selective medium to be used in a spore trap. *Neth. J. Plant Pathol.* 96:247-250.

Kerssies, A. 1993, Influence of environmental conditions on dispersal of *Botrytis cinerea* conidia and on postharvest infection of gerbera flowers grown under galss. *Plant Pathol.* 42:754-762.

Kerssies, A., Bosker-van Zessen, A.I. and Frinking, H.D. 1995, Influence of environmental conditions in a glasshouse on conidia of *Botrytis cinerea* and on post-harvest infection of rose flowers. *Eur. J. Plant Pathol.* 101:201-216.

Klijnstra, J.W., Brueren, G.J. and de Vlieger, J.J. 1993, Formulation research for biocontrol preparations, pp. 93-98 In: Lepoivre, P. (ed.) *Proc. Eur. Com. Workshop "Biological control of fruit and foliar diseases"*, Gembloux, Belgium.

Köhl, J., Molhoek, W.M.L., van der Plas, C.H. and Fokkema, N.J. 1995a, Effect of *Ulocladium atrum* and other antagonists on sporulation of *Botrytis cinerea* on dead lily leaves exposed to field conditions, *Phytopathology* 85:393-401.

Köhl, J., Molhoek, W.M.L., van der Plas, C.H. and Fokkema, N.J. 1995b, Suppression of sporulation of *Botrytis* spp as a valid biocontrol strategy, *Eur. J. Plant Pathol.* 101:251-259.

Köhl, J., Molhoek, W.M.L., van der Plas, C.H., Kessel, G.J.T. and Fokkema, N.J. 1992, Biological control of *Botrytis* leaf blight of onions: significance of sporulation suppression, pp. 192-196 In: Verhoeff, K., Malathrakis, N.E. and Williamson, B. (eds.) *Recent Advances in* Botrytis *Research,* Pudoc, Wageningen.

Koning, G.P. and Köhl, J. 1994, Wound protection by antagonists against *Botrytis* stem rot in cucumber and tomato, In: *Environmental Biotic Factors in Integrated Plant Disease Control, Proc. 3rd IFPP Conference,* Poznan, Poland.

Kraeutler, E. 1995, Production et échanges de tomates, *Infos CTIFL* 111: 16-21.

Lagier, J., Henry, B., Reich, P. and Baille, A. 1995, "Serriste", système d'aide à la décision. *Culture Légumière* 29:21-24.

Lagier, J., Mermier, M. and Reich, P. 1996, "Serriste", un logiciel d'aide au choix des consignes climatiques pour la tomate sous serre: présentation et résultats d'essais, *PHM-Revue horticole* (in press).

Laterrot, H. 1990, Situation de la lutte génétique contre les parasites de la tomate dans les pays méditerranéens, *P.H.M. Revue Horticole* 303:53-56.

Leifert, C., Li, H., Chidburee, S., Hampson, S., Workman, S., Sigee, D., Epton, H.A.S. and Harbour, A. 1995, Antibiotic production and biocontrol activity by bacillus subtilis CL27 and bacillus pumilus CL45, *J. Appl. Bacteriol.* 78:97-108.

Leroux, P. 1994, Effect of pH, amino acids and various organic compounds on the fungitoxicity of pyrimethanil, glufosinate, captafol, cymoxanil and fenpiclonil in *Botrytis cinerea*, *Agronomie* 14:541-554.

Leroux, P. and Clerjeau, M, 1985, Resistance of *Botrytis cinerea* Pers. and *Plasmopara viticola* (Berk. & Curt.) Berl. and de Toni to fungicides in French vineyards, *Crop Protect.* 4:137-160.

Leroux, P. and Gredt, M. 1981, Méthode de détection de la résistance de *B. cinerea* Pers. aux fongicides, à partir d'échantillons prélevés dans le vignoble, *Phytiatrie-Phytopharmacie* 30:57-68.

Lin, M.W., Watson, J.F. and Baggett, J.R. 1995, Inheritance of resistance to neck-rot disease incited by *Botrytis allii* in bulb onions, *J. Am. Soc. Hort. Sci.* 120:297-299.

Löcher, F.J., Lorenz, G. and Beetz, K.J. 1987, Resistance management strategies for dicarboximides in grapes: Results of six year's trial work. *Crop Protect..* 6:139-147.

Lorenz, G. 1988, Dicarboximide fungicides: History of resistance development and monitoring methods, pp. 45-51 In: Delp, C.J. (ed.) *Fungicide Resistance in North America.* American Phytopathological Society, St Paul, MN.

Luck, J.E. and Gillings, M.R. 1995, Rapid identification of benomyl resistant strains of *Botrytis cinerea* using the polymerase chain reaction, *Mycol. Res.* 99:1483-1488.

Malathrakis, N.E. and Klironomou, E.J. 1992, Control of grey mould of tomatoes in greenhouses wih fungicides and antagonists, pp. 282-286 In: Verhoeff, K., Malathrakis, N.E. and Williamson, B. (eds.) *Recent Advances in* Botrytis *Research,* Pudoc, Wageningen.

Malathrakis, N.E., Markellou, E. and Goumas, D. 1995, Biological control of *Botrytis cinerea* in greenhouse crops, *IOBC WPRS Bull.* 18(3)91-97.

Marois, J.J. 1992, Biological control of *Botrytis cinerea*, pp. 109-111 In: Tjamos, E.C., Papavizas, G.C. and Cook, R.J. (eds.) *Biological control of plant diseases*, Plenum Press, New York.

Marois, J.J., Redmond, J.C. and MacDonald, J.D. 1988, Quantification of the impact of environment on the susceptibility of *Rosa hybrida* flowers to *Botrytis cinerea*, *J. Amer. Soc. Hort. Sci.* 113:842-845.

Martin, C., Ravetti, F., Decognet, V. and Nicot, P.C. 1994, Perspectives de lutte intégrée contre la pourriture grise de la tomate sous abri, pp. 861-868 In: *Proc. Third Internat. A.N.P.P.Conf. on Plant Diseases (Volume 2)*, Bordeaux, France.

Meneses, J.F. and Monteiro, A.A. 1990, Permanent ventilation in non heated greenhouses to reduce *Botrytis* on tomatoes, *Proc. International Seminar and British-Israel Workshop on Greenhouse Technology*, Tel-Aviv, Israel.

Meneses, J.F., Monteiro, A.A. and Abreu, P.E. 1994, Influence of two differential ventilation methods on greenhouse climate, tomato production and *Botrytis* control, *Plasticulture* 101:3-12.

Mermier, M. and Fauvel, C. 1996, Capteur d'humectation, In: *Proc. Séminaire AIP "SERRES"*, INRA, Alénya, France.

Migheli, Q., Herrera-Estrella, A., Avataneo, M. and Gullino, M.L. 1994, Fate of transformed *Trichoderma harzianum* in the phyllosphere of tomato plants, *Mol. Ecol.* 3:153-159.

Milling, R.J.and Richardson, C.J. 1995, Mode of action of the anilino-pyrimidine fungicide pyrimethanil .2. effects on enzyme secretion in botrytis cinerea, *Pestic. Sci.* 45:43-48.

Milner, J.L., Raffel, S.J., Lethbridge, B.J. and Handelsman, J. 1995, Culture conditions that influence accumulation of swittermicin A by *Bacillus cereus* UW85, *Appl. Microbiol. and Biotech.* 43:685-691.

Moorman, G.W. and Lease, R.J. 1992, Benzimidazole- and dicarboximide-resistant in *Botrytis cinerea* from Pennsylvania greenhouses, *Plant Dis.* 76:477-480.

Moorman, G.W. and Lease, R.J. 1995, Incidence of dicarboximide fungicide resistance in *Botrytis cinerea* monitored in two greenhouses, *Plant Dis.* 79:319.

Moorman, G.W., Lease, R.J. and Vali, R.J. 1994, Bioassay for dicarboximide resistance in *Botrytis cinerea*, *Plant Dis.* 78:890-891.

Nicot, P.C. 1992, Inhibition of sporulation of *Botrytis cinerea* on plant tissue under polyethylene films filtering near ulta-violet light: potential for control of grey mold of greenhouse-grown vegetables. *Proc. Xth Botrytis Symposium*, Heraklio, Greece.

Nicot, P.C. and Allex, D. 1991, Grey mold of greenhouse-grown tomatoes: disease control by climate management ?, *IOBC WPRS Bull.* 14(5):200-210.

Nicot, P.C., Mermier, M., Vaissière, B.E. and Lagier, J. 1996, Differential spore production by *Botrytis cinerea* on agar medium and plant tissue under near-ultraviolet light-absorbing polyethylene film, *Plant Dis.* 80: 555-558.

Nicot P.C., Morison, N., Guinebretière, M.H. and Nguyen Thé, C. 1993a, Evaluation of the effectiveness of selected microbial strains for the control of grey mold of tomato. pp. 9-15 In: Lepoivre, P. (ed.) *Proc. Eur. Com. Workshop "Biological control of fruit and foliar diseases"*, Gembloux, Belgium.

Nicot P.C., Saltzis, V. and Guinebretière, M.H. 1993b, A miniaturized *in vivo* assay for the screening of potential antagonists of *Botrytis cinerea* on tomato plants. *IOBC WRS Bull.* 16(11):30-33.

Nicot, P.C., Zimmer, N., Jacques, M.A. and Morris, C. 1994, Effect of nutrient amendments on the protection of pruning wounds on tomatoes against *Botrytis cinerea* by bacterial antagonists, *Mol. Ecol.* 3:611.

Northover, J. 1988, Persistance of dicarboximide resistant *Botrytis cinerea* in Ontario vineyards, *Can. J. Plant Pathol.* 10:123-132.

Pak, H.A., Beever, R.E. and Laracy, E.P. 1990, Population dynamics of dicarboximide-resistant strains of *Botrytis cinerea* on grapevine in New Zealand, *Plant Pathol.* 39:501-509.

Peng, G., Sutton, J.C. and Kevan, P.G. 1992, Effectiveness of honeybees for applying the biocontrol agent *Gliocladium roseum* to strawberry flowers to suppress *Botrytis cinerea*, *Can. J. Phytopathol.* 14:117-129.

Ragsdale, N.N. and Sisler, H.D. 1994, Social and political implications of managing plant diseases with decreased availability of fungicides in the United States, *Annu. Rev. Phytopathol.* 32:545-557.

Raposo R., Colgan R., Delcan J. and Melgarejo P. 1995, Application of an automated quantitative method to determine fungicide resistance in *Botrytis cinerea*, *Plant Dis.* 79:294-296.

Reglinski, T., Lyon, G.D. and Newton, A.C. 1995, The control of *Botrytis cinerea* and *Rhizoctonia solani* on lettuce using elicitors extracted from yeast cell walls, *Z. PflKrankh. PflSchutz*, 102:257-266.

Reuveni, R. and Raviv, M. 1992, The effect of spectrally-modified polyethylene films on the development of *Botrytis cinerea* in greenhouse-grown tomato plants, *Biol. Agric. and Hort.* 9:77-86.

Reuveni, R., Raviv, M. and Bar, R. 1989, Sporulation of *Botrytis cinerea* as affected by photoselective polyethylene sheets and filters, *Ann. Appl. Biol.* 115:417-424.

Salinas, J., Glandorf, D.C.M., Picavet, E.D. and Verhoeff, K. 1989, Effect of temperatures, relative humidity and age of conidia on the incidence of spotting on gerbera flowers caused by *Botrytis cinerea*, *Neth. J. Plant Pathol.* 95:51-64.

Sasaki, T., Honda, Y., Umekawa, M. and Nemoto, M. 1985, Control of certain diseases of greenhouse vegetables with ultraviolet-absorbing vinyl film, *Plant Dis.* 69:530-533.

Scherm, H., Koike, S.T., Laemmlen and van Bruggen, A.H.C. 1995, Field evaluation of fungicide spray advisories against lettuce downy mildew (*Bremia lactucae*) based on measured or forecast morning leaf wetness, *Plant Dis.* 79:511-516.

Sirjusingh, C. and Sutton, J.C. 1996, Effects of wetness duration and temperature on infection of geranium by *Botrytis cinerea*, *Plant Dis.* 80:160-165.

Smith, C.M. 1998. History of benzimidazole use and resistance, pp. 23-24 In: Delp, C.J. (ed.) *Fungicide Resistance in North America*. American Phytopathological Society, St Paul, MN.

Stall, R.E. 1963, Effects of lime on incidence of *Botrytis* gray mold of Tomato, *Phytopathology* 53:149-151.

Sutton, J.C., Gillespie, T.J. and James, T.D. 1988, Electronic monitoring and use of microprocessors in the field, pp. 99-113 In: Kranz, J. and Rotem, J. (eds.) *Experimental Techniques in Plant Disease Epidemiology*, Springer Verlag, Berlin.

Sutton, J.C., James, T.D. and Rowell, P.M. 1986, BOTCAST: a forecasting system to time the initial fungicide spray for managing Botrytis leaf blight of onions, *Agric. Ecosystems Environ.* 18:123-143.

Sutton, J.C. and Peng, G. 1993a, Biosuppression of inoculum production by *Botrytis cinerea* in strawberry leaves, *IOBC WRS Bull.* 16(11):47-52.

Sutton, J.C. and Peng, G. 1993b, Biocontrol of *Botrytis cinerea* in strawberry leaves, *Phytopathology* 83:615-621.

Tan, K.K. 1974, Blue-light inhibition of sporulation in *Botrytis cinerea*, *J. Gen. Microbiol.* 82:191-200.

Tan, K.K 1975, Interaction of near-ultraviolet, blue, red and far-red light in sporulation of *Botrytis cinerea*, *Trans. Br. Mycol. Soc.* 64:215-222.

Terrentroy, A. 1994, Tomate Serre: enquête sur le *Botrytis* dans les cultures de tomate précoce, *APREL Bull.* N° S-641, Chamber of Agriculture of Bouches du Rhône, France.

Thonard, P., Jacques, P., Cornelius, C., Gregoire, J., Zgoulli, S., Destain, J., Meurisse, E., Tossut, P., Gilsoul, J.J., Hbid, C and Moukoum, J.B. 1993, Technological aspects of biopesticide production, pp. 87-92 In: Lepoivre, P. (ed.) *Proc. Eur. Com. Workshop "Biological control of fruit and foliar diseases"*, Gembloux, Belgium.

Trivellas, A.E. 1988, Benzimidazole resistance monitoring techniques and the use of monitoring studies to guise benomyl marketing, pp. 28-30 In: Delp, C.J. (ed.) *Fungicide Resistance in North America.*, American Phytopathological Society, St Paul, MN.

Vakalounakis, D.J. 1991, Control of early blight of greenhouse tomato, caused by *Alternaria solani*, by inhibiting sporulation with ultraviolet-absorbing vinyl film, *Plant Dis.* 75:795-797.

Vakalounakis, D.J. 1992, Control of fungal diseases of greenhouse tomato under long-wave infrared-absorbing plastic film, *Plant Dis.* 76:43-46.

Vali, R.J. and Moorman, G.W. 1992, Influence of selected fungicide regimes on frequency of dicarboximide-resistant and dicarboximide-sensitive strains of *Botrytis cinerea*, *Plant Dis.* 76:919-924.

Vincelli, P.C. and Lorbeer, J.W. 1989, BLIGHT-ALERT: a weather-based predictive system for timing fungicide applications on onion before infection periods of *Botrytis squamosa*, *Phytopathology* 79:493-498.

Volpin, H. and Elad, Y. 1991, Influence of calcium nutrition on susceptibility of rose flowers to *Botrytis* blight, *Phytopathology* 81:1390-1394.

Wang, Z.N., Coley-Smith, J.R. and Wareing, P.W. 1986, Dicarboximide resistance of *Botrytis cinerea* in protected lettuce, *Plant Pathol.* 35:427-433.

Wilcox, W.F. and Seem, R.C., 1994, Relationship between strawberry gray mold incidence, environmental variables and fungicide applications during different periods of the fruiting season, *Phytopathology* 84:264-270.

Winspear, K.W., Postlethwaite, J.D. and Cotton, R.F. 1970, The restriction of *Cladosporium fulvum* and *Botrytis cinerea* attacking glasshouse tomatoes, by automatic humidity control. *Ann. Appl. Biol.* 65:75-83.

Yunis, H., Shtienberg, D., Elad, Y. and Mahrer, Y. 1994, Quallitative approach for modelling outbreaks of grey mould epidemics in non-heated cucumber greenhouses, *Crop Protect.* 13:99-104.

12

THE ROLE OF PLANT SURFACE BACTERIA IN THE HYGIENIC AND MARKET QUALITY OF MINIMALLY PROCESSED VEGETABLES

Cindy E. Morris[1] and Christophe Nguyen-The[2]

INRA
[1] Station de Pathologie Végétale
[2] Station de Technologie des Produits Végétaux
Centre de Recherches d'Avignon
France

INTRODUCTION

In the past 20 years fresh and "fresh-like" vegetables have assumed a greater place in the diet of Americans and western Europeans. In the US, for example, per capita consumption of fresh vegetables increased by 26% from 1978 to 1988 (Schlimme, 1995). The introduction of minimally processed (MP) - pre-cut, washed, ready-to-use - products to the retail market responded to the growing demand for "fresh-like" vegetables. Since 1992, the retail sales of MP salads, for example, grew by 95% a year in the US, reaching a total sales volume of $600 million in 1994 (Shapiro, 1995). In France, Europe's largest market for MP salads, 35,000 tonnes of MP salads were sold in 1993 representing 5% of the total fresh salad consumption (Harzig, 1994). It has been estimated that, by the year 2000, 50% of the sales volume of supermarket produce sections will be generated by MP products (Graziano, 1993).

Vegetables that are marketed as MP products include broccoli, celery, cucumbers, tomatoes and numerous leafy vegetables (lettuce, chicories, spinach, etc.) (Schlimme, 1995). Some root crops are also available in MP forms, but their microbiology is outside the scope of this chapter. Salads are the principal MP vegetable, constituting 85% of the market in France, for example. Whereas species of *Cichorium* are the most popular salad vegetables in France, lettuce is the main component of MP salads in the US.

Vegetables destined for minimal processing are generally produced under conditions identical to those used for fresh market vegetables. Hence, the microflora of the surface of the aerial parts of these plants is as diverse and dense as that of typical field- or greenhouse-grown crops. Occasionally, these vegetables are contaminated with human pathogens during cultivation; however, industrial processing provides another opportunity for introduction of human pathogens if strict hygiene conditions are not respected in the factory. Subsequently, packaging of these vegetables may provide favourable conditions for bacterial proliferation

Aerial Plant Surface Microbiology, edited by Morris et al.
Plenum Press, New York, 1996

if the packs are not maintained at sufficiently cool temperatures during storage. This bacterial proliferation may lead to important losses in market quality of these vegetable products before the sell-by date. Furthermore, psychrotrophic human pathogens such as *Listeria monocytogenes* may also survive and multiply on the surfaces of these vegetables. Studies of the microbiology of MP vegetables aim to understand the factors controlling microbial colonisation of vegetable surfaces, mechanisms by which micro-organisms degrade plant tissues and interactions among the various members of the microflora in view of assuring the market and hygienic quality of these products.

In this chapter we will describe the variety of known and potential mechanisms by which the diverse bacterial flora of MP vegetables may influence market quality of these products. Furthermore, we will evaluate the risk of installation of human pathogens on MP vegetable surfaces based on their ability to grow in this environment and to co-exist with saprophytic bacteria. We will focus on MP products made from aerial plant parts - leafy vegetable salads in particular. Finally, in light of the diversity and ecology of this community, we will consider possible strategies for management of the bacteria associated with MP vegetables.

SAPROPHYTIC AND PHYTOPATHOGENIC BACTERIA ON SURFACES OF MP VEGETABLES

The types and numbers of bacteria, fungi and yeasts on MP vegetables have been described in detail by Nguyen-the and Carlin (1994) in their review of the literature on the microbiology of this product. They note that bacteria tend to be dominant on most leafy vegetable products and, among the bacteria, members of the Pseudomonadaceae and Entero-bacteriaceae are the most frequent. This general pattern of composition of the microbial community was also observed in a study of MP broad-leaved endive (Jacques, 1994). Although the flora on MP vegetables tends to be dominated by two bacterial families, it is nevertheless very diverse. For example, Jacques (1994) has tentatively identified over 100 species of Gram (-) and Gram (+) bacteria from MP broad-leaved endive. Among these, the most frequently isolated was *Pseudomonas fluorescens* which is one of the principal organisms responsible for decay of this product (see below).

Another essential feature of the bacterial populations on MP vegetables and on vegetables destined for minimal processing is their variability in composition and size among plants, leaves, leaf pieces, etc. This notion of variability is not unique to MP vegetables and has been well-developed for bacterial populations on other crops (Hirano and Upper, 1991). For broad-leaved endives destined for minimal processing, the size of their bacterial populations may be significantly different among different plants cultivated at the same time in the same field (Morris and Lucotte, 1993), among plants sown at different dates in the same season (Morris and Lucotte, 1993) or cultivated in different years in the same location (Jacques and Morris, 1995) and among leaves of different ages on the same plant (Jacques et al., 1995). This latter point is described in detail in the chapter by Jacques (this volume). Furthermore, the dynamics of bacterial population sizes during storage of MP salads is significantly effected by leaf age (Jacques and Morris, 1995) and the species composition of these populations may vary greatly among the leaf pieces in the same pack (Morris et al., 1990). This variability can greatly complicate efforts to evaluate the importance of a given micro-organism in alteration of MP salads or to predict market quality.

Vegetables destined for minimal processing are highly contaminated with bacteria having total mesophile populations as large as 10^5 colony forming units (cfu) g^{-1} for lettuce (King et al., 1991) and 10^7 cfu cm^{-2} for broad-leaved endive (Jacques and Morris, 1995) at harvest. Although the disinfection step of processing may reduce the bacterial population sizes by about two log units, during storage these populations may increase to sizes equal to or greater than the pre-processing population size (Jacques and Morris, 1995). For MP broad-leaved endive, decay of leaf pieces in packs is infrequent when populations of fluorescent pseudomonads, for example, are less than 10^7 cfu cm^{-2} (Nguyen-the and Prunier, 1989; Jacques and Morris, 1995). Storage of MP vegetable packs at temperatures around 4°C is generally the most widely used, effective method of limiting bacterial proliferation; however, in practice, this temperature is not respected throughout distribution and storage (Scandella, 1989) favouring more rapid microbial growth.

MECHANISMS BY WHICH BACTERIA CAN ALTER THE MARKET QUALITY OF MP VEGETABLES

Negative Effects on Quality

After processing, MP vegetables can suffer from several types of alteration of market quality before the sell-by date. The principal types of alterations are browning of cut edges and wounds, soft rot and the development of silage-like odours. In fact, any type of transformation leading to the loss of fresh-like appearance or taste should be considered a loss in market quality.

Although bacteria are not always responsible for changes in market quality of MP vegetables (cf. Nguyen-the and Carlin, 1994), their role in soft rot of these products has been well documented. Furthermore, MP vegetables represent wounded tissues excised from the plant and incubated under conditions ideal for the growth and activity of psychrotrophic bacteria. Hence, it is likely that bacteria contribute in numerous other, perhaps subtle, ways to the market quality of these stressed plant tissues. In this section we will describe the known mechanisms by which bacteria alter the quality of MP vegetables and suggest additional mechanisms that should be investigated in more detail.

Induction of soft rot, due to the action of pectolytic enzymes, is the chief mechanism by which bacteria attack MP vegetables during storage. It is possible that soft rot is induced by other types of microbial enzymes or that the macerating enzymes are produced by the plant. However, these latter possibilities have not yet been demonstrated for MP vegetables. The most abundant pectolytic bacteria on MP salads during storage are the oxidase (+), arginine dihydrolase (+) fluorescent pseudomonads (Brocklehurst et al., 1987; Nguyen-the and Prunier, 1989) (Pseudomonas fluorescens, P. putida, P. chlororaphis, etc.). Pectolytic strains in this group of pseudomonads are also involved in post harvest decay of a wide range of fresh market vegetables including pepper, tomato, cucumber, broccoli, spinach and lettuce (Liao and Wells, 1987a) and are capable of multiplying and macerating plant tissues at 4°C (Brocklehurst and Lund, 1981), the recommended storage temperature for many MP vegetables. The capacity to liquefy a calcium pectate gel was perfectly correlated with the capacity to macerate MP chicory salads for fluorescent pseudomonads isolated by Nguyen-the and Prunier (1989). However, in a study of over 250 isolates of fluorescent pseudomonads from MP broad-leaved endive, the production of pectolytic enzymes could not be detected in in vitro assays for 41% of the isolates that were mildly or strongly aggressive in their maceration of this MP product (Jacques, 1994). It is possible that the pectolytic enzymes of these isolates were not induced under the experimental conditions employed or that other types of enzymes were responsible for the maceration. Wang and Kelman (1982) have reported that a protease was apparently responsible for maceration of potato tuber slices by a strain of Pseudomonas fluorescens for which no pectate lyase, polygalacturonase nor cellulase production could be detected. However, in a further study Schlemmer *et al.* (1987) determined that the protease activity resolved by ion-exchange chromatography by Wang and Kelman (1982) contained a pectin lyase responsible for the maceration.

The role of pectolytic bacteria other than the fluorescent pseudomonads in soft rot of MP vegetables has been examined less intensively. Numerous species of bacteria found on vegetable surfaces are known to produce pectolytic enzymes capable of softening plant tissues. These include species of *Erwinia, Xanthomonas, Cytophaga* and *Bacillus* as well as *Aeromonas liquefaciens* and *Leuconostoc mesenteroides* (Lund, 1992). Pectolytic strains of *Erwinia* spp. were frequently isolated from MP salads by Brocklehurst *et al.* (1987) but were not detected by Nguyen-the and Prunier (1989). *Cytophaga johnsonae* and *Xanthomonas campestris* have been isolated from soft rots of fresh market vegetables (cucumber, pepper, squash, tomato, etc.) and induce maceration upon re-inoculation of the plant tissues (Liao and Wells, 1986; Liao and Wells, 1987b). However, these pathogenicity tests were conducted at 20°C and may not be representative of the degradative capacity of these bacteria at typical storage temperatures of MP vegetables. For example, the optimal conditions for maceration of stored plant tissues by certain strains of *Bacillus* spp. have been reported to be at ambient temperatures (Fernando and Stevenson, 1952; Ono, 1969), whereas strains of *Bacillus* spp. isolated from Chinese cabbage were able to cause soft rot at 4° or warmer (Chiu *et al.*, 1964). Preliminary results from our laboratory indicate that pectolytic strains of bacteria tentatively identified as *Xanthomonas maltophilia* (Jacques and Morris, 1994) and non-fluorescent pseudomonads (unpublished data) can cause substantial soft rot of MP broad-leaved endive in experimental packs stored at 10°C, but the frequency of occurrence of these bacteria on MP salads has not been determined.

The action of pectolytic enzymes in macerating plant tissues may be facilitated by biosurfactants. Biosurfactants are produced by a wide range of plant-associated bacteria including species of *Bacillus, Corynebacterium, Rhodococcus* and the fluorescent pseudo-monads (Laycock, 1991; Bunster *et al.*, 1989; Cooper, 1986). Production of biosurfactant compounds is also frequent among the fluorescent pseudomonads isolated from MP broad-leaved endive (Jacques, 1994). Inoculation of broccoli heads with mixed populations of a pectolytic fluorescent pseudomonad and an isolate that produced biosurfactants resulted in significantly more soft rot of the broccoli heads than when only the pectolytic strain was used as inoculum (Hildebrand, 1989). The authors concluded that the biosurfactants in-creased wetting of the broccoli surface, hence fostering better contact of the pectolytic enzymes with their substrates. Biosurfactants also induce leakage of nutrients from plant tissues (Laycock, 1991). In a previous chapter, Schönherr and Baur (this volume) have suggested that the induction of leakage by biosurfactants may create greater nutrient reserves on leaf surfaces. This in turn may permit bacteria to establish population sizes necessary for effective attack of host tissues.

Although soft rot of MP vegetables can cause dramatic changes in market quality if its development is not restrained, tissue browning is probably the most frequent type of decay, particularly for MP salads. Browning occurs when products of phenylpropanoid metabolism (phenolics, anthocyanins, etc.) are oxidised in reactions catalysed by phenolases (*cf.* Brecht, 1995) often in response to wounding. The occurrence and severity of browning have been attributed principally to the physiological condition of the plant independent of the action of any micro-organisms. The degree of browning of different varieties of chicories, for example, after minimal processing is well correlated with the variety's polyphenol content (Varoquaux, 1990). However, plant pathogenic bacteria - and in particular those producing pectolytic enzymes - can induce plant defence responses (*cf.* Alghisi and Favaron, 1995; van Gijsegem *et al.*, 1995; Lojkowska and Holubowska, 1992). Inoculation of potato slices with a suspension of *Erwinia carotovora* ssp. *carotovora* caused significant increases in polyphenol oxidase and peroxidase activity when compared to slices that were simply wounded (Lojkowska and Holubowska, 1992). This enzymatic activity was coupled to intense browning of tissues in the hybrids of potato with the highest levels of these enzymes. Vantomme *et al.* (1989) have also reported that many strains of pectolytic fluorescent pseudomonads induce a brown rot of whitloof chicory leaves but do not cause soft rot of this host.

Is it possible that the bacteria associated with the surfaces of MP vegetables contribute to browning of tissues during storage? The role of bacteria in the browning of tissue discussed in the previous paragraph may be the result of elicitation of host defence mechanisms by fragments of the plant cell wall released by the action of pectolytic enzymes as described in the review by Alghisi and Favaron (1995). This function of pectolytic enzymes has been described for endo-polygalacturonases and endo-pectate lyases from plant pathogenic bacteria (*cf.* Alghisi and Favaron, 1995). It would be interesting to determine if pectolytic enzymes from the so-called saprophytic bacteria associated with MP vegetables also induce plant defence mechanisms. Another potential elicitor of plant defence mechanisms produced by bacteria associated with MP vegetables is indole acetic acid (IAA). Bacteria that are ubiquitous epiphytes such as *Erwinia herbicola* and fluorescent pseudomonads are known to produce IAA (Clark and Lindow, 1989; Brandl *et al.*, 1996). On pear these bacteria have been shown to contribute to russetting of fruits (Clark and Lindow, 1989). IAA is known to induce production of ethylene which in turn may induce phenylpropanoid metabolism (*cf.* Brecht, 1995) or lipid peroxidation (Stanley, 1991). In lettuce, ethylene induces phenolic metabolism which leads to russet spotting (Hyodo *et al.*, 1978). Although wounding of plant tissues increases ethylene production, rates of ethylene production in lettuce, for example, return to normal levels within 24 hr after wounding (Ke and Saltveit, 1989). It would be interesting to determine if IAA-producing bacteria associated with plant surfaces contribute to the accumulation of ethylene in packs of MP vegetables.

Do bacteria contribute in other ways to the loss of market quality of MP vegetables? Are they involved in the development of odours or anaerobic conditions in packs? If they induce plant defence mechanisms, are the resulting products dangerous for the consumer? These are interesting questions that need to be explored.

Beneficial Effects on Quality

The diverse bacterial species associated with MP vegetables represent a wide range of metabolic capacities. While some of the metabolic capacities of the bacteria on MP vegetables may lead to reductions in the market quality of these products, it is very likely that some of the bacteria in this community have beneficial effects for market quality. Among these effects may be those that directly intervene in the physiology of the host thereby altering its sensitivity to decay or those that act on other micro-organisms in the community (via antagonism, for example). A component of the bacterial flora of MP vegetables that may be beneficial for market quality is the non symbiotic nitrogen fixers. *Rahnella aquatilis* is a nitrogen fixing enteric bacterium (Berge *et al.*, 1991) that has been frequently isolated from MP salads (Geiges *et al.*, 1990; Marchetti *et al.*, 1992; Jacques, 1994). If the *R. aquatilis* strains on MP vegetables fix nitrogen they could help to delay senescence of these vegetable tissues. In other systems it has been observed that the nitrogen fixed by phyllosphere micro-organisms can be directly absorbed in the form of amino acids by the leaf (Jones, 1970). Preliminary results from our laboratory suggest that packs of MP broad-leaved endive inoculated with certain strains of *R. aquatilis* maintain a fresh appearance longer than non inoculated packs (Jacques and Morris, 1994). Such observations give impetus for studies of the importance of nitrogen fixation - or other metabolic pathways by which epiphytic bacteria could provide nutrients for plant tissues - for the shelf life of MP vegetables.

Another component of the bacterial flora of MP vegetables that may have beneficial effects are lactic acid bacteria which are found on many types of plants in the field and at harvest (Mundt and Hammer, 1968; Mundt *et al.*, 1969) and have been found on MP vegetables (Denis and Picoche, 1986; Brocklehurst, 1987; Barriga *et al.*, 1991). Some lactic acid bacteria are known to produce inhibitory compounds such as bacteriocins and antibiotics (Klaenhammer, 1988). In a test of the effect of lactic acid bacteria on quality of MP vegetables, Vescovo *et al.* (1995) inoculated packs of MP salads with strains of *Lactobacillus* spp. and *Pediococcus* spp. able to produce antimicrobial compounds. They noted that, after 8 days of storage of the packs at 8°C, the population sizes of total mesophilic bacteria in packs inoculated with the lactic acid bacteria were 0.1% to 1% of those in packs that were not inoculated. Although naturally occurring lactic acid bacteria are usually present on MP vegetables at population sizes smaller than those used in the experiments by Vescovo *et al.* (1995), it should be determined if they play a role in limiting the growth of certain components of the bacterial community on this product. In the work mentioned above, no reference was made of the quality of MP vegetables treated with lactic acid bacteria. Beside bacteriocins which act in very specific conditions, the antimicrobial effect of lactic acid bacteria relies mostly on acidification and production of organic acids (Piard and Desmazeau, 1991). To which extent MP vegetables withstand organic acid accumulation is not known. The prevailing lactic acid bacteria identified in MP vegetables are *Leuconostoc* spp. (Denis *et al.*, 1988; Nguyen-the and Carlin, 1994). In some products such as shredded carrots, growth of *Leuconostoc* spp. was identified as an important cause of spoilage (Carlin *et al.*, 1989). It is obviously not advisable to promote the indigenous lactic acid bacteria to control other spoilage bacteria in MP vegetables. However, *Leuconostoc* spp. are heterofermentative bacteria producing, in addition to lactic acid, acetic acid, ethanol and carbon dioxide, compounds which could be more phytotoxic and more detrimental to quality than lactic acid alone. The potential use of introducing homofermentative lactic acid bacteria to limit microbial development in MP vegetables deserves more attention.

BACTERIA RESPONSIBLE FOR FOODBORNE DISEASES

Occurrence of Bacterial Foodborne Pathogens on Vegetables and Consequences for Human Health

Contamination of vegetables with human pathogens is largely documented (Lund, 1988; Doyle, 1990; Nguyen-the and Carlin, 1994). Most of the work has focused on pathogenic bacteria such as *Clostridium botulinum* (Lund, 1988; Notermans, 1993), *Clostridium perfringens* (Lund, 1988; Roberts *et al.*, 1982), *Bacillus cereus* (Roberts *et al.*, 1982; Kramer and Gilbert, 1989), *Salmonella* spp.(Garcia-Villanova Ruiz *et al.*, 1987; Roberts *et al.*, 1982; Ercolani, 1976; Tamminga *et al.*, 1978), coagulase positive *Staphylococcus aureus* (Stewart *et al.*, 1978), *Listeria monocytogenes* (Ryser and Marth, 1991), and thermotolerant *Campylobacter* spp. (Park and Sanders, 1992). Bacterial foodborne pathogens have always been found in low numbers on fresh vegetables, usually after a step of selective enrichment and in all cases represented less than 0.1% of the total microflora. Incidence of foodborne pathogens in published works may be high (30% to 80% of the samples analysed were contaminated), but incidences of a few percent or below one percent are more frequent.

Vegetables are a possible source of foodborne disease. Examples of pathogens that have caused foodborne diseases involving raw or MP fresh vegetables include: *L. monocytogenes* (Schlech *et al.*, 1983), *Shigella sonnei* (Davis *et al.*, 1988; Kapperud *et al.*, 1995), *Vibrio cholera* (Shuval *et al.*, 1984), *Salmonella* spp. (O'Mahoney *et al.*, 1990; Doyle, 1990),

C. botulinum (Solomon *et al.*, 1990), *B. cereus* (Portnoy *et al.*, 1976), enterotoxigenic *Escherichia coli* (MMWR, 1994). In some cases (Schlech *et al.*, 1983; Doyle, 1990; Shuval *et al.*, 1984), contamination of the vegetables implicated in the outbreaks was traced to the use of manure as a fertiliser in the field or the use of polluted water for irrigation. Bacteria which are not normal inhabitants of the phyllosphere can therefore survive on plant surfaces in the field and during storage until consumption. However, vegetables are a minor source of foodborne diseases and accounted for less than 5% of cases in North America from 1973 to 1987 (Bean and Griffin, 1990; Todd, 1992).

Bacterial Foodborne Pathogens on MP Vegetables

Micro-organisms present on MP vegetables are those normally found on raw vegetables (Nguyen-the and Carlin, 1994). Most surveys have focused on *L. monocytogenes* which was found present in frequencies varying from 3% to 19% of samples analysed (Nguyen-the and Carlin, 1994), although recent work in France reported an absence (*i.e.* less than 1 per 25g) of *L. monocytogenes* in over 100 (Beaufort *et al.*, 1992) and 195 (DGCCRF, 1994) samples of MP vegetables. In a few cases, the origin of contamination of MP vegetables with foodborne pathogens was identified: contamination in the field through the use of organic fertilisation (Schlech *et al.*, 1983), contamination through the processing equipment (Lainé and Michard, 1988) and contamination from an employee during processing (Davis *et al.*, 1988). The recent implementation of good manufacturing practices for minimal processing of vegetables in various countries (Anon, 1988; Brown and Oscroft, 1989) should reduce the occurrence of such contamination.

Minimally processed vegetables undergo treatments such as peeling, shredding and cutting, which create large surfaces of wounded tissues and presumably release nutrients from the vegetable cells. In addition, MP vegetables are sealed in polymeric materials which maintain a high relative humidity. Both factors are presumably favourable to the growth of foodborne pathogens, and for this reason, minimal processing was suspected to increase the risk of foodborne diseases, as compared to raw vegetables. Refrigeration combined with the choice of appropriate packaging conditions permits a shelf-life of 7 days or more for most MP vegetables, a delay theoretically amply sufficient for the multiplication of psychrotrophic pathogens like *L. monocytogenes* (Ryser and Marth, 1991).

The Fate of Foodborne Pathogens on MP Vegetables: The Case of *Listeria monocytogenes*

Most bacterial pathogens can multiply or produce toxin on the surface of vegetables stored at temperatures above 10°C (Nguyen-the and Carlin, 1994). At temperatures of refrigeration, *L. monocytogenes* is one of the few human pathogens capable of significant growth (Nguyen-the and Carlin, 1994) on fresh vegetables, together with *Aeromonas hydrophila,* but the status of the latter as a foodborne pathogen is still an object of debate. Recently, the International Commission for Microbial Specification for Foods considered that the ubiquitous nature of *L. monocytogenes* makes it impossible to totally preclude its presence in raw foods and, although the infectious dose remains unknown, proposed a tolerance limit of 100 *L. monocytogenes* per g (Farber, 1993). Several countries, such as France and England, implemented these specifications in their legislation (Anon, 1992 a, b). It is therefore necessary to quantify the growth potential of *L. monocytogenes* in MP vegetables to assess the risk of reaching numbers above the tolerance limit during the shelf-life. The initial numbers of *L. monocytogenes* on MP vegetables cannot be known precisely, but all surveys found numbers below 50 per g and in France it was estimated to between 1 and 0.1 per g (Lainé and Michard, 1988). Another key point is the definition of shelf-life. Rather than the use-by-date at the recommended storage temperature specified by the manufacturer, or the legislator, it is rather the period before the appearance of noticeable decay, at any storage temperature. Growth of *L. monocytogenes* must therefore be followed together with spoilage development.

Growth of *L. monocytogenes* on MP vegetables was tested by inoculating strains grown in laboratory media on various vegetables. It grew on shredded iceberg lettuce (Beuchat and Brackett, 1990a; Steinbruegge *et al.*, 1988), on shredded cabbage (Kallender *et al.*, 1991; Beuchat *et al.*, 1986), on shredded endive (Carlin *et al.*, 1995) and on shredded butterhead lettuce (Carlin *et al.*, 1994). In contrast, MP lambs lettuce did not support growth of *L. monocytogenes* (Carlin *et al.*, 1994) and the bacterium was killed upon contact with cut or shredded carrot tissues (Nguyen-the and Lund, 1991 and 1992; Beuchat and Brackett 1990a). Carlin *et al.* (1995) and Steinbruegge *et al.* (1988) pointed out the variability of results obtained among various samples of a same product. For instance, on shredded broad-leaved endive, the carrying capacity of MP leaves was influenced by their physiological age (Carlin *et al.*, 1995), the bacterium reaching higher numbers on young etiolated leaves than on green leaves. As discussed by Jacques (this volume), this is the inverse of the trend observed for saprophytic and plant pathogenic bacteria on this plant. For leaves of the same physiological age, the carrying capacity for *L. monocytogenes* was positively correlated with the extend of spoilage after 7 days at 10°C (r=0.72 over 195 results, Carlin *et al.*, 1995). The authors found that *L. monocytogenes* would hardly grow above 10^6 cfu g^{-1}, for an initial number of 10^4 cfu g^{-1}, on leaves without noticeable spoilage. For a comparison, saprophytic bacteria reached 10^8 to 10^9 cfu g^{-1} in the same time, for an initial number of 10^4 to 10^5 cfu g^{-1}. On leaves showing decay, growth of *L. monocytogenes* could reach 10^8 to 10^9 cfu g^{-1} after 7 days at 10°C. Decay was probably the result of vegetable cell death and caused a release in nutrients that could foster *L. monocytogenes* growth. A similar relation between decay, cell death, release of soluble materials from the vegetable tissues and microbial development was observed for shredded carrots and the lactic acid bacterium *Leuconostoc mesenteroides* (Carlin *et al.*, 1990). This would imply that *L. monocytogenes* is nutrient limited on the surface of MP endive leaves, an assumption supported by the observation that most of the essential amino acids required by *L. monocytogenes* (Premaratne *et al.*, 1991) were not found, or in concentrations far below the minimum requirements, in exudates of endive leaves (Carlin *et al.*, 1996).

Listeria monocytogenes is capable of growth on MP vegetables at temperatures of 3°C and above. At temperatures from 3°C to 6 °C, the maximum growth rate of *L. monocytogenes* recorded on green salads was markedly lower than that of saprophytic bacteria. At 10°C and above, maximum growth rates of both populations tended to be similar (Steinbruegge *et al.*, 1988; Beuchat and Brackett, 1990a; Carlin *et al.*, 1995). This is illustrated in Figure 1 for MP green endive by the doubling times of *L. monocytogenes* and saprophytic bacteria plotted versus storage temperature. It is possible that saprophytic bacteria from green salad leaves are better adapted to low temperatures than *L. monocytogenes*: for instance a fluorescent pseudomonad from vegetables had a doubling time of 15 h at 0.2°C (Lund, 1983), whereas *L. monocytogenes* hardly grows at this temperature. However, in lettuce juice at 5°C, *L. monocytogenes* and saprophytic bacteria from lettuce leaves grew similarly (Steinbruegge *et al.*, 1988), indicating that the lower fitness of *L. monocytogenes* on green salad leaves at low temperatures may not only be explained by a lower psychrotrophy. Practically, these results imply that refrigeration storage of MP vegetables does not increase the risk of listeriosis by delaying the development of a competitive microflora as suspected for other foods. Refrigeration does not either increase risk by delaying spoilage and increasing the time available for growth of listeria: growth achieved

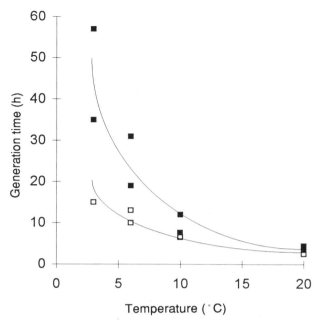

Figure 1. Generation time of *Listeria monocytogenes* (closed symbols) and of the saprophytic microflora (open symbols) on minimally processed broad-leaved endive as a function of storage temperature. Adapted from Carlin *et al.* (1995).

by *L. monocytogenes* when spoilage became noticeable was similar for the storage temperatures between 3 and 20°C (Table 1).

Testing growth of bacteria by inoculating foods with laboratory grown strains may cause important experimental bias. One relies on the nature of the inoculum which is often prepared from collection strains. However, no variations were observed among collection strains of *L. monocytogenes* and strains isolated from cabbage on either shredded lettuce (Beuchat and Brackett, 1990a) or MP endive (Carlin *et al.*, 1995), and strains from various origins behaved similarly on vegetables (Breer and Baumgartner, 1992). In most work, inoculum was prepared from strains grown in liquid, laboratory media, whereas MP vege-

Table 1. Growth of *Listeria monocytogenes* in log units achieved on minimally processed broad-leaved endive stored at different temperatures, just before spoilage became noticeable and at the time when spoilage became noticeable. Adapted from Carlin *et al.* (1995)

Storage temperature (°C)	growth of listeria (log units) before noticeable spoilage	growth of listeria (log units) when spoilage was first noticeable
3	1.24	1.72
	0.59	0.46
6	1.91	2.25
	1.11	1.38
10	1.90	2.66
	1.25	1.80
20	1.46	2.95
	1.64	1.75

tables are solid surfaces. The influence of the preparation conditions of inoculum on the adaptation of *L. monocytogenes* to low temperatures and to low pH has been demonstrated (Buchanan and Klatwitter, 1991; George and Lund, 1992), but there is no indication that such effects exist for the adaptation of *L. monocytogenes* to the conditions prevailing on minimally processed vegetables. Fate of micro-organisms is usually studied after inoculation of sufficiently high numbers to measure either an increase or a decline with conventional methods for enumeration of micro-organisms. In the case of *L. monocytogenes* most studies have been performed after inoculating between 10^4 and 10^5 cfu g^{-1}, whereas fresh vegetables are probably contaminated with less than 50 cfu g^{-1}. In laboratory media, growth parameters such as lag time, growth rate and maximum number, are independent of the inoculum level (Buchanan and Philipps, 1990). In contrast, the initial number of *L. monocytogenes* had a significant influence on its fate on MP endive (Carlin *et al.*, 1995): growth rates were slightly increased by low inoculum levels during the first days of storage at 10°C but the population tended toward a lower maximum level (Figure 2a). Similarly, at 12°C enterotoxigenic *E. coli* grew on various MP vegetables for high inoculum level, but declined for low inoculum level (Abdu Raouf *et al.*, 1993) (Figure 2b). However, at 20°C, a temperature closer than 12°C to the optimum for *E. coli*, growth was observed for both inoculum levels (Abdu Raouf *et al.*, 1993). Growth parameters of pathogenic bacteria on the surface of fresh vegetables obtained experimentally from high inoculum may therefore over-estimate the risk which could arise from a natural contamination. It is possible that low inoculum levels reduce the ratio of introduced bacteria to indigenous microflora, and thus increase competition. The importance of population density on the magnitude of competition among bacteria on plant surfaces is discussed in the chapter by Kinkel (this volume). Competition may also be a consequence of the heterogeneity of the vegetable surface where there are possibly only a discrete number of favourable sites. In this case a low inoculum level reduces the probability for the bacterial population to reach and to colonise such sites. A similar assumption was proposed by Wilson

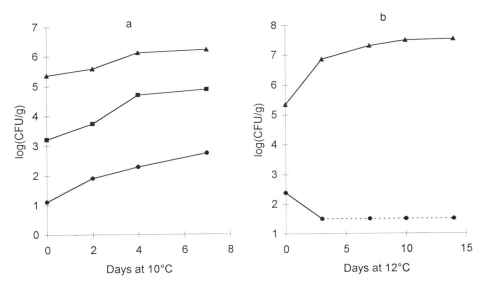

Figure 2. Fate of *Listeria monocytogenes* on minimally processed broad-leaved endive (a) and fate of *E. coli* O157:H7 on minimally processed iceberg lettuce (b) as influenced by inoculum level. Adapted from Carlin *et al.* (1995) (a) and from Abdu Raouf *et al.* (1993) (b). The dashed line indicates that *E. coli* was detected by enrichment but was below the limit for direct enumeration.

and Lindow (1994) to explain the lower survival of pseudomonads on bean leaves for low inoculum levels.

INTERACTIONS BETWEEN SAPROPHYTES AND HUMAN PATHOGENS

Fresh vegetables carry large microbial populations which multiply rapidly on the MP products. The antagonistic properties of indigenous micro-organisms against spoilage agents of fruits and vegetables have been described and are the subject of active research and practical applications (Droby *et al.*, this volume). In several countries a low microbial population is an indication of a good hygiene and a marker of quality, and processing aims at reducing the saprophytic microflora by washing procedures or sanitation (Anon, 1988). Therefore, it is important to know whether the saprophytic microflora limits the development of human pathogens on MP vegetables and what is the impact of sanitation.

Listeria monocytogenes requires a few essential amino acids and is not proteolytic. In milk, fluorescent pseudomonads stimulated growth of *L. monocytogenes*, presumably because proteolysis of milk proteins by pseudomonads released free amino-acids (Marshall and Smith, 1990 and 1991). However, other authors (Farrag and Marth, 1989) found little effect of pseudomonads on growth of *L. monocytogenes* in milk. On raw meat, growth of *L. monocytogenes* was stimulated by the presence of pseudomonads, according to Gouet *et al.*. (1978), but Mallila-Sandholm and Skyttä (1991) observed a competition between these micro-organisms for nutrients. It was also proposed that fluorescent pseudomonads could inhibit *L. monocytogenes* by chelating iron (Freedman *et al.*, 1989), although Simon *et al.* (1995) found that iron chelated by siderophores of fluorescent pseudomonads could be used by *L. monocytogenes*. The relation between *L. monocytogenes* and saprophytic bacteria from foods is variable and probably depends on the nutrient status of the bacterium.

On MP green endive, sanitation may achieve a 100-fold reduction in the population size of saprophytic bacteria. Inoculation of *L. monocytogenes* on disinfected endive leaves led to a higher growth of the bacterium than on non-disinfected controls (Carlin *et al.*, 1996). Increasing the number of the saprophytes of endive leaves 10 to 100-fold by introducing selected components of this microflora (fluorescent pseudomonads and enterobacteria) reduced growth of *L. monocytogenes* (Carlin *et al.*, 1996). However, the results presented by the authors varied with the saprophytic strains introduced, and more surprisingly, the effect of a strain was not consistent between replicate experiments. In a liquid medium prepared with exudates from endive leaves, *L. monocytogenes* was outgrown by saprophytic bacteria, but was only partially inhibited by single strains. Only a mixed population freshly isolated from the leaves completely suppressed the development of *L. monocytogenes* (Carlin *et al.* 1996). The relation between *L. monocytogenes* and saprophytes on MP vegetables is obviously complex. In practice, both populations grow simultaneously on MP vegetables but the saprophytic microflora probably contribute in limiting the growth potential of *L. monocytogenes*. In all cases, sanitation of minimally processed vegetables must be followed by strict hygiene conditions to prevent recontamination with foodborne pathogens.

Interactions between saprophytes and human pathogens may also occur during processing and may have implications for the colonisation of equipment and surfaces of the processing line. Bacteria are known to form biofilms on industrial surfaces. These biofilms are particularly resistant to disinfectants and may also break off of these surfaces to contaminate foods in the processing chain (Zottola and Sasahara, 1994). Normally, biofilm formation on industrial surfaces requires several days to several weeks; hence, they would probably be eliminated in processing plants practicing daily cleaning and sanitising (Zottola and Sasahara, 1994). However, biofilms of saprophytic and plant pathogenic bacteria exist on vegetables destined for minimal processing, on broad-leaved endive in particular (Gras *et al.*, 1994; Morris *et al.*, 1994). These biofilms may pose a hazard if they adhere to equipment in minimal processing plants and provide sites were cells of human pathogens can rapidly integrate into biofilms.. For instance, the integration of *L. monocytogenes* into a biofilm of pseudomonads was demonstrated by Sasahara and Zottola (1993). Conversely, the multiplication of *L. monocytogenes* cells adherent to a surface would probably be reduced by the competition of saprophytic bacteria from vegetables. This was shown by Jeong and Frank (1994) with saprophytic bacteria from meat and dairy products.

STRATEGIES FOR MANAGEMENT OF BACTERIA ASSOCIATED WITH MP VEGETABLES

At present, most strategies for limiting the occurrence of human pathogens and the proliferation of saprophytic and phytopathogenic bacteria associated with MP vegetables are employed during processing and storage of these products, and a few are employed during field cultivation. These strategies include:

- strict hygiene in the processing plants (Hurst, 1995) and avoidance of use of manure during field cultivation.
- maintenance of a cold chain throughout production, distribution and sales (Scandella, 1989).
- partial disinfection of vegetable surfaces during processing (Adams *et al.*, 1989.; Garg *et al.*, 1989).
- packaging under modified atmospheres (Farber, 1991).
- incorporation of antimicrobial compounds into the plastic films used for packaging (Baldwin *et al.*, 1995).

As discussed previously in this chapter, there is also research directed at the possibility of introducing micro-organisms that are antagonistic to the development of the normal bacterial flora of MP salads (Vescovo *et al.*, 1995).

Although contamination of MP vegetable surfaces by saprophytic and phytopathogenic bacteria occurs principally before harvest of the raw plant material, little research effort has been focused on managing bacterial populations during vegetable cultivation as a strategy to improve MP product quality. Criteria for choice (or breeding) of cultivars of vegetables destined for minimal processing has traditionally included their tolerance to wounding and modified atmospheres, sensory qualities and resistance to diseases and pests important during cultivation. Might it also be possible to develop vegetable cultivars that inhibit the growth of bacteria such as *P. fluorescens* or human pathogens such as *L. monocytogenes*? Certain plants are known to produce compounds that are toxic to *L. monocytogenes*, for example lambs lettuce does not seem to provide conditions favourable for growth of *L. monocytogenes* (Carlin *et al.*, 1994) and carrot tissues are known to release an antimicrobial compound upon shredding or cutting (Beuchat and Brackett, 1990b; Nguyen-the and Lund, 1991, 1992). The genetic determinants for the production of these compounds should be identified. As Andrews has described in a subsequent chapter (this volume), it may also become possible to genetically modify plants to secrete nutritional compounds that provide selective pressure against certain components of epiphytic microflora. These technologies may foster creation of vegetable varieties with a microbial flora that enhances shelf life. Another possible approach is to develop more effective bactericides for use during cultivation or to modify bacterial populations through applications of nutrients that induce competition among indigenous micro-organisms as has been attempted with other crops (Morris and Rouse, 1984; Wilson and Lindow, 1994). We are currently exploring these possibilities in our laboratory.

Modification of other cultural practices may also be a means to manage the bacterial flora of vegetables destined for minimal processing. In a previous chapter (this volume) Lindow has illustrated why effective cultural control methods for epiphytic plant pathogenic bacteria should be based on knowledge of the role of either immigration/emigration or growth/death of cells in the colonisation process of aerial plant surfaces. Jacques *et al.* (1995), have attempted to determine which of these processes governs the colonisation of broad-leaved endive destined for minimal processing. Their results suggest that immigration of bacteria, particularly the fluorescent pseudomonads, may sporadically lead to 10 - 100 fold increases in the size of populations of these bacteria on newly emerging leaves. Before cultural techniques are developed consequent to these observations we must determine the source of the bacterial immigrants and the environmental conditions that favour a shift in the balance between bacterial growth and immigration. Hence, development of strategies for management of the bacterial flora associated with MP vegetables will depend on continued investment in research on the ecology of these bacteria.

CONCLUSION

The studies published on the microbiology of minimally processed vegetables illustrate the difficulties in ascribing spoilage to a specific component of their microflora. Spoilage probably involves several microbial species and several mechanisms. Stress caused by minimal processing also triggers biochemical and physiological reactions in the vegetables tissues such as oxidation of phenolic compounds, acceleration of senescence, decompartmentalization at the cell and at the tissue levels. Such processes contribute either directly to spoilage by causing brown discoloration or vitrescence, or indirectly by reducing the resistance of vegetable tissues to microbial attack. Minimally processed vegetables are packaged in a sealed polymeric films in which carbon dioxide accumulates rapidly and oxygen is rapidly depleted. These atmosphere modifications may extend shelf-life but levels of carbon dioxide (of oxygen) above (below) tolerance limits are a cause of spoilage. A delicate balance between respiration of the product and permeability of the film must be achieved. The contribution of the rapidly growing microbial population of minimally processed vegetables in gas exchanges has not been determined, but it is not unlikely that it could accelerate spoilage simply by accelerating oxygen depletion or carbon dioxide production. The identification of components of the microbial population that are not involved in spoilage is probably as important as the identification of spoilage micro-organisms. This offers means of managing the epiphytic microflora towards a reduction of spoilage, and perhaps, toward a reduction of the carrying capacity for human pathogens as suggested by the case of *L. monocytogenes* on broad-leaved endive. Future studies should take into consideration the following trait which clearly distinguishes minimally processed vegetables from either unprocessed vegetables or traditional processed products: a rapid multiplication of the microbial population on a plant tissue that is still alive but on the verge of senescence.

REFERENCES

Abdul-Raouf, U.M., Beuchat, L.R. and Ammar, M.S. 1993, Survival and growth of *Escherichia coli* O157:H7 on salad vegetables, *Appl. Environ. Microbiol.* 59:1999-2006.

Adams, M.R., Hartley, A.D. and Cox, L.J. 1989, Factors affecting the efficacy of washing procedures used in the production of prepared salads, *Food Microbiol.* 6:69-77.

Alghisi, P. and Favaron, F. 1995, Pectin-degrading enzymes and plant-parasite interactions, *Eur. J. Plant Pathol.* 101:365-375.

Anon. 1988, Guide des bonnes pratiques hygièniques concernant les produits végétaux dits de la 4ème gamme, *Bull. Offic. Concurr. Consomm. Repress. Fraud.* 48:221-232.

Anon. 1992a, Provisional microbiological guidelines for some ready-to-eat foods samples at point of sale, *PHLS Microbiology Digest* 9(3):98-99.

Anon. 1992b, Conseil Supérieur d'Hygiène Publique de France, Section Alimentation, Séance du 8 Septembre 1992, Avis relatif à *Listeria monocytogenes* et l'alimentation, *Cah. Nutr. Diét* 27(6):325.

Baldwin, E.A., Nisperos-Carriedo, M.O. and Baker, R.A. 1995, Edible coatings for lightly processed fruits and vegetables, *HortScience* 30:35-38.

Barriga, M.I., Trachy, G., Willemot, C. and Simard, R.E. 1991, Microbial changes in shredded iceberg lettuce stored under controlled atmospheres, *J. Food Sci.* 56:1586-1599.

Bean, N.H. and Griffin, P.M. 1990, Foodborne disease outbreaks in the United States, 1973-1987: pathognes, vehicles, and trends, *J. Food Prot.* 53:804-817.

Beaufort, A., Poumeyrol, G. and Rudelle, S. 1992, Fréquence de contamination par listeria et yersinia d'une gamme de produits de 4ème gamme, *Rev. Gen. Froid* 82(3):28-31.

Berge, O., Heulin, T., Achouak, W., Richard, R. and Balandreau, J. 1991, *Rahnella aquatilis*, a nitrogen fixing enteric bacterium associated with the rhizosphere of wheat and maize, *Can. J. Microbiol.* 37:195-203.

Beuchat, L.R. and Brackett, R.E. 1990a, Survival and growth of Listeria monocytogenes on lettuce as influenced by shredding, chlorine treatment, modified atmosphere packaging and temperature *J. Food Sci.* 55:755-758,870.

Beuchat, L.R. and Brackett, R.E. 1990b, Inhibitory effect of raw carrots on *Listeria monocytogenes*, *Appl. Environ. Microbiol.* 56:1734-1742.

Beuchat, L.R., Brackett, R.E., Hao, D.Y.-Y. and Conner, D.E. 1986, Growth and thermal inactivation of *Listeria monocytogenes* in cabbage and cabbage juice, *Can. J. Microbiol.* 32:791-795.

Brandl, M., Clark, E.M. and Lindow, S.E. 1996, Characterization of the indole-3-acetic acid biosynthesis pathway in an epiphytic strain of *Erwinia herbicola* and IAA production *in vitro*, *Can. J. Microbiol.* 42: 586-592.

Brecht, J.K. 1995, Physiology of lightly processed fruits and vegetables, *HortScience* 30:18-22

Breer, C. and Baumgartner, A. 1992, Vorkomen und Verhalten von *Listeria monocytogenes* auf Salaten und Gemüsen sowie in Frischgepressten Gemüsesäften, *Archiv. für Lebensmittelhygiene* 43:108-110.

Brocklehurst, T.F. and Lund, B.M. 1981, Properties of pseudomonads causing spoilage of vegetables stored at low temperature, *J. Appl. Bacteriol.* 50:259-266.

Brocklehurst, T.F., Zaman-Wong, C.M. and Lund, B.M. 1987, A note on the microbiology of retail packs of prepared salad vegetables, *J. Appl. Bacteriol.* 63:409-415.

Brown, K.L. and Oscroft, C.A. 1989, Guidelines for the hygienic manufacture, distribution and retail sale of sprouted seeds with particular reference to mung beans, *Technical Manual n° 25*, Campden Food and Drink Research Association, Chipping Campden, 25 pp.

Buchanan, R.L. and Klatwitter, L.A. 1991, Effect of temperature history on the growth of *Listeria monocytogenes* Scott A at refrigeration temperatures, *Int. J. Food Microbiol.* 12:235-246.

Buchanan, R.L. and Phillips, J.G. 1990, Response surface model for predicting the effects of temperature, pH, sodium chloride content, sodium nitrite concentration and atmosphere on the growth of *Listeria monocytogenes. J. Food Prot.* 53:370-376,381.

Bunster, L., Fokkema, N.J. and Schippers, B. 1989, Effect of surface-active *Pseudomonas* spp. on leaf wetability, *Appl. Environ. Microbiol.* 55:1340-1345.

Carlin, F. and Nguyen-the, C. 1994, Fate of *Listeria monocytogenes* on four types of minimally processed green salads, *Lett. Appl. Microbiol.* 18:222-226.

Carlin, F., Nguyen-the, C. and Abreu da Silva, A. 1995, Factors affecting the growth of *Listeria monocytogenes* on minimally processed fresh endive, *J. Appl. Bacteriol.* 78:636-646.

Carlin, F., Nguyen-the, C., Chambroy, Y. and Reich, M. 1990, Effects of controlled atmospheres on microbial spoilage, electrolyte leakage and sugar content of fresh " ready-to-use " grated carrots. *Int. J. Food Sci. Technol.* 25:110-119.

Carlin, F., Nguyen-the, C., Cudennec P. and Reich, M. 1989, Microbial spoilage of fresh ready-to-use grated carrots. *Sci. Alim.* 9:371-386.

Carlin, F., Nguyen-the, C. and Morris, C.E. 1996, The influence of the background microflora on the fate of *Listeria monocytogenes* on minimally processed fresh broad leaved endive (*Cichorium endivia* var. *latifolia*), *J. Food Prot.* 59: 698-703.

Clark, E. and Lindow, S.E. 1989, Indoleacetic acid production by epiphytic bacteria associated with pear fruit russetting, *Phytopathology* 79:1191.

Cooper, D.G. 1986, Biosurfactants, *Microbiol. Sci.* 3:145-149.

Davis, H., Taylor, P., Perdue, J.N., Stelma, G.N., Humphrey, J.M., Rowntree, R. and Greene K.D. 1988, A shigellosis outbreak traced to commercially distributed shredded lettuce, *Am. J. Epidemiol.* 128:1312-1321.

Denis, C. and Picoche, B. 1986, Microbiologie des légumes frais prédécoupés, *Ind. Agr. Alim.* 103:547-553.

Denis, F., Veillet, L. and Michel, M. 1988, Détermination des *Leuconsostocs* dans les produits alimentaires: comparaison de milieux pour leur identification - Microbiologie des légumes de la quatrième gamme, *Microbiol. Alim. Nutr.* 6:185-192.

DGCCRF. 1993, Qualité microbiologique des produits végétaux dits de 4ème gamme, p. 54. In: *Rapport d'activité 1993 des laboratoires de la Direction Générale de la Consommation de la Concurrence et des Fraudes*, Paris.

Doyle, M.P. 1990, Fruit and vegetable safety - Microbiological considerations, *HortScience* 25:1478-1482.

Ercolani, G.L. 1976, Bacteriological quality assesment of fresh marketed lettuce and fennel, *Appl. Environ. Microbiol.*31: 847-852.

Farrag, S.A. and Marth, E.H. 1989, Growth of *Listeria monocytogenes* in the presence of *Pseudomonas fluorescens* at 7 or 13°C in skim milk. *J. Food Prot.* 52:852-855.

Farber, J.M. 1991, Microbiological aspects of modified-atmosphere packaging technology - a review, *J. Food Protect.* 54:58-70.

Farber, J.M. 1993, Current research on *Listeria monocytogenes*: an overview. *J. Food Prot.* 56:640-646.

Fernando, M. and Stevenson, G. 1952, Studies in the physiology of parasitism. XVI. Effects of the condition of potato tissue, as modified by temperature and water content, upon attack by certain organisms and their pectinanse enzymes, *Ann. Bot. Lond. N.S.* 16:103-114.

Freedman, D.J., Kondo, J.K. and Willrett, D.L. 1989, Antagonism of foodborne bacteria by *Pseudomonas* spp.: a possible role for iron, *J. Food Prot.* 52:484-489.

Garcia-Villanova Ruiz, B., Galvez Vargas, R. and Garcia-Villanova R. 1987, Contamination of fresh vegetables during cultivation and marketing, *Int. J. Food Microbiol.* 4: 285-291.

Garg, N., Churey, J.J. and Splittstoesser, D.F. 1990, Effect of processing conditions on the microflora of fresh-cut vegetables, *J. Food Prot.* 53:701-703.

Geiges, O., Stählin, B. and Baumann, B. 1990, Mikobiologische Beurteilung von Schnittsalat und Spross-gemüse, *Mitt. Gebiete Lebensm. Hyg.* 81:684-721.

George, S.M. and Lund, B.M. 1992, The effect of culture medium and aeration on growth of *Listeria monocytogenes* at pH 4.5. *Lett. Appl. Microbiol.* 15:49-52.

Gouet, P., Labadie, J. and Serratore, C. 1978, Development of *Listeria monocytogenes* in monoxenic and polyxenic beef minces, *Zbl. Bakt. Hyg., I. Abt. Orig. B.* 166:87-94.

Gras, M. H., Druet-Michaud, C. and Cerf, O. 1994, La flore bactérienne des feuilles de salade fraîche, *Sciences des Aliments* 14:173-188.

Graziano, J. 1993, Fresh-cut at a glance, *Produce Business* 9(10):42-48.

Harzig, J. 1994, 4ème gamme: l'optimisme est de retour, *L'echo des MIN.* Feb. 1994. pp. 52-54.

Hirano, S.S. and Upper, C.D. 1991, Bacterial community dynamics, pp. 271-294. In: Andrews, J.H. and Hirano, S.S. (eds.) *Microbial Ecology of Leaves*. Springer Verlag, New York, 499 pp.

Hurst, W.C. 1995, Sanitation of lightly processed fruits and vegetables, *HortScience* 30:22-24.

Hyodo, H., Kuroda, H. and Yang, S.F. 1978, Induction of phenylalanine ammonium-lyase and increase in phenolics in lettuce in relation to the development of russet spotting caused by ethylene, *Plant Physiol.* 62:31-35.

Jacques, M.-A. 1994, Ecologie quantitative et physiologie de la communauté bactérienne épiphylle de *Cichorium endivia* var. *latifolia* L, *Thesis (Docteur en Sciences)*, Université Paris XI Orsay, 125 pp.

Jacques, M.-A., Kinkel, L.L. and Morris, C.E. 1995, Population sizes, immigration, and growth of epiphytic bacteria on leaves of different ages and positions of field-grown endive (*Cichorium endivia* var. *latifolia*), *Appl. Environ. Microbiol.* 61:899-906.

Jacques, M.A. and Morris, C.E. 1994, Diversity of bacteria contributing to the decay of ready- to-use salads, *Proc. 8th Int. Conf. on Plant Pathogenic Bacteria.* 9-12 June 1992, Versailles, pp. 165-171.

Jacques, M.-A. and Morris, C.E. 1995, Bacterial population dynamics and decay on leaves of different ages of ready-to-use broad-leaved endive, *Int. J. Food Sci. Technol.* 30:221-236.

Jeong, D.K. and Frank, J.F. 1994. Growth of *Listeria monocytogenes* at 10°C in biofilms with microorganisms from meat and dairy processing environments, *J. Food Prot.* 57:576-586.

Jones, K. 1970, Nitrogen fixation in the phyllosphere of Douglas fir *Pseudotsuga douglasii*, *Annals of Botany* 34:239-244.

Kallender, K.D., Hitchins, A.D., Lancette, G.A., Schmieg, J.A., Garcia, G.R., Solomon, H.M. and Sofos, J.N. 1991, Fate of *Listeria monocytogenes* in shredded cabbage stored at 5°C and 25°C under a modified atmosphere *J. Food Prot.* 54:302-304.

Kapperud, G., Rorvik, L.M., Hasseltvedt, V., Hoiby, E.A., Iversen, B.G., Staveland, K., Johnsen, G., Leitao, J., Heriskstad, H., Andersson, Y., Langeland, G., Gondrosen, B. and Lassen, J. 1995, Outbreak *of Shigella sonnei* infection traced to imported iceberg lettuce, *J. Clin. Microbiol.* 33:609-614.

Ke, D. and Saltveit, M.E. Jr. 1989, Wound-induced ethylene production, phenolic metabolism and susceptibilty to russet spotting in iceberg lettuce, *Physiol. Plant.* 76:412-418.

King, A.D.Jr., Magnuson, J.A., Török, T. and Goodman, N. 1991, Microbial flora and storage quality of partially processed lettuce, *J. Food Sci.* 56:459-461.

Klaenhammer, T.R. 1988, Bacteriocins of lactic acid bacteria, *Biochimie* 70:337-349.

Kramer, J.M. and Gilberts, R.J. 1989, *Bacillus cereus* and other *Bacillus* species, pp. 21-70, In: *Foodborne Bacterial Pathogens,* Doyle, M.P. (ed), Marcel Dekker Inc., New York.

Lainé, K, and Michard, J. 1988, Fréquence et abondance des *Listeria* dans les légumes frais découpés prêts à l'emploi. *Microbiol. Alim. Nutr.* 6:329-335.

Laycock, M.V., Hildebrand, P.D., Thibault, P., Walter, J.A. and Wright, J. L. C. 1991, Viscosin, a potent peptidolipid biosurfactant and phytopathogenic mediator produced by a pectolytic strain of *Pseudomonas fluorescens*, *J. Agric. Food Chem.* 39:483-489.

Liao, C.H. and Wells, J.M. 1987a, Diversity of pectolytic, fluorescent pseudomonads causing soft rots of fresh vegetables at produce markets, *Phytopathology* 77:673-677.

Liao, C.H. and Wells, J.M. 1987b, Association of pectolytic stains of *Xanthomonas campestris* with soft rots of fruits and vegetables at retail markets, *Phytopathology* 77:418-422.

Liao, C.H. and Wells, J.M. 1986, Properties of *Cytophaga johnsonae* strains causing spoilage of fresh produce at food markets, *Appl. Environ. Microbiol.* 52:1261-1265.

Lojkowska, E. and Holubowska, M. 1992, The role of polyphenol oxidase and peroxidase in potato tuber resistance to soft rot caused by *Erwinia carotovora*, *J. Phytopathol.* 136:319-328.

Lund, B.M. 1983, Bacterial spoilage, pp. 219-257, In: Dennis, C. (ed.) *Post-Harvest Pathology of Fruits and Vegetables*, Academic press, London.

Lund, B.M. 1988, Bacterial contamination of food crops, *Aspects of Appl. Biol.* 17:71-81.

Lund, B.M. 1992, Ecosystems in vegetable foods, *J. Appl. Bacteriol. Symposium Suppl.* 73:115S-126S.

Marchetti, R., Casadei, M.A. and Guerzoni, M.E. 1992, Microbial population dynamics in ready-to-use vegetable salads, *Ital. J. Food Sci.* 2:97-108.

Marshall, D.L. and Schmidt, R.H. 1989, Growth of *Listeria monocytogenes* at 10°C in milk preincubated with selected pseudomonads. *J. Food Prot.* 51:277-282.

Marshall, D.L. and Schmidt, R.H. 1991, Physiological evaluation of stimulated growth of *Listeria monocytogenes* by *Pseudomonas* species in milk, *Can J. Microbiol.* 37:594-599.

Mattila-Sandhom, T. and Skyttä, E. 1991, The effect of spoilage flora on the growth of food pathogens in minced meat stored at chilled temperature. *Lebensm.-Wiss.-Technol.* 24:116-121.

MMWR. 1994, Foodborne outbreaks of enterotoxigenic *Escherichia coli* - Rhode Island and New Hampshire, 1993, *JAMA* 271:652-654.

Morris, C.E., Guinebretiere, M.H. and Mazollier, J. 1990, Identity and distribution of bacteria contributing to the decay of ready-to-use salads, In : SFP (eds), *Compte-rendus du 2ème congrès de la Société Française de Phytopathologie,* Montpellier, France. Thème 5, Posters.

Morris, C.E., Jacques, M.-A. and Nicot, P.C. 1994, Microbial aggregates on leaf surfaces: characterization and implications for the ecology of epiphytic bacteria, *Molec. Ecol.* 3:613.

Morris, C.E. and Lucotte, T. 1993, Dynamics and variability of bacterial population density on leaves of field-grown endive destined for ready-to-use processing, *Int. J. Food Sci. Technol.* 28:201-209.

Morris, C.E. and Rouse, D.I. 1984, The role of nutrients in regulating epiphytic bacterial populations, pp. 63-82 In: Windels, C.E. and Lindow, S.E. (eds.) *Biological Control in the Phylloplane*, American Phytopathological Society, St. Paul, MN.

Mundt, J.O. and Hammer, J.L. 1968, Lactobacilli on plants, *Appl. Microbiol.* 16:1326-1330.

Mundt, J.O., Beattie, W.G. and Wieland, F.R. 1969, Pediococci residing on plants, *J. Bacteriol.* 98:938-942.

Nguyen-the, C. and Carlin, F. 1994, The microbiology of minimally processed fresh fruits and vegetables, *Crit. Rev. Food Sci. Nutr.* 34:371-401.

Nguyen-the, C. and Lund, B. 1991, The lethal effect of carrot on *Listeria* species, *J. Appl. Bacteriol.* 70:479-788.

Nguyen-the, C. and Lund, B. 1992, An investigation of the lethal effect of carrot on *Listeria monocytogenes,* *J. Appl. Bacteriol.* 73:23-30.

Nguyen-the, C. and Prunier, J.P. 1989, Involvement of pseudomonads in deterioration of "ready-to-use" salads, *Int. J. Food Sci. Technol.* 24:47-58.

Notermans, S.H.W. 1993, Control in fruits and vegetables, pp. 233-260, In: Hauschild, A.H.W. and Dodds, K.L. (eds.) *Clostridium botulinum, Ecology and Control in Foods,* Marcel Dekker Inc., New York.

O'Mahony, M., Cowden, J., Smyth, B., Lynch, D., Hall, M., Rowe, B., Teare, E.L., Tettmar, R.E., Rampling, A.M., Coles, M., Gilbert, R.J., Kingcott, E., and Bartlett, C.L.R. 1990, An outbreak of *Salmonella Saint-Paul* infection associated with beansprouts, *Epidemiol. Infect.* 104:229-235.

Ono, K. 1969, Occurrence of *Bacillus polymyxa* (Praz.) Migula causing rot of harvested tobacco leaves, *Bull. Morioka Tob. Exp. Stn.* 4:77-83.

Park, C.E. and Sanders, G.W. 1992, Occurrence of thermotolerant camplylobacters in fresh vegetables sold at farmers'outdoor markets and supermarkets, *Can. J. Microbiol.* 38: 313-316.

Piard, J.C. and Desmazeaud, M. 1991, Inhibiting factors produced by lactic acid bacteria. 1. Oxygen metabolites and catabolism end-products, *Lait* 71:525-541.

Portnoy, B.J., Goepfert, J.M. and Harmon, S.M. 1976, An outbreak of *Bacillus cereus* food poisoning resulting from contaminated vegetables sprouts, *Am. J. Epidempiol.* 103:589-594.

Premaratne, R.J.,Wei-Jen Lin and Johnson, E.A. 1991, Development of an improved chemically defined minimal medium for *Listeria monocytogenes. Appl. Environ. Microbiol.* 57:3046-3048.

Roberts, D., Watson G.N. and Gilbert, R.J. 1982, Contamination of food plants and plant products with bacteria of public health significance, pp. 169-195, In: Rhodes-Roberts, M. and Skinner, F.A. (eds.) *Bacteria and Plants,* Academic Press, London.

Ryser, E.T. and Marth, E.A., 1991, *Listeria, listeriosis and food safety,* Marcel Dekker Inc., New York 632 pp.

Sasahara, K.C. and Zottola, E.A. 1993, Biofilm formation by *Listeria monocytogenes* utilizes a primary colonizing microorganism in flowing systems, *J. Food Prot.* 56:1022-1028.

Scandella, D. 1989, Maîtrise de la qualité des produits de IVe gamme dans la filière de production et de distribution, *Rev. Gén. Froid* 1989 n°3: 94-101.

Schlech, W.F., Lavigne, P.M., Bortolussi, R.A., Allen, A.C., Haldane, E.V., Wort, A.J., Hightower, A.W., Johnson, S.E., King, S.H., Nicholls, E.S. and Broome, C.V. 1983, Epidemic listeriosis - Evidence for transmission by food. *New Engl. J. Med.* 308:203-206.

Schlemmer, A.F., Ware, C.F. and Keen, N.T. 1987, Purification and characterization of a pectin lyase produced by *Pseudomonas fluorescens* W51, *J. Bacteriol.* 169:4493-4498.

Schlimme, D.V. 1995, Marketing lightly processed fruits and vegetables, *HortScience* 30:15-17.

Shapiro, L. 1995, The lazy man's leafy greens, *Newsweek* 75(25):46.

Shuval, H.I., Yekutiel, P. and Fattal, B. 1984, Epidemiological evidence for helminth and cholera transmission by vegetables irrigated with wastewater: Jerusalem - a case study, *Wat. Sci. Tech.* 17:433-442.

Simon, N., Coulanges, V., André, P. and Vidon, D.J.-M. 1995, Utilization of exogenous siderophores and natural catechols by *Listeria monocytogenes, Appl. Environ. Microbiol.* 61:1643-1645.

Solomon, H.M., Kauter, D.A., Lilly, T. and Rhodehamel, E.J. 1990, Outgrowth of *Clostridium botulinum* in shredded cabbage at room temperature under a modified atmosphere, *J. Food Prot.* 53:831-833.

Stanley, D.W. 1991, Biological membrane deterioration and associated quality losses in food tissues, *Crit. Rev. Food Sci. Nutr.* 30:487-553.

Steinbruegge, E.G., Maxcy, R.B. and Liewen, M.B. 1988, Fate of *Listeria monocytogenes* on ready to serve lettuce, *J. Food Prot.* 51:596-599.

Stewart, A.W., Langford, A.F., Hall, C.and Johnson, M.G. 1978, Bacteriological survey of raw " soulfoods " available in South Carolina, *J. Food Prot.* 41:364-366.

Tamminga, S.K., Beumer, R.R. and Kampelmacher, E.H. 1978, The hygienic quality of vegetables grown in or imported into the Netherlands: a tentative survey, *J. Hyg., Camb.* 80:143-154.

Todd, E.C.D. 1992, Foodborne disease in Canada - a 10 years summary from 1975 to 1984, *J. Food Prot.* 55:123-132.

van Gijsegem, F., Somssich, I.E. and Scheel, D. 1995, Activation of defense-related genes in parsley leaves by infection with *Erwinia chrysanthemi, Eur. J. Plant Pathol.* 101:549-559.

Vantomme, R., Sarrazyn, R., Goor, M., Verdonck, L., Kersters, K. and De Ley, J. 1989, Bacterial rot of whitloof chicory caused by strains of *Erwinia* and *Pseudomonas*: symptoms, isolation and charaterization, *J. Phytopathology* 124:337-365.

Varoquaux, P. 1990, Connaissances de la matière première et transformation des légumes de quatrième gamme, *Agoral,* Nantes, 7p.

Vescovo, M., Orsi, C., Scolari, G. and Torriana, S. 1995, Inhibitory effect of selected lactic acid bacteria on microflora associated with ready-to-use vegetables, *Lett. Appl. Microbiol.* 21:121-125.

Wang, J.S. and Kelman, A. 1982, Injury to potato tissue by protease of *Pseudomonas fluorescens* (Biotype A), *Phytopathology* 72:936.

Wilson, M. and Lindow, S.E. 1994, Inoculum density-dependent mortality and colonization of the phyllosphere by *Pseudomonas syringae, Appl. Environ. Microbiol.* 60:2232-2237.

Wilson, M. and Lindow, S.E. 1994, Coexistence among epiphytic bacterial populations mediated through nutritional resource partitioning, *Appl. Environ. Microbiol.* 60:4468-4477.

Zottola, E.A. and Sasahara, K.C. 1994, Microbial biofilms in the food processing industry - Should they be a concern?, *Int. J. Food Microbiol. 23:125-148.*

13

MICROBIOLOGY OF GRAPE SURFACES AND ITS IMPORTANCE FOR WINE QUALITY

Tomás G. Villa and Elisa Longo[1]

Department of Microbiology
Faculty of Pharmacy
University of Santiago de Compostela
[1] Department of Microbiology
Faculty of Sciences
University of Vigo
Spain

INTRODUCTION

Although humankind has been involved in winemaking since the very beginning of civilization, it was not until the second half of the last century that Pasteur (1857) demonstrated the microbial origin of spontaneous fermentations. Years later, Hansen (1903) established the taxonomic basis for a rational approach to the study of yeasts associated with grapes and to those actually involved in the fermentation process. By the middle of the present century studies in microbial ecology have shown that the vinification process involves a sequential proliferation of different types of yeast and bacteria coming either from the vineyard itself or from the cellar (Mrak and McClung, 1940; Castelli, 1954 and Galzy, 1958)

The study of the microorganisms that colonize the surface of grapes has been and still is the central research topic of many groups studying microbial taxonomy. There are several reasons for this; among others: i) certain fungi cause severe diseases of grapes leading to important economic losses; ii) yeasts and bacteria are the natural inoculum for winemaking and the acceptability of the final product to a large extent depends on such microbiota (Suomalainen and Lehtonen, 1979; Gillian and Fleet, 1985); iii) the vineyard as an ecosystem may be a good source of microorganisms with potential applications in biotechnology. Some of the hydrolytic enzymes currently used in the food processing industry are produced by microorganisms first isolated from vineyards (Fogarty and Kelly, 1983; Gacto et al., 1992).

Colonization of Grape Berries

Using scanning electron microscopy Belin (1972) demonstrated the presence of yeasts on the upper part of the berries. These yeasts show particularly high proliferation rates

Aerial Plant Surface Microbiology, edited by Morris et al.
Plenum Press, New York, 1996

around the pedicels and stomata as well as on microwounds causing juice leakages. It was then held that yeast cells must somehow become embedded in waxy substances or exudates from the grapes themselves (van Zyl and Du Plessis, 1961). Starting from these points, yeast populations would disseminate due to the activity of insects, this notably increasing as the berries ripen and their aroma increases (Benda, 1970; Phaff, 1978).

Epiphytic yeasts were tentatively quantified by Domerq(1956) and Galzy (1958), who reported values of 10^3-10^5 yeast cells per square centimeter. Later Rossini et al. (1982) and Farris et al. (1987) described higher values (up to 10^6). In the middle 1960's and beginning of the 1970's most attention was given to the fast-fermenting strains of S. cerevisiae (Kunkee and Amerine, 1970). Although apiculate, more weakly fermenting yeasts were observed in fresh musts and were rapidly substituted by strains of S. cerevisiae with the onset of fermentation, it was concluded that S. cerevisiae belonged to the indigenous vineyard microbiota. However, with time this conclusion proved to be incorrect.

Of particular relevance is the work of Davenport (1976). Contrary to what was currently accepted, after elaboration of a complex sampling calendar including all stages of grape maturation, this author recovered Hanseniospora, Candida, Pichia and Hansenula, all of low or null enological potential, whereas yeasts with high fermentative potential such as Saccharomyces cerevisiae were rarely found. Also, dark mucoid and carotenoid pigmented yeasts and some rough ones (Sporobolomyces, Cryptococcus and Rhodotorula) were always isolated during grape ripening. Davenport therefore concluded that the yeast microbiota associated with the vineyard could be classifed in two categories, resident and transient. The study of the first category was evidently most important. The main conclusion from this type of ecological study is that the principal yeast microflora associated with vineyards almost everywhere comprises non-fermentative species that are entirely different from those found during spontaneous grape juice fermentation. Rossini and co-workers (1982) clearly showed that the grape colonization process is intimately linked to berry ripening and that it is inversely proportional to the distance to the principal vine. As indicated above, the upper part of the grapes and the zones around the pedicel contain up to 100 times more yeast than the lower face. Yeast proliferation begins two weeks before harvest as only 5% of the berries show a significant number of yeast three weeks before harvest. Species belonging to Sporobolomyces, Rhodotorula, Cryptococcus, and to a lesser extent, Candida and Pichia were always isolated. In another study, regional differences were found in the transient population, which included species of Debaryomyces, Zygosaccharomyces, Lodderomyces, Trichosporon and Saccharomyces (Ribereau-Gayon, 1985).

Climatic conditions, including both temperature and rainfall, are among the most important external factors affecting the berry-associated yeast microflora. Thus, Hanseniaspora has been described on grapes in warm climates by many authors, while the imperfect form Kloeckera is always associated with cold temperatures. Addtionally, rain has been shown to reduce yeast population causing a "washing effect" on the grapes (Longo et al., 1991). Other factors such as mechanical harvesting or the use of antifungal compounds may have a negative effect on yeast strain equilibrium by favouring strains with oxidative metabolism (Marchetti et al., 1984; Monteil et al., 1986). Indeed, it has been demonstrated that the different species considered to form part of the normal grape microbiota do respond in different ways to commercial antifungal compounds, which tend to inhibit fermentative species favouring proliferation of oxidative or apiculate species. This aspect, initially reported by Sapis-Domerq and co-workers (1977), has been more recently reported by Girond et al (1989). The conclusion drawn from all the above is that S. cerevisiae is quite sensitive to antifungal compounds whereas species belonging to *Rhodotorula, Cryptotoccus and Sporobolomyces* show remarkable polyresistance to such agents.

It is also possible that synergistic and antagonistic relationships exist among yeasts, filamentous fungi and even bacteria on the grape surface. Thus, the development of the fungus *Botrytis cinerea* slows the growth of *S. cerevisiae* and favours the appearance of yeasts with oxidative metabolism. Other vineyard-associated fungi, such as species of *Aspergillus, Mucor,* and *Fusarium*, are also reported to be involved in imbalanced yeast growth. Another type of antagonistic relationship among yeasts would result from the existence of intergeneric killer reactions, such as those described by Starmer and co-workers (1987) for species of *Cryptococcus* and *Pichia* in decaying stems and fruits of cacti as well as in the slime fluxes of trees. These authors clearly showed that yeasts found in decaying fruits have a far higher incidence of killer activity than those isolated from tree exudates or from cactus necroses.

Other factors, directly linked to the physiological properties of the species, might also be involved in the predominance of certain yeasts. For example, carotenoid-containing yeasts might be more resistant to radiation than carotenoidless ones. Also, the ability of certain yeasts to secrete mucoid substances would result in their being protected against water loss caused by high temperatures. Finally, a phytopathologic role has been invoked for some yeast genera as agents of the disease known as "sour rot", characterized by high levels of acetic acid and general peel weakness of the berries. Species of *Candida, Hanseniaspora,* and *Pichia* are systematically recovered from such grapes and, more important, they are able to reproduce the disease *in vitro* (Marchetti, 1984; Bisiach *et al.*, 1986; Guerzoni and Marchetti, 1987). Hydrolytic enzymes such as pectinases have been proposed as important colonizing and virulence factors for pathogens attacking grape berries. These types of enzyme are produced by both phytopathogenic filamentous fungi and bacteria (and as such accepted as virulence factors) and may play a significant role in grape surface colonization by saprophytic yeast microflora. In fact, this kind of pectic enzyme has been demonstrated in *Cryptococcus* (Federici, 1985) and some enological *S. cerevisiae* strains (McKay, 1990; Blanco *et al.*, 1994)

Apart from certain deviations caused by the above external factors, it may be concluded that the normal yeast microflora in the vineyards is characterized by a very low abundance, if not the total absence, of the main species able to conduct fermentation: *S. cerevisiae.*

Species Distribution during Must Fermentation

From the foregoing, it may be proposed that the different strains of *S. cerevisiae* as well as of *Pichia* spp. must be incorporated into the musts in the wine cellar itself from materials, or during manipulations or undesirable insect activity; at least, this is the conclusion to be drawn from the work of Cuinier(1980), Bouix *et al.*(1981), Fleet (1990) and Frezier and Dubordieau (1992). Moreover, the number of species may vary from year to year for a same wine region as shown by Longo *et al.* (1991)(see Table 1).

The distribution of species and strains during the different stages of wine elaboration has been studied in many countries and always seems to follow a common ecological pattern (Table 2).

As indicated, the grape juice microflora comes mainly from the grape surface and from cellar utensils. The yeast species responsible for the onset of fermentation depend on the physicochemical properties of the must and on the different must treatments. Once the grapes have been transformed into must, apiculate yeasts (*Hanseniaspora* or *Kloeckera)* proliferate very rapidly. It has been observed that sulfite addition tends to halt the growth of

Table 1. Differences in the yeast populations of two spontaneous fermentations
in an Atlantic region of Northwest Spain

Species	Class	1986	1987
Kloeckera apis	II	-	+
Candida guillermondii	III	-	+
C. glabrata	III	-	+
C. lusitaneae	III	+	-
C. rugosa	III	+	-
C. stellata	III	+	-
Pichia etchellsii	III	+	-
Rhodotorula mucilaginosa	III	+	-

* 1986 had a mean temperature of 20°C and a rainfall of 178 mm; in 1987 the temperature
was 23.5 °C and rainfall 111mm. Roman numbers represent apiculate species (II) and
species with mainly oxidative metabolism (III). Fermentative species (class I: *S. cerevisiae*,
S. kluyveri, *Kluyveromyces marxianus* and *Torulaspora globosa*), the apiculate *K.apiculata*
and the oxidative species *C.pulcherrima*, *Pichia anomala*, *Hansenula silvicola*, *P.
membranefaciens*, *P.farinosa*, *Debaryomyces hansenii* and *Cryptococcus albidus* were
isolated in both years.

these species, thus favouring fermenting yeast species, including several strains of *S.
cerevisiae* (Goto, 1980; Beech and Thomas, 1985). After a few days, the entire microbial
population increases to as much as 10^8 cells per ml, after which the succession of *S. cerevisiae*
strains begins. Accordingly, the strains responsible for producing the first 4-5 degrees of
ethanol are substituted by others that are more resistant to high ethanol concentrations. In
turn, the latter bring the fermentation to completion (Ribereau-Gayon *et al.*, 1978; Fleet *et
al.*, 1984). However, this aparently simple scheme may have quite a different outcome in the
case of grapes infected with *Botrytis cinerea*, as demonstrated by Goto (1984). Other genera
such as *Zygosaccharomyces*, *Torulaspora* or *Kluyveromyces* have been reported in spontaneous grape juice fermentations together with *S. cerevisiae*.

Table 2. Isolation of yeasts during three fermentation stages

Yeast	Stage		
	IP	AF	SF
S. cerevisiae	+	++	+++
Torulaspora (2 species)	++	+	++
Kluyveromyces marxianus	++	++	+
Debaryomyces(2 species)	+++	-	-
Kloeckera (2 species)	+++	-	-
Candida (9 species)	+++	+	-
Pichia (4 species)	++	+/-	-
Saccharomycodes ludwigii	+	-	-
Dekkera intermedia	+	+	-
Rhodotorula (2 species)	+	+/-	-
Cryptococcus albidus	+	-	-

IP: Induction period; AF: Active fermentation; SF: Slow fermentation.

IS THE KILLER PHENOTYPE OF *SACCHAROMYCES CEREVISIAE* A RELEVANT FEATURE IN WINEMAKING?

General Considerations

As indicated before in this chapter, whatever the yeast population at the moment of grape harvest, alcoholic fermentation is carried out and finished by different strains belonging to the species *S. cerevisiae*. Many such strains harbor intracellular viral particles that contain linear doubled-stranded RNA (dsRNA) of basically two sizes (L and M), and they form the new viral *totiviridae* group. Additionally, many of these strains are able to inhibit the growth of other strains (by killing them) through a property known as the *killer* phenotype. Three killer phenotypes (K1, K2 and K28) have been clearly defined and characterized in *S. cerevisiae* (Tipper and Schmitt, 1991), K2 and K1 having received the most attention. K1 was the first to be described (Bevan and McKover, 1963), whereas K2 was shown to be a different phenotype for certain strains of *S. cerevisiae* isolated from natural and spontaneous wine fermentations (Naumova and Naumov, 1973).

Among the killer systems, the K1 phenotype is the best known (for a review see Wickner, 1992). dsRNAs include L-A (4.6 Kb), which encodes the major capsid-forming coat protein and also a minor, although larger, protein (Fujimura and Wickner, 1988; Icho and Wickner, 1989; Esteban and Fujimura, 1992). dsRNAs are transcribed and replicated conservatively and sequentially (Fujimura, 1986) where messenger polarity (+) strands are formed via virion-associated RNA-dependent RNA polymerase using the minus strand of the dsRNA as the template (Welsh and Leibowitz, 1980; Bruenn *et al.*, 1980).

Killer toxins are typical eukaryotic N-glycosidic glycoproteins, encoded by M-type dsRNA, like the immune system. Each killer phenotype has a toxin-specific immune system (Cansado *et al.*, 1992); thus, K2 is active against K1 and *vice versa*. The K1 toxin was first purified in 1979 by Palfree and Bussey, who characterized it as a 11.5 kd glycoprotein with pI and pH values of 4.5 and 4.7, respectively. The mature toxin contains two subunits (alpha and beta) of 103 and 83 aminoacids each. Current data support the notion that it is the alpha subunit which interacts and somehow disorganizes the plasma membrane of sensitive cells. Although no specific protein responsible for immunity has been found, it has been inferred from mutagenesis experiments carried out on the appropriate cDNA region located within the alpha region (Sturley *et al.*, 1986). K1 toxin enters sensitive cells by interacting with 1,6-ß-D-glucans, which are virtual receptors at the cell wall level (Zhu and Bussey, 1991), and is later transferred to a plasma membrana receptor (yet to be found, but genetically hinted at; Al-Adrioss and Bussey, 1977). Cell death would result from an energy imbalance (dela Peña *et al.*, 1981).

K2 toxin is less well known than K1; It is also a glycoprotein but with a lower optimum pH (4.3). The coding genes for K2 and K1 toxins must be different since the respective amino acid sequences of the peptides differ substantially.

For many years, it has been assumed that killer activity of the K2 type would represent one of the mechanisms of yeast antagonism during spontaneous wine fermentations (provided there is no pH effect). The use of killer wine strains might therefore be of industrial interest in fermentations requiring strict microbiological control, such as for preventing contamination by undesirable *S. cerevisiae* strains (Seki *et al.*, 1985; Longo *et al.*, 1990). Several killer strains have been successfuly constructed for use in wine (Seki *et al.*, 1985), beer (Young, 1980) and saké (Ouchi *et al.*,1979) fermentations.

Some factors such as the pH value of the musts (Ciolfi, 1984), the size of the inoculum of the killer strain (Heard and Fleet, 1987), and certain wine manipulations (Radler and Schmitt, 1987) are also critically involved. There is apparently no general consensus on the effect of pH; Barre (1984) and Benda (1985) demonstrated killer activity at pH values of 3.4 whereas Rossini (1986), working under the same conditions, found only traces of such activity. In relation to inoculum size, there seems to be a clear relationship between the initial proportions of killer/killer sensitive strains and predominance of either of the strains. Both aspects will be further dealt with in this chapter.

Occurrence of Killer, Resistant, and Sensitive Phenotypes during Spontaneous Wine Fermentations

After several years of sampling in Galicia (Northwest Spain) we have isolated 392 *S. cerevisiae* wine strains. Of these, 71% are killer sensitive; 6.6% are killer-resistant, and 22.4% K2 killers (Cansado *et al.*, 1991). Higher (Barre, 1980) or lower (Tiourina *et al*, 1980) percentages for killer strains have been reported in other countries. All the yeast species and genera other than *S. cerevisiae* reported here are resistant to both K1 and K2 toxins.

In spontaneous fermentations the highest percentages of isolation of killer strains occur in the active phase of yeast growth; i.e. in the phase when maximum killer toxin activity is detected (Bussey, 1981). Additionally, the percentages of killer strains decrease during late fermentation, showing that the killer sub-population is not the predominant one. By contrast, the percentages of isolation of killer-resistant *S. cerevisiae* strains increase steadily, this being independent of the vinification tank sampled, the cellar or even the region studied.

All killer-positive strains (88) could be grouped into 13 different karyotypes on the basis of their differences in the number, position, and intensity of visible chromosomal bands when the strains were subjected to pulsed-field electrophoresis. Two of the 13 karyotypes accounted for 73% of the killer strains. All were K2 killers. Rapid dsRNA extractions revealed the presence of L and M bands. Relevant data on the 88 strains are shown in Table 3.

Northern blot hybridization employing *in vitro* ^{32}P-labeled transcripts from L2 and M2 of laboratory strains revealed strong homology with the isolates, thus confirming that we were dealing with true K2 killers.

While L2 dsRNA exhibited the same size in all strains, this was not the case for M2 dsRNA. Variations of up to 0,3 Kb are observed even for strains within the same karyotype. An explanation for this may be found in the actual structure of this dsRNA, where 200-300 unpaired base pairs define a highly alterable region.

We also observed important variations in the killer activity shown by the isolates, possibly due to differences in the genetic background. More important than the actual amount

Table 3. Main features of the 88 K2 killer strains of *S. cerevisiae* isolated in different spontaneous fermentations

Number of strains	M2 dsRNA (Kb)			Activity		Optimum pH	
	Max.	Min.	VM(%)*	Max.	Min	Max	Min
55	1.77	1,51	15	100	10	4.5	4.1
15	1.77	1.58	11	87	12	4.5	4.1
3	1.70	1.48	13	37	10	4.4	4.2
2	1.73	1.71	1	25	25	4.3	4.2
3	1.71	1.63	5	75	62	4.4	4.0
4	1.69	1.64	3	75	25	4.4	4.2
4	1.62	1.58	2.5	62	37	4.4	4.2
2	1.63	1.63	0	62	62	4.3	4.3

*Percentage of M2 dsRNA size variation

of toxin (which may be explained by the number of copies of M2 dsRNA) is the fact that some strains showed optimum killer activity at pH 4.5 instead of 4.3. This would indicate the existence of different animo acids, suggesting the presence of different genes within the K2 killer strains of *S. cerevisiae*. It has also been suggested that differences in the glycosylation pattern of K2 toxins may affect the optimum pH of activity. This latter notion, however, does not fit in with the present state of knowledge of the enzymic activities of glycoproteins, where under- or deglycosylated forms may exhibit lower pH stability but where the optimum pH is always the same.

Two hypotheses may be advanced to explain the increased presence of killer-resistant strains of *S. cerevisiae* during spontaneous fermentations: i) the killer-resistant phenotype is induced by the presence of a killer population, or ii) these strains in fact derive from previously killer-positive strains. Recently, Cansado *et al.* (1992) have shown that in 92% of the cases the second hypothesis is correct. By karyotype analysis they even found evidence that all killer-resistant strains can be grouped into 4 different classes, that are also found among killer-positive strains. Of them, all but two have L and M dsRNA. It is thus plausible that non-operative toxin genes would emerge and at the same time the toxin-resistance phenotype would remain active (probably due to the accumulation of spontaneous mutations occurring during fermentation). The other two killer-resistant strains were probably modified in their toxin receptors since no L or M dsRNAs were observed.

The K2 killer toxin shows a narrow pH profile for activity. In all cases so far tested, the optimum pH lies between 4.0 and 4.5. This fact alone would throw doubt on the role of killer phenotypes in the dominance of strains harboring K2 dsRNA plasmids over non-killer strains in spontaneous fermentations. Indeed, taxonomic studies clearly show that such dominance does not exist. Controlled fermentations were conducted in order to ascertain the involvement of cell numbers (killer-positive *vs.* killer-sensitive) as well as the effect of pH. Population dynamics were studied in fermentation media buffered at pH 3,4 and inoculated with a K2 killer wine strain and its isogenic killer-cured counterpart (Fig.1)

When the killer strain constituted 50% of the total inoculum it did dominate the fermentations. However, when it represented only 20% or less, the fermentations were brought about by the isogenic killer-cured (and sensitive) strain. As reported by Cansado *et al.*(1992), there are two possible reasons for these unexpected results (if compared with those published by Tiourina *et al.*, 1980): either the killer strains produce low levels of the toxin (which does not seem to be the case) or the use of non-isogenic strains involves other factors that depend on the genetic background of the strain, which would explain the findings of Tiourina *et al.*

Competition assays at different inital pH values were carried out with killer-positive and killer-resistant isogenic strains. As shown in Fig.2, in all cases the killer-resistant strain reached higher numbers of viable cells than the killer-positive strain.

The killer phenotype exhibited by many *S. cerevisiae* strains when first isolated from grapes may have a hitherto undemonstrated role in that particular ecosystem. Its role in the context of wine-making, however, is doubtful.

New Enological Strains of *S. cerevisiae* Obtained by Biotechnological Approaches

The development of genetic engineering techniques has stimulated researchers to construct new tailor-made *S. cerevisiae* strains with improved or even new properties for wine-making. After the demonstration in industrial brewer's yeast strains by Suihko *et al.* (1991) that it was possible to maintain and express an endo-ß-1,4-glucanase gene (*egl1*) from

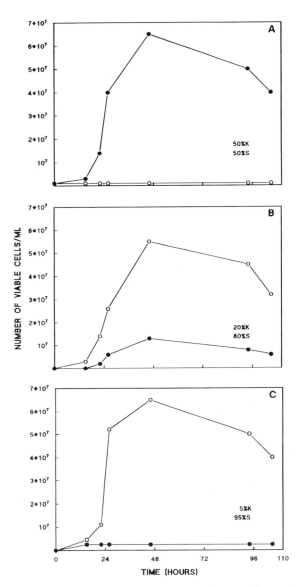

Figure 1. Controlled fermentations at pH 3.4 conducted by a K2 killer-positive strain (closed circles) of *S. cerevisiae* together with its isogenic killer-cured and sensitive counterpart (open circles) and inoculated in different proportions.

Trichoderma reesei , the idea was also applied for wine strains of *S. cerevisiae*. Perez-González *et al.*,(1993) also used an *egl*1 gene, this time from *T.longibranchiatum*, which when cloned in *S. cerevisiae* str.73 as pTLEGY3 (cycloheximide and ampicilin resistances) is expressed under natural fermentation conditions. The resulting endo-ß-1,4-glucanase secreted into the medium is able to hydrolyze terpenoid glucosides, thus giving a fruity aroma to the final product.

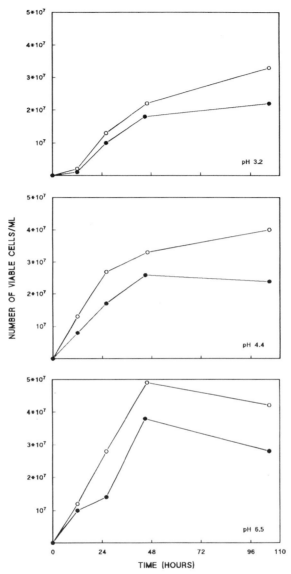

Figure 2. Population dynamics of controlled fermentations conducted by a K2 killer-positive strain (closed circles) of *S. cerevisie* and a killer-resistant (open circles) isogenic strain at different pHs. Ratio of the strains = 1:1.

Grape juices or musts are particularly rich in low molecular weight pectins and polygalacturonic acid. Winemakers have to eliminate these before the wine can be considered totally elaborated. This is normally accomplished by adding exogenous polygalacturonases and/or pectinases isolated from filamentous fungi. Since these enzymes may be considered as food additives, recent research has focused on the use of selected natural *S. cerevisiae* strains able to form quite large amounts of polyglacturonase in quasi-anaerobic media, thus reducing must viscosity (Blanco *et al.*, 1994). A pectin lyase gene from *Glomerella cingulata* (Templeton *et al.*, 1994) and a pectate lyase gene from *Fusarium solani* (González Candelas *et al.*, 1995) have been successfully cloned and expressed in wine strains of *S. cerevisiae*. Preliminary reports have shown that the wines produced by such engineered yeasts do not differ in their main physico-chemical parameters from controls.

Since the vinification proccess is a quasi-open microbiological system, the use of starters obtained by DNA recombinant technologies in the wine industry may encounter serious legal hurdles. Meanwhile, the use of "natural" starters prepared by the appropriate blending of "natural" strains isolated either from the vineyard, the grape surface or cellars, and which exhibit high levels of aryl-ß-glucosidases and polygalacturonases should certainly not be overlooked.

This work was supported by the CICYT (Grant Bio-668/92), the Spanish Ministry of Education and the Xunta of Galicia (Grant XUGA 20309B94)

REFERENCES

Al-Aidroos, K. and Bussey H. 1977, Chromosomal mutants of *Saccharomyces cerevisiae* affecting the cell wall binding site for killer factor. *Can. J. Microbiol.* 24:228-237.

Barre, P. 1980, Role du facteur killer dans la concurrence entre souches de levures. *Bull. O.I.V.* 52:560-567.

Barre, P. 1984, Le mecanisme killer dans la concurrence entre souches de levures. Evaluation et prise en compte. *Bull. O.I.V.* 641:635-643.

Belin, J.M. 1972, Reserches sur la repartition des levures a la surface de la grappe de raisin. *Vitis* 11:135-145.

Beech, F.W. and Thomas, S. 1985, Action antimicrobienne de l'anhidride sulfureux. *Bull. O.I.V.* 58:564-581.

Benda, J. 1970, Natural and controlled microbial processes in grape must in young wine. *Bayer Ladwisch. Jahrh.* 47:19-29.

Benda, J. 1985, Yeast in winemaking-investigating so-called killer yeasts during fermentation. *Der Deutsche Weinbau. Wisbaden.* 40:166-171.

Bevan, E. A. and MacKover, M. 1963, The physiologycal basis of the killer character in yeast. In: Geerts, S.J. (ed.) *Proc.Intern.Congress of Genet.* Le Hage. 11:127.

Bisiach, M., Minervini, G. and Zerbetto, F. 1986, Possible integrated control of grapevine sour rot. *Vitis* 25:118-128.

Blanco, P., Sieiro, C., Diaz, A. and Villa, T.G. 1994, Production and Partial Characterization of an Endopoly-galacturonase from *Saccharomyces cerevisiae. Can. J. Microbiol.* 40:974-977.

Bouix, M., Leveau, J.Y, et Cuinier C. 1981, Determination de l'origine de levures de vinification par une methode de differenciation fine des souches. *Conn. Vigne Vin.* 15:41-52.

Bruenn, J., Bobek, L., Brennan, V. and Held, W. 1980, Yeast viral RNA polymerase is a transcriptase. *Nucleic Acid Res.* 8:2985-2997.

Bussey, H. 1981, Physiology of killer factor in yeast. *Adv. Microbiol. Physiol.* 23:93-122.

Cansado, J., Velazquez, J.B., Sieiro, C., Calo, P., Longo, E. and Villa, T.G. 1992, Role of killer strains of *Saccharomyces cerevisiae* in spontaneous fermentations, pp. 127-142 In: Villa, T.G. and Abalde, J. (eds.) *Profiles on Biotechnology* Santiago University Press.Spain.

Cansado, J., Longo, E., Calo, P., Sieiro, C., Velazquez, J.B. and Villa, T.G. 1991, Role of killer character in spontaneous fermentations from NW Spain: ecology, distribution and significance *Appl. Microbiol. Biotechnol.*34:643-647.

Castelli, T. 1954, Les agents de la fermentation vinaire. *Arch. Microbiol.* 20:323-342.

Ciolfi, G. 1984, Effect of pH on the killer phenotype in *Saccharomyces cerevisiae. Riv. Vitic. Enol.* 37:76-79.

Cuinier, C. 1980, Origine des levures assurant l'elaboration d'un vin blanc de Tourraine. Identification des species. *Connaiss. Vigne Vin* 14:111-126.

Davenport, R.R. 1976, Distribution of yeasts and yeast-like organisms from aerial surfaces of developing apples and grapes, pp.325-359 In: Dickinson, C.H. and Preecee, T.F. (eds.) *Microbiology of Aerial Plant Surfaces,* Academic Press. London.

dela Peña, P., Barros, F., Gascón, S., Lazo, P.S. and Ramos, S. 1981, Effect of yeast killer toxin on sensitive cells of *S. cerevisiae. J. Biol. Chem.* 256:10420-10425.

Domerq, S. 1956, Etude et classification des levures de vin de la Gironde. *These de Docteur-Ingenieur.* Universite de Bordeaux

Esteban, R. and Fujimura, T. 1992, RNA expresion vectors based on the killer dsRNA viruses from *Saccharomyces* cerevisiae, pp. 33-42 In: Villa, T.G. and Abalde, J. (eds.) *Profiles on Biotechnology* Santiago University Press, Spain.

Farris, G.A., Forteleoni, M. and Fatichenti, F. 1987, Effects microbiologique de la recolte mecanique. *Rev. Fr. Oenol.* 110:42-45.

Federici, F. 1985, Production, purification and partial characterization of an endopolygalacturonase from *Cryptococcus albidus* var. *albidus*. *Antonie van Leeuwenhoek* 51:139-150.

Fleet, G.H., Lafon-Lafourcade, S. and Ribereau-Gayon ,1984, Evolution of yeast and lactic acid bacteria during fermentation and storage of Bordeaux wines. *Appl. Env. Microbiol.* 48:1034-1038.

Fleet, G.H. 1990, Which yeast species really conduct the fermentation?, pp. 153-156 In: Williams, P.I., Davison, D. and Lee, T.H. (eds.) *Proceedings of the Seventh Australian Wine Industry,* Australian Industrial Publishers, Adelaide, S.A.

Fogarty, W.M. and Kelly, C.T. 1983, Pectic enzymes, pp. 131-182 In: Fogarty, W.M. (ed.) *Microbial Enzymes and Biotechnology*, Applied Science Publishers,London.

Frezier, V. and Dubourdieu, D. 1992, Ecology of yeast strain *S. cerevisiae* during a spontaneous fermentation in a Bordeaux winery. *Am. J. Enol. Vitic.* 43:375-380.

Fujimura, T., Esteban, R. and Wickner, R.B. 1986, *In vitro* L-A double-stranded RNA synthesis in virus-like particles from *Saccharomyces cerevisiae. Proc. Natl. Acad. Sci.* USA. 83:4433-4437.

Fujimura, T. and Wickner, R.B. 1988, Gene overlap results in a viral protein having an RNA binding domain and a major coat protein domain. *Cell.* 55:663-671.

Gacto, M., Vicente Soler, J and Pardo, C. 1992, Production and characterization of pectic enzymes from yeasts, pp. 173-181 In: Villa, T.G. and Abalde, J. (eds.) *Profiles on Biotechnology*, Santiago University Press, Spain.

Galzy, P. 1958, Origine de la microflore des vins. *Proc. Agric. Vitic.* 75:58-65.

Gillian, M.H. and Fleet, G.H. 1985, Growth of natural flora during fermentation of inoculated wines. *Appl. Env. Microbiol.* 50:727-728.

Girond, S., Blazy-Mougen, F. and Michell, G 1989, Influence de quelque pesticides viticoles sur les levures et la fermentation. *Rev. Fr. Oenol.* 119:14-22.

Gonzalez-Candelas,L., Cortell,A. and Ramon D. 1995, Construction of a recombinant wine yeast strain expressing a fungal pectate lyase gene. *FEMS Microbiol. Let.* 126:263-270.

Goto, S. 1980, Changes in the wild yeast flora of sulfited grape musts. *J. Inst. Enol. Vitic.* 15:29-32.

Goto, S., Oguri, H. and Yamazaki, M. 1984, Yeast population in botrytised grapes. *J. Inst. Enol. Vitic.* 19:1-5.

Guerzoni, E. and Marchetti, R. 1987, Analysis of the yeast flora associated with grape sour rot and the chemical disease markers. *Appl. Env. Microbiol.* 53:571-576.

Hansen, E.C. 1903, Neue Untersuchungen uber den Kreislanf der Hefernanten in der natur. *Zentr. Bakteriol.* 10:1-8.

Heard, G.M. and Fleet, G.H. 1987, Occurrence and growth of killer yeasts during wine fermentation. *Appl. Environ. Microbiol.* 53:2171-2174.

Icho, T. and Wickner,R.B. 1989, The double-stranded RNA genome of yeast virus L-A encodes its own putative RNA polymerase by fusing two open reading frames *J. Biol. Chem.* 252:6716-6723.

Kunkee, R.E. and Amerine, M.H. 1970, Yeast in winemaking, pp. 5-71 In: Rose, A.H. and Harrison, J.S. (eds.) *The Yeasts,* Academic Press, New York.

Longo, E., Velazquez, J.B., Cansado, J., Calo, P. and Villa, T.G. 1990, Role of killer effect in fermentations conducted by mixed strains of *Saccharomyces cerevisiae. FEMS Microbiol. Let.* 71:331-336.

Longo, E., Cansado, J., Agrelo, D. and Villa, T.G. 1991, Effect of climatic conditions on yeast diversity in grape musts from northwest Spain. *Am. J. Enol. Vitic.* 42:141-144.

Longo, E., Cansado, J., Sieiro, C., Calo, P., Velazquez, J.B. and Villa T.G. 1992, Influence of the curing of the killer phenotype in *Saccharomyces cerevisiae* wine strains on their fermentative behaviour *World J. Microbiol. Biotechnol.* 8:147-150.

McKay, A.M. 1990, Degradation of polygalacturonic acid by *Saccharomyces cerevisiae, Let. Appl. Microbiol.* 11:41-44.

Marchetti, R., Guerzoni, E. and Gentile, M. 1984, Reserche sur l'étiologie d'une nouvelle maladie de la grappe:la pourriture acide. *Vitis.* 23:55-65

Mrak, E.M. and McClung, S. 1940, Yeasts occurring on grapes and grape products in California. *J. Bacteriol.* 40:395-399.

Monteil, H., Blazy-Mangen, F. and Michel, G. 1986, Influence des pesticides sur la croissance des levures des raisins et des vins. *Science des Aliments.* 6:349-360.

Naumova, G.I. and Naumov T.I. 1973, Comparative genetics of yeasts. XIII. Study of antagonistic interrelations in *Saccharomyces cerevisiae. Genetika* 9:85-90.

Ouchi, K., Wickner, R.B., Tohe, A. and Akiyama H. 1979, Breeding of killer yeast for saké brewing by cytoduction *J. Ferment. Technol.* 57:483-487.

Palfree, R.G.E. and Bussey, H. 1979, Yeast killer toxin: purification and characterization of the protein from *S. cerevisiae. Eur. J. Biochem.* 93:487-493.

Pasteur, L. 1857, Memoires sur la fermentation alcoolique. *CR Acad. Scs.* Paris. XLV:1032-1036.

Perez-Gonzalez, J.A., Gonzalez, R., Querol, A., Sendra, J. and Ramon, D. 1993, Construction of a recombinant wine yeast strain expressing ß-(1,4)-endoglucanase and its use in microvinification processes *Appl. Environ. Microbiol.* 59:2801-2806.

Phaff, H.J., Miller, M.W. and Mrak, E.M. 1978, *The Life of Yeasts.* Harvard University Press. Cambridge. Mass. USA.

Radler, F. and Schmitt, M. 1987, Killer toxins of yeasts: inhibitors of fermentation and their adsorption. *J. Food. Protect.* 50:234-238.

Ribereau-Gayon, P., Peynaud, E., et Sudraud, P. 1978. *Sciences et Techniques du Vin* (vol. 2) Dunod. Paris.

Ribereau-Gayon, P. 1985, New developments in wine microbiology. *Am. J. Enol. Vitic.* 36:1-10.

Rossini, G., Federici, F. and Martini, A. 1982, Yeast flora of grape berries during ripening. *Microb. Ecol.* 8:83-89.

Rossini, G. 1986, Effect of killer strains on winemaking yeast starters. *Ann. Fc. Agrar. Univ. Stud. Perugia.* 38:345-354.

Sapis-Domerq, S., Bertrand, A., Mur, F. and Sarre, C. 1977, Influence des produits de traitement de la vigne sur la microflore des raisins et des vins. *Conn. Vigne Vin.* 11:227-242.

Seki, T., Choi, E. and Ryu, D. 1985, Construction of killer wine yeast strain. *Appl. Environ. Microbiol.* 49:1211-1215

Soumalainen, H. and Lehtonen,M. 1979, The production of aroma compounds by yeasts. *J. Inst. Brewing* 85:149-156.

Starmer, W.T., Ganter, P.F., Aberdeen, V., Lachance, M.A. and Phaff, H.J. 1987, The ecological role of killer yeast in natural communities of yeasts. *Can. J. Microbiol.* 33:783-796.

Sturley, S.L., Elliot, Q., Le Vietre, J., Tipper, D.J. and Bostian, K.A. 1986, Mapping of functional domains within the *S. cerevisiae* type I killer preprotoxin. *EMBO J.* 5:3381-3389.

Suihko, M.L.,Lehtinen, U., Zurbriggen, B., Kilpola, A., Nowles, J. and Penttla, M. 1991, Construction and analysis of recombinant glucanolytic brewer's yeast strains. *Appl. Microbiol. Biotechnol.* 35:781-787.

Templeton, M.D., Sarrock, K.R., Bowen J.K., Crowhurst, R.N. and Rikkerink, E.H.A. 1994, The pectin lyase gene (*pnl*) family from *Glomerella cingulata*: characterization of *pnlA* and its expression in yeast. *Gene* 142:141-146.

Tiourina, L.V., Bourjan, N.I. and Skarikova T.K. 1980, Emploi des cultures pures du phenotype killer dans la fermentation des mouts de raisin. *Bull. O.I.V.* 593:573-576.

Tipper, D.J. and Schmitt, M.J. 1991, Yeast ds-RNA viruses replication and killer phenotypes. *Mol. Microbiol.* 5:2331-2338.

van Zyl, J.A. and Du Plessis, L. 1961, The microbiology of South African winemaking. Part 1. The yeasts occuring in vineyards, musts and wines. *S. Afr. J. Agric. Sci.* (Pretoria) 4:393-401.

Welsh, D.J. and Leibowitz, M.J. 1980, Transcription of killer virion double-stranded RNA *in vitro. Nucleic Acid Res.* 8:2365-2375.

Wickner, R.B. 1992, Yeast RNA virologyer systems, pp. 263-269 In: Broach, J.R., Pringle, J.R. and Jones, E.W. (eds.) *The Molecular Biology of the Yeast Saccharomyces,* vol. 1, Cold Spring Harbor Laboratory, New York, USA.

Young, T.W. 1980, The genetic manipulation of killer character into brewing yeast. *J. Inst. Brew.* 87:292-295.

Zhu, H. and Bussey, H. 1991, Mutational analysis of the functional domains of yeast K1 killer toxin. *Mol. Cell. Biol.* 11:175-181.

14

PLANT SURFACE MICRO-ORGANISMS AS SOURCES OF COMPOUNDS TOXIC FOR HUMANS AND DOMESTIC ANIMALS

Peter G. Mantle

Biochemistry Department
Imperial College of Science
Technology and Medicine
London SW7 2AY
United Kingdom

INTRODUCTION

At first sight this topic may evoke a variety of expectations, or even cause surprise that toxinogenic micro-organisms might be of any consequence on the surface of plants. Generally most non-senescent plant surfaces do not support populations of potentially saprophytic micro-organisms that are significant in the local generation of toxins. Sometimes pathogenic fungi may establish sub-cuticular, or deeper, infections in particular host plant leaves, stems, inflorescences and fruits, but most of these fungi are not recognised sources of toxins for human and/or domestic animals. Populations of bacteria and yeasts may be numerically large on plant surfaces such as fruits, through which sugars and amino acids exude or where exudates from insect damage may accumulate. Again, these micro-organisms are not notably toxinogenic except indirectly as sources of acids and alcohol in fermentation processes. However, microbial propagules, often in the form of spores, readily impact on aerial plant surfaces, so that plant material used as human food or animal feed is often inoculated with an array of micro-organisms that may proliferate in appropriate conditions of moisture and temperature if optimal storage conditions are not maintained. Then, super- ficial latent micro-organisms, mainly fungi, will proliferate superficially on, or penetrate deeply into, food commodities and cause spoilage. More moisture and heat will be generated through such metabolic activity and an even more complex environment, nutritionally and physically, will arise within which regulatory influences on fungal metabolism may switch on pathways for secondary metabolite biosynthesis. These pathways are widespread in their distribution amongst opportunist fungi and are extremely diverse in their expression, leading to a vast array of end-product compounds. Many of these compounds have been shown to exhibit biological activity against other organisms including man and domestic animals, and some of these activities are adverse to health usually through chronically-developing pathologies. Such mycotoxic expressions are the principal concern of the present review,

Aerial Plant Surface Microbiology, edited by Morris et al.
Plenum Press, New York, 1996

and are illustrated by reference to examples chosen to show how activities range from those that are virtually superficial, organ-specific and caused by ecologically obligate parasites to those that arise from indiscriminate deposition of saprophytic propagules on aerial plant surfaces.

PLANT PARASITIC TOXINOGENIC ERGOT FUNGI

A specific example of plant surface micro-organisms that lead directly to toxic compounds concerns the ergot fungi (*Claviceps* spp.). As it happens, these are some of the oldest recognised toxinogenic fungi and there is a long history of their involvement in human and animal toxicoses (Bové, 1970; Mantle, 1978). Airborne propagules of these fungi become impacted on plant surfaces randomly in the form of ascospores ejected from the teleomorph that is usually located near the soil surface, arising from sclerotia that fell from ripened grass or cereal inflorescences of a previous crop. Random dissemination of asexually produced conidia of the anamorphic state (*Sphacelia* sp.) occurs from infected inflorescences a few days after flowering by means of rain splash and head-to-head contact in stands of monocultured cereals or grasses, but more focused dissemination occurs when insects fly from flower to flower. By whatever means of distribution, most ascospores and conidia impact on plant surfaces with absolutely no further consequence. However, a very small proportion are deposited on a surface within a cereal or grass floret that is gaping for perhaps only a few hours at the time of anthesis. In the extreme modern artificial circumstances, the gaping of florets of male-sterile lines in F_1 hybrid breeding systems may last for several days if pollination from another source is not perfectly synchronised and adequately abundant. In some Graminae, stigmas remain exserted after floret gaping, in which case the period of consequential impaction of a spore is extended. Probably, a viable spore must land either on a stigmatic hair or close to the ovary wall for germination to take place. It has not been experimentally demonstrated whether spores applied only on the inner surface of a glume can give rise to ovary infection but it is perhaps not surprising that, for the very specialised ergot parasites, only the most succulent floral apices (ovaries) provide access to an only-delicately aggressive pathogen. Direct penetration through the wall of an unfertilised ovary may occur, but the most sophisticatedly adapted scenario is when a spore impacts on a stigmatic hair and, within a few hours, germinates to form a germ tube that quickly penetrates between the cells of the stigmatic hair. Then the hypha follows a transmission tract designed primarily for the passage of a pollen tube, thereby mimicking the pollination route. Pollination is several-fold quicker than ergot infection but, given an equal start, or a few hours advantage, the pathogen may reach the ovule before the rapid ovarian histological changes following pollination makes penetration through the ovule wall impossible. Commonly, therefore, ergot fungal spores landing on a receptive gynoecium must immediately focus on a directional race for the target that is only a few parenchymatous cells below the ovary surface. As pollen rehydrates within seconds of landing on a suitable stigmatic hair, so ergot spores, whether the 4-celled needle-shaped ascospores of *Claviceps purpurea* or the single-celled ovoid conidia, must immediately initiate organisation of a penetrative germ tube that can seek and follow a transmission tract where nutrients are available to fuel pollen tube or ergot hyphal growth. Subsequent parasitic development has been described for several of the ergot fungi and in most infected florets this leads to the development of a sclerotium in place of a seed and the hard mass of fungal tissue itself becomes an extended part of the aerial plant surface. In a strict sense, the sclerotium, having completely disrupted the ovary, proliferates on the plant surface that is the fractured vascular system that was to have translocated nutrients to a developing seed. Thus ergot fungi go through a transient sub-surface phase, lasting a few days, but then make all subsequent development on a lesioned plant surface. There is no penetration further than between the most distal cells of the rachilla. Sclerotial tissues usually elaborate the complex ergoline alkaloids and these, in some *Claviceps* spp., have profound pharmacological activity that may constitute toxicity in humans or domestic animals.

The most significant, and as it happens temperate-zone, species is *Claviceps purpurea*, classically associated with ergotism in humans and agricultural ruminants since Mediaeval times. The principal toxins are ergotamine and the ergotoxine group of related alkaloids, but it is important to recognise that toxic pharmacology may, at appropriate dosage, become therapeutant pharmacology. Consequently, ergot alkaloids have long had an important place in medicine for the treatment of, for example, post-partum haemorrhage and migraine headache by which the oxytocic and vasopressor properties of the alkaloids are exploited at sub-toxic doses, though of course the potential problems of over-dose remain.

Claviceps paspali parasitises the sub-tropical *Paspalum* grasses and is unusual amongst ergot fungi in elaborating tremorgenic indole-diterpenoid metabolites, in addition to accumulating very small amounts of rather typical ergot-type alkaloids. These tremorgenic mycotoxins are the cause of *Paspalum* Staggers, a neurological disorder of cattle grazing rank infected *Paspalum* pasture (Mantle *et al.*, 1978). *Claviceps fusiformis* is the most significantly toxic tropical ergot fungus, parasitising pearl millet in the Indian sub-continent and in parts of Africa. It is perhaps best known as the cause of reproductive failure in sows, through agalactia, in Zimbabwe where the poisoning was quite widespread in the 1960's (Shone *et al.*, 1959). The pig just happens to be sensitive to the suppression of prolactin caused by a cascade of consequences of central nervous stimulation by the clavine alkaloids that occur in the sclerotia of *C. fusiformis* (Mantle, 1968).

Periodically, a significant reduction in milk yield occurs in the Cape region of South Africa in dairy cattle fed a ration supplement derived in part from the cleanings from barley destined for brewing (Schneider *et al.*, 1996). It has recently been shown that the cleanings contain grass ergots (*C. purpurea*) (Mantle, 1996) and even the rather low intake of ergot alkaloids is sufficient to impose just enough general peripheral vasopressor action to reduce some sub-cutaneous capillary function, although not to cause gangrene of the extremities as is a classical effect of ergot alkaloids in cattle. In hot summer conditions the consequential impairment of thermoregulation causes loss of appetite in the affected dairy cows, and milk yield declines. Therefore, brief superficial growth of *C. purpurea* on the gynoecia of weed grasses in cereal crops may subsequently have a significant impact on dairy productivity through an unexpectedly complex sequence of events.

SAPROPHYTIC FUNGI

Pithomyces chartarum Producing Hepatotoxins in Spores

Aerial plant surfaces are not always in the peak of physiological condition but may eventually become senescent without actually being shed from the plant. Progress of fungal association from superficial to sub-epidermal makes fungi the main initial colonisers of ageing leaves, so that most leaves of deciduous plants are already partly colonised before they are shed. A significant opportunist coloniser of senescing pasture herbage in New Zealand is the fungus *Pithomyces chartarum*. As its specific epithet implies, this fungus readily grows on damp cellulose, and large amounts of this substrate in the form of fibrous senescent leaves of pasture grasses are often available in summer months for ingestion during the close-grazing of high grazing pressure. Following subsequent rain and high day and night temperatures this endemic fungus rapidly colonises the vegetation and produces masses of black spores that contain the hepatotoxic sporidesmins. Ruminants put to graze the new grass growth that accompanies the flush of moisture may also consume large numbers of toxic spores and the cryptic insult to the liver and associated bile ducts may be so severe as to become fatal (Mortimer *et al.*, 1978). Animals that subsequently develop the typical facial eczema syndrome, caused by the photosensitisation arising from the abnormal high circulating concentration of unexcreted phylloerythrin in blood, will not only be demonstrating indirectly that severe liver damage has occurred but will also be experiencing more discomfort than that caused probably by any other mycotoxicosis. Consequently, losses may be by mortality or by morbidities affecting productivity of meat, milk or wool in agricultural ruminants. The potential particularly in New Zealand agriculture for the development of outbreaks of *Pithomyces*-toxicosis is partly a function of topography and climate but also a consequence of high intensity pastoral agriculture.

Oestrogenic Mycotoxins of *Fusarium* Fungi

The heightened awareness of mycotoxicoses in New Zealand has recognised that there may be chronic sub-clinical expressions of toxinogenic fungi growing on plant surfaces. One such is the recently established connections between the *Fusarium* metabolite zearalenone detected in forage (di Menna *et al.*, 1987) and impaired reproduction performance in ewes (Smith *et al.*, 1990). Even silage supplementing dairy cattle rations may increase the rate of return to service (McDougall *et al.*, 1995), and zearalenone metabolites have been found in the urine of sheep with a history of poor lambing percentages (Towers and Sprosen, 1992). Where the effects are so subtle, and seen only when the performance of large groups of animals is analysed, the actual origin of trace amounts of zearalenone is difficult to determine. However, in New Zealand zearalenone is produced by fungi of genera other than *Fusarium*, also observed elsewhere by El-Kady *et al.* (1988), and there is still the possibility that the full range of fungi able to produce small amounts of zearalenone on forage in the field is not yet ascertained. In some cases the association between fungus and host may verge on parasitism, but not in the ecologically obligate way that it is for the association between ergot fungi and plants.

A less subtle effect of zearalenone - producing fungi, particularly *Fusarium graminearum*, growing superficially on maize or barley is the overt porcine oestrogenism which lead to zearalenone being first isolated in the US (Kurtz and Mirocha, 1978). The superficial moulding before harvest of ripe corn in the Mid-West corn belt put the fungus in an ideal position to exploit sub-optimal bulk storage that provided conditions of high humidity, thereby promoting further fungal growth. In falling autumn temperatures the octaketide biosynthetic pathway lead to the production and accumulation of zearalenone. A rationale for the production of many fungal toxins having bizarre chronic effects on some animals is not easy to perceive, but zearalenone seems to be unusual in having a function as a sex hormone for the producing fungus (Wolf and Mirocha, 1977). Presumably, such biological activity occurs by a mechanism different from that involved in zearalenone's mimicking of oestrogen in mammals.

The quite widespread natural incidence of small amounts of zearalenone in pasture forage may on occasion result in unwitting positive detection of zearalenone and/ or its structurally analogous metabolites in meat or body fluids by ultra-sensitive ELISA techniques, thereby giving the false impression in foreign markets that artificial growth promoting substances such as zearalanol may have been deliberately used. One of the consequences of modern improvements in sensitive analysis of xenobiotics is that unreasonable suspicions may be generated. Particularly where a microbial toxin may pass along a food chain to man, and even more so if a toxin with a feminising pharmacology might be involved, it soon becomes apparent how difficult it is to estimate a safe dosage, because dose-response experiments have not been, or can not be, done. Consequently, any trace recognition of toxins can lead to alarmist responses, and trends to legislate, quite understandably, against undesirable contaminants in the environment or in food chains are increasing. Risk assessment is becoming a fashionable branch of toxicology, and is subject to as wide a diversity of opinion as is economics. However, it is wise to be cautious when sub-clinical effects of toxins may occur and toxins may have totally different effects in different animals. For example, the fumonisins, produced by *Fusarium moniliforme* mainly in, rather than on, maize in many parts of the world, can cause leukoencephalomalacia (holes-in-the-brain) in the horse (Marasas *et al.*, 1988), pulmonary oedema in the pig (Harrison *et al.*, 1990) or hepatic carcinoma in the rat (Gelderblom *et al.*, 1991). The first was the natural disease; the effects in pig and rat were found experimentally. No wonder, therefore, that there is current debate on the etiological involvement of fumonisins in oesophageal cancer in parts of Southern Africa and whether or not these compounds are also of any wider significance in human health in countries where maize is a staple food.

Mycotoxinogenic *Aspergillus* and *Penicillium* spp.

Fungi of the genera *Aspergillus* and *Penicillium* readily produce abundant small spores that can frequently be found in the complex populations of air-spora world-wide. By impaction from air, or contamination in the dusty process of harvesting, spores of these relatively quick growing fungi are on most aerial plant surfaces and therefore account for the diversity of these fungi that are readily isolated from unsterilised plant surfaces or from washings therefrom. However, several *Aspergillus* and *Penicillium* fungi can also be isolated from surface sterilised food commodities of plant origin, such as legumes, cereals and coffee. The range of types is less diverse than that endemic in soil and on other niches of the particular locality and, of course, varies considerably between tropical and cold temperate regions. However, many of the fungi that achieve sub-epidermal colonisation of aerial plant surfaces, often without evoking any visible plant tissue response, are able to synthesise curious secondary metabolites and some of these compounds are potential toxins for man and other animals. The best known is aflatoxin B_1, one of a family of hepatotoxic and carcinogenic metabolites of, mainly, *Aspergillus flavus* and *A. parasiticus*. Until more recently, the aflatoxins have been regarded as products of moulding occurring in poor storage conditions. While this is mainly so, it is important to recognise that the mould spoilage does not just arise from surface contamination of, for example, sub-terranean groundnuts during harvesting, but can also arise from invasion of an aerial plant surface (the inflorescence) at a much earlier stage in the crop's growth cycle (Wells *et al.*, 1972; Griffin and Garren, 1976). Latent infection in the developing seeds can proliferate later, with concomitant synthesis of aflatoxins even without obvious moulding, if moist warm storage conditions occur (Cole *et al.*, 1986). This is the reason why groundnuts are such a potent, frequent source of aflatoxins; the causal fungus may have had an advantageous short period of growth on an aerial plant surface.

Probably, the aflatoxins have been the focus of the largest amount of scientific attention assigned to any of the fungal toxins that constituted the new field of mycotoxicology that arose during the 1960s. This is mainly due to their extreme potency as primary liver carcinogens in the rat (Heathcote and Hibbert, 1978). In the model rat system these compounds need to be metabolically activated before binding to DNA and initiating chronic events leading to tumourigenesis. Acute hepatic toxicity in the human has been diagnosed (Shank, 1978) but there is still much debate as to the actual nature and extent of the chronic toxicological significance of trace amounts of aflatoxins in human diet. Consequently, the uncertainty of toxicological significance of aflatoxins has led to the routine sampling and sophisticated analysis of certain food commodities in many of the developed countries, particularly of commodities imported from sub-tropical or tropical regions. A more recent example concerns dried figs amongst which occasional highly contaminated fruits, that may even be recognised fluorescing blue under ultraviolet light, can theoretically give a significant aflatoxin component to fig pastes destined for use in confectionery.

Ochratoxin and Ochratoxicosis

If the precise significance of aflatoxins in human and animal health is unclear, even though aflatoxin B_1 was discovered specifically in response to a need to find the cause of serious diseases in poultry and trout, so much more debatable is that of ochratoxin. Discovered at about the same time as aflatoxin, in the early 1960s in South Africa, but not in response to a specific natural toxicological problem, ochratoxin is of less general concern. Nevertheless, since its significant synthesis occurs mainly on or in plant commodities that were once aerial crop plant surfaces, it is appropriate to discuss the significance in the present chapter.

Ochratoxin's claim to fame is mainly threefold. First, it is generally recognised as the principal cause, possibly in conjunction with the related isocoumarin citrinin, of the porcine nephropathy that for many years prior to the 1970s had been a chronic, low-incidence problem in the extensive bacon industry in Denmark (Krogh, 1978). The problem was due to storage moulding of cereal by the fungus *Penicillium verrucosum*. This grows particularly well at low temperatures and is an effective competitor under temperature conditions that are below optimum for most spoilage fungi. So, under moist cool or cold storage conditions a numerically minor population component of this fungus can become more dominant and cause spoilage.

Secondly, as a result of an 800-rat study in the U.S. National Institutes of Health ochratoxin must now be recognised as a carcinogen (Boorman, 1989). This has considerably elevated the interest in its toxicological significance, especially since the toxin occurs in some food commodities world-wide (Speijers and van Egmond, 1993) and is from time to time detected in human serum (Hult *et al.*, 1982), and even in human milk (Micco *et al.*, 1991).

The third feature is the assumption that, because ochratoxin may naturally cause nephropathy that may readily be recognised in a 6 month bacon pig and is a renal carcinogen in the rat, it has a role in human Balkan endemic nephropathy (Castegnaro *et al.*, 1990). This disease, which in Bulgaria is restricted to the Vratza region, is an idiopathic progressive bilateral renal atrophy which may only become apparent in middle age and which happens to be associated with a much increased incidence of tumours of the urinary tract. There are, however, considerable difficulties with this idea of an etiological role, not the least in reconciling the evidence concerning the fungi that may produce the toxin, the commodities in which it occurs and the timing of initial colonisation of the plant tissue substrate.

Mycological studies in the Balkans agree that ochratoxinogenic *P. verrucosum* has not been found and therefore direct extrapolation from the mycobiota around latitude 56^0 N can not reasonably be made to the 43-45^0 N region. At first sight this presents no problem because ochratoxin, as its name implies, was first found as a metabolite of *Aspergillus ochraceus* isolated in South Africa, no part of which is further than 35^0 South of the Equator. Thus, presumably the ochratoxin that has been found in human sera, and in some food commodities in the Balkan region, could come from *A. ochraceus,* or some similar persistently yellow-sporing *Aspergillus* of the *ochraceus* group. As it happens, the evidence for ochratoxinogenic *A. ochraceus* is rather poor. The survey by Cvetnic and Pepeljnjak (1990) in Yugoslavia is the most positive. However, only one tenth of the isolates, all obtained from smoked meat and grain, produced ochratoxin in laboratory culture. Most isolates produced none, or only a trace. Subsequently, studies on material taken from foodstuffs and the environment of nephropathy households in Yugoslavia and Bulgaria found *A. ochraceus* to be rare, and isolates usually failed to produce ochratoxin in laboratory culture (Macgeorge and Mantle, 1990; Mantle and McHugh, 1993).

A porcine nephropathy, histologically similar to classical Danish mycotoxic porcine nephropathy, has been described from the Stara Zagora area of central Bulgaria where human Balkan endemic nephropathy does not occur (Stoev and Stoikov, 1993). The present author has also recently seen the evidence at first hand. Sera from the affected animals contained small amounts of ochratoxin. A very recent mycological study of 14 compounded pig feeds, visually of very variable quality, from the Stara Zagora area again showed *A. ochraceus* to be rather rare. Isolates of all seven of the *A. ochraceus* colonies observed on spread plates of only four of the 14 pig feed samples failed to produce ochratoxin in laboratory culture (P.G. Mantle and L. Collinson, unpublished).

Detailed mycological study is exceptionally time consuming, especially if toxic metabolite production by other than the expected fungi is to be considered. The poor fit between the incidence of ochratoxinogenic *A. ochraceus* in the Balkans and from time to time the rather widespread finding of ochratoxin in pig sera poses the question as to where the ochratoxin is coming from. It also fuels the debate over whether it is reasonable to extrapolate from the cause of porcine nephropathy in central Bulgaria to the human Balkan endemic nephropathy that is restricted to the Vratza region 200 Km distant.

A threat to an important national beverage may induce perhaps disproportionate allocation of scientific resource, and a recent example may be the piece in the New Scientist (18th Feb, 1995) concerning ochratoxin in coffee and the reported analyses that were to be made in official laboratories. Irrespective of the outcome of such analyses, a small pilot mycological survey was made in the author's laboratory, seeking the fungi growing after surface sterilisation from green beans of good quality coffees from the main producing countries in Central America, Eastern Africa, India and the Yemen. The samples were from the current commodities used by a long-established coffee merchant in Scotland. The extent of fungal infection varied from virtually none to a rather prominent incidence of Aspergilli. *A. ochraceus* was only found as a minor component in samples from Colombia, Honduras, Uganda and the Yemen. While these broadly covered the main geographical coffee-producing regions, none of the fungi produced ochratoxin in laboratory culture on shredded wheat, but they did all produce a novel alkaloid ($C_{31}H_{27}NO$) the structure of which is being investigated (Harris, 1996). These findings show that not only can coffee seeds become infected by several fungi that are typical of tropical latitudes, but that also a plant commodity that has never been implicated in any human or animal nephropathy can give a glimpse of the indigenous *A. ochraceus* population and again demonstrate that ochratoxin production is an uncommon attribute of this fungus. It may be some comfort that this brief survey of good quality coffees showed that there were no acquired alkaloids more toxic than caffeine.

It seems therefore that it is by no means clear what indigenous fungi produce the ochratoxin that enters the food chain in other than cool temperate climates. Attention to this detail was first paid by Mantle and McHugh (1993) in seeking nephrotoxinogenic fungi in and on food plant materials from nephropathy households. Contrary to expectation, several fungi, best assigned at the time to the concepts of *P. solitum, P. viridicatum* and *P. griseofulvum,* all produced ochratoxin on shredded wheat. Some of these were isolated from surface sterilised Bulgarian maize that had a rather large ochratoxin content in spite of appearing to be entirely healthy seed. During the incubation necessary to isolate the fungi on a nutrient medium, the ochratoxin content of the seeds increased further several-fold, implying active biosynthesis. It is assumed that the fungi responsible were isolated and it would appear that they are notably toxinogenic when hidden below the surface of the seed. Commonly, laboratory cultural methods for testing biosynthetic potential of penicillia, growing stationary on solid substrates, results in abundant superficial mycelium and sporulation and could be an abnormally luxuriant experimental expression of fungal growth.

Recent experiments in the author's laboratory (Harris, 1996), expanding and refining the principle of shaken solid substrate fermentation that Hesseltine (1972) first used successfully for the production of ochratoxin by *A. ochraceus*, have given outstanding yields of up to 12 mg of ochratoxin A per gram of shredded wheat breakfast cereal. This is approximately an order of magnitude greater than that achieved by Hesseltine (1972). The shaken conditions, 200 rpm and 10 cm eccentric throw in 500 ml conical flasks prevented any perceptible superficial growth on the extruded wheat shreds and consequently there was no sporulation. The shreds darkened to a chocolate brown as the fermentation proceeded and it is assumed that *A. ochraceus* hyphae had no option but to grow within the substrate. The conditions were mimicking more closely the cryptic growth that may occur under a substrate water activity that first permits slow fungal growth in a protected environment, where water released from metabolic activity may be retained locally to facilitate further cryptic colonisation of the substrate. These rather harsh conditions were associated with a massive diversion to secondary metabolism. It remains to be tested whether the ochratoxinogenic penicillia from Bulgaria could flourish better with respect to ochratoxin biosynthesis in shaken solid substrate fermentation. If so, the techniques may provide the optimum test for ochratoxin biosynthetic potential amongst fungi isolated from cereal and legume commodities.

Necessary Rigour in Diagnosing Ochratoxicosis

Correct attribution of human and animal disease causation to microbial metabolites relies on maintaining rigorous standards of proof. The interdisciplinary nature of the subject, ranging from the medical or veterinary diagnosis of idiopathic syndromes to the rational understanding of toxin synthesis by known or unknown micro-organisms in the environment, presents a complex challenge to researchers.

A recent example of an apparent failure of scientific rigour concerns the claim in a very reputable journal that a case of acute renal failure was undeniably due to inhalation of ochratoxin of *Aspergillus ochraceus* (Di Paolo *et al.*, 1993). The report describes a most interesting clinical case of transitory acute renal disease in a farm worker presenting five days after spending 8 hours working in a granary of somewhat moulded wheat. Biopsy of an enlarged kidney showed histopathological changes consistent with the haematological and urinary evidence of acute renal failure. Therapy, including diuretics and a special diet, allowed the patient to re-establish normal renal function within six weeks. In addressing a cause, a reasonable hypothesis of kidney damage due to inhalation of substances present in the granary was initially made. Mycological study of mouldy grains on a nutrient medium led to the isolation of *Aspergillus ochraceus*. No mention was made of other fungi and there was no indication on the relative abundance of *A. ochraceus* within what would be expected to be a complex mycobiota (probably rich in *A. flavus*) in this ecological situation. Thin layer chromatography of a simple chloroform extract of the wheat revealed ochratoxin, but no quantitative measurement was made.

Four guinea pigs and four rabbits were exposed for 8 hours to the air from venting 8 Kg of the moulding wheat by a continuous current of air. Two rabbits and a guinea pig died within 2 days and post-mortem examination of these, and the others killed at 5 days, showed marked macroscopic and microscopic changes in liver and/or kidney of five of the animals. In discussing the reported findings it was concluded that isolation of *A. ochraceus*, and the recognition of ochratoxin in the wheat and the evidence of acute inhalation toxicity in experimental animals constituted sufficient proof that the acute kidney failure in the patient was due to ochratoxin. The discussion was taken a step further to suggest that inhaled mycotoxins were responsible for mysterious diseases, formerly ascribed to curses, of archaeologists of ancient tombs and scholars of old books. This all makes romantic reading, but initially strays from the scientific rigour necessary to establish cause and effect as tightly as is reasonable, given the current state of scientific knowledge and analytical techniques.

The report was essentially repeated last year in another journal (Di Paolo *et al.*, 1994) and, although not specifically stated, there could be the implication that inhalation of "dust full of toxins" includes spores of *A. ochraceus*. Thus it is important to comment that there is no evidence that spores of *A. ochraceus* contain ochratoxin; indeed, there is plenty of evidence to imply that ochratoxin biosynthesis is a feature of vegetative mycelium only. Further, the abstract states that a strain of *A. ochraceus* producing ochratoxin was isolated. No evidence of that was presented and indeed the authors admit that not all strains of this fungus produce toxic secondary metabolites. This of course allows the additional observation that there was no proof that the ochratoxin in the wheat had been produced by *A. ochraceus*. It would not be surprising if other fungi were involved. Therefore there appears to be the flawed assumption that ochratoxin is a volatile substance, and/ or that fine particles of the spoiled wheat, bearing toxin, were inhaled by both human and animal subjects. No evidence of airborne particles in the animal experiment was presented.

An alternative conclusion on the clinical case is that there is absolutely no compelling evidence of significant involvement of ochratoxin. Further, the unreasonable diagnosis of ochratoxicosis obscures the interesting nature of the case that was more likely caused by volatile products of microbial spoilage of material that had once been an aerial plant surface. Unfortunately, the diagnosis of ochratoxicosis may be perpetuated through uncritical literature referencing and proper study of important toxic volatiles may thus be neglected.

Ochratoxin is currently being suggested as having a role in the etiology of karyomegalic interstitial nephritis, proposed as a new disease entity (Spoendlin *et al.*, 1995; Godin *et al.*, 1995). Unfortunately, karyomegaly was not illustrated when it was described as a consistent feature in the U.S. National Institutes of Health study establishing ochratoxin as a renal carcinogen in rats, and therefore its relative magnitude is unclear. However, experimental production of karyomegaly in renal proximal tubules of Sprague-Dawley rats is much more readily and dramatically achieved by a toxin of a common *Penicillium* than by relatively large doses of ochratoxins (Mantle *et al.*, 1991). The fungus, more recently assigned to *Penicillium polonicum* in the revision of the broad species *P. aurantiogriseum* (Lund and Frisvad, 1994), superficially readily colonises the surface of cereal seed and, with sufficient moisture, sporulates profusely. Insect damage of grain still on its source plant will expose surfaces that are particularly readily colonised. The fungus has been shown to be the most common *Penicillium* isolated from maize grain and other plant products in Southern Europe (Mantle and McHugh, 1993). The nephrotoxin, that has as yet been only partially characterised and for which no analytical method is available, occurs in the spores (Macgeorge and Mantle, 1990), in contrast to the apparent absence of nephrotoxic ochratoxins from the spores of *A. ochraceus*. The question arises, therefore, as to the relative significance of the common *Penicillium polonicum*, most isolates of which are nephrotoxic, as a renal disease determinant in Europe with respect to the significance of ochratoxin producing fungi. It is interesting to note that Spoendlin *et al.* (1995) assumed that, because work on *P. polonicum* has been with isolates from Bulgaria and former Yugoslavia, such fungi are hardly relevant to people living in Switzerland and Sicily and could not be a causal factor of their described karyomegalic interstitial nephritis; at least there is a weak appreciation of fungal ecology here.

CONCLUSIONS

In determining that evidence for a risk exists concerning toxigenicity of a plant-associated micro-organism it is important that any diagnosis that clinical disease is caused by that agent is subject to the degree of rigour of proof enshrined, for example, in Koch's postulates. While these were designed for pathogens, the principles are readily transposed to apply to toxins. Microbiologists should therefore beware of occasional excessive claims in toxicological literature.

Advances in the sensitivity of analytical methodology now result in an increasing incidence of positive findings of xenobiotics in plant-derived food, or indirectly in serum. Where these compounds are toxins of microbial origin the full range of their biological effects in animals may not be fully defined; immunological impairment is one of the more recently recognised aspects of mycotoxicology. Consequently, there is the danger of too readily and uncritically invoking a homeopathic philosophy when ascribing significance to trace amounts of toxin in food or serum, perceiving that even vanishingly small amounts could still be important.

Conversely, biological effects of toxins described from laboratory experiments on isolated organs, tissue slices, cultured cells or other experimental systems have often arisen from amounts of toxin that exceed those actually available to such organs or tissues in vivo under natural conditions. Again, risk assessment based on laboratory findings is set the challenge of correctly extrapolating from effects seen in high dose laboratory experiments to the reality of natural exposures.

Unfortunately there is a shortage of whole animal experimental data on low dosages and nil effects. Nil effect experiments usually give low personal intellectual satisfaction to researchers and may appear unattractive to funding agencies, but would be very valuable in assisting realistic risk assessment.

From the point of view of a microbiologist, mycotoxins in plant, fruit, or seed commodities are seen as the products of fungal biochemistry in the particular physiological states associated with colonising plant substrates. There is a need to understand the factors affecting toxin production by cryptic sub-epidermal fungal hyphae within seeds that are not visibly spoiled by this microbial activity and are consequently accepted for human food or animal feed. There is also a need to explore further the range of micro-organisms that can elaborate certain toxins. Further, it should be recognised that the range of toxic expressions is only as comprehensive as can be deduced from known involvement in natural syndromes and positive responses in laboratory assays for particular pharmacologies. Correct and balanced assessment of the significance of microbial activity in plant-derived commodities poses a particular challenge with respect to chronic toxicities with a long sub-clinical phase.

It is at this point that the incomplete knowledge of the fungi that actually produce the ochratoxin found particularly in human food, serum and urine should be emphasised. The deficit urgently needs to be addressed. The fungi concerned are likely to have begun their important role as colonisers of aerial plant surfaces and, if for no other reason, are relevant to participants of a conference on the microbiology of this ecological niche.

REFERENCES

Boorman, P. 1989. Toxicology and carcinogenesis studies of ochratoxin A. *U.S. National Institutes of Health Publication* No. 89-2813. Washington. U.S.A.

Bové, F.J. 1970, *The Story of Ergot.* S. Karger, Basel and New York.

Castegnaro, N., Chernozemsky, I.N., Hietanen, E. and Bartsch, H. 1990, Are mycotoxins risk factors for endmic nephropathy and associated urothelial cancers? *Arch. Geschwulstforsch.* 60: 295-303.

Cole, R.J., Hill, R.A., Blankenship, P.D. and Sanders, T.H. 1986, Color mutants of *Aspergillus flavus* and *Aspergillus parasiticus* in a study of preharvest invasion of peanuts. *Appl. Environ. Microbiol.* 52: 1128-1131.

Cvetnic, Z. and Pepeljnjak, S. 1990, Ochratoxinogenicity of *Aspergillus ochraceus* strains from nephropathic and non-nephropathic areas of Yugoslavia. *Mycopathologia* 110: 93-99.

di Menna, M.E., Lauren, D.R., Poole, P.R., Mortimer, P.H., Hill, R.A. and Agnew, M.P. 1987, Zearalenone in New Zealand pasture herbage and the mycotoxin-producing potential of *Fusarium* species from pasture, *N.Z.J. Agric. Res.* 30: 499-504.

Di Paolo, N., Guarnieri, A., Loi, F., Sacchi, G., Mangiarotti, A.M. and Di Paolo, M. 1993, Acute renal failure from inhalation of mycotoxins. *Nephron*, 64: 621-625.

Di Paolo, N., Guarnieri, A., Garosi, G., Sacchi, G., Mangiarotti, A.M. and Di Paolo, M. 1994, Inhaled mycotoxins lead to acute renal failure, *Nephrol. Dial. Transplant* 9 (suppl 4): 116-120.

El-Kady, I.A., Moubasher, A.H. and Mohamed El-Maraghy, S.S. 1988, Zearalenone production by several genera of fungi other than *Fusarium. Egypt J. of Bot.* 31: 99-108.

Gelderblom, W.C.A., Kriek, N.P.J., Marasas, W.F.O. and Thiel, P.G. 1991, Toxicity and carcinogenicity of the *Fusarium moniliforme* metabolite, fumonisin B$_1$, in rats. *Carcinogenesis* 12: 1247-1251.

Godin, M., Francois, A., Le Roy, F., Morin, J.-P., Creppy, E., Hemet, J. and Fillastre, J.-P. 1996, Karyomegalic interstitial nephritis, *Amer. J. Kidney Dis.* 27: 166.

Griffin, G. and Garren, K.H. 1976, Colonisation of aerial peanut pegs by *Aspergillus flavus* and *A. niger* - group fungi under field conditions. *Phytopathology.* 66: 1161-1162.

Harris, J.P. 1996, Biosynthesis of ochratoxins and structurally related polyketides by *Aspergillus ochraceus*, *Ph.D. thesis*. University of London.

Harrison, L.R., Colvin, B.M., Greene, J.T., Newman, L.E. and Cole, J.R. 1990, Pulmonary edema and hydrothorax in swine produced by fumonisin B$_1$, a toxic metabolite of *Fusarium moniliforme, J. Vet. Diagnostic Investigations.* 2: 217-221.

Heathcote, J.G. and Hibbert, J.R. 1978, *Aflatoxins: Chemical and Biological aspects. Developments in Food Science, 1.* Elsevier, Amsterdam.

Hesseltine, C.W. 1972, Solid state fermentations. *Biotech. and Bioeng.* 14: 517-532.

Hult, K., Plestina, R., Habazin-Novak, V., Radic, B. and Ceovic, S. 1982, Ochratoxin A in human blood and Balkan endemic nephropathy, *Arch. Toxicol.* 51: 313-321.

Krogh, P. 1978, Causal associations of mycotoxic nephropathy. *Acta Path. et Microbiol. Scand.* A, supplement No. 269.

Kurtz, H.J. and Mirocha, C.J. 1978, Zearalenone (F-2) induced estrogenic syndrome in swine, pp. 256-268 In: Wyllie, T.D. and Morehouse, L.G. (eds.) *Mycotoxic Fungi, Mycotoxins, Mycotoxicoses*, Marcel Dekker, New York.

Lund, F. and Frisvad, J.C. 1994, Chemotaxonomy of *Penicillium aurantiogriseum* and related species. *Mycol. Res.,* 98: 481-492.

Macgeorge, K.M. and Mantle, P.G. 1990, Nephrotoxicity of *Penicillium aurantiogriseum* and *P. commune* from an endemic nephropathy area of Yugoslavia. *Mycopathologia.* 112: 139-145.

Mantle, P.G. 1968, Inhibition of lactation in mice following feeding with ergot sclerotia (*Claviceps fusiformis* Loveless) from the bulrush millet (*Pennisetum typhoides* Staph and Hubbard) and an alkaloid component. *Proc. Roy. Soc. Lond.* B. 170: 423-434.

Mantle, P.G. 1978, Ergotism in cattle and sheep, pp. 145-151; 207-213 In: Wyllie, T.D. and Morehouse, L.G. (eds.) *Mycotoxic Fungi, Mycotoxins, Mycotoxicoses*, Marcel Dekker, New York.

Mantle, P.G. 1996, Detection of ergot (*Claviceps purpurea*) in a dairy feed component by gas chromatography and mass spectrometry. *J. Dairy Sci.* (in press).

Mantle, P.G., Mortimer, P.H. and White, E.P., 1978, Mycotoxic tremorgens of *Claviceps paspali* and *Penicillium cyclopium:* a comparative study of effects on sheep and cattle in relation to natural staggers syndromes. *Res. Vet. Sci.* 24: 49-56.

Mantle, P.G., McHugh, K.M., Adatia, R., Gray, T. and Turner, D.R. 1991, Persistent karyomegaly caused by *Penicillium* nephrotoxins in the rat. *Proc. Roy. Soc. Lond.* series B., 246: 251-259.

Mantle, P.G. and McHugh, K.M., 1993, Nephrotoxic fungi in foods from nephropathy households in Bulgaria. *Mycol. Res.* 97:205-212.

Marasas, W.F.O., Kellerman, T.S., Gelderblom, W.C.A., Coetzer, J.A.W., Thiel, P.G. and Van der Lugt, J.J. 1988, Leukoencephalomalacia in a horse induced by fumonisin B$_1$, isolated from *Fusarium moniliforme, Onderstepoort J. Vet. Res.* 55: 197-203.

McDougall, S., Clark, D.A., Macmillan, K.L. and Williamson, N.B. 1995, Some effects of feeding pasture silage as a supplement to pasture on reproductive performance in lactating dairy cows, *N.Z. Vet. J.* 43: 4-9.

Micco, C., Ambruzzi, M.A., Miraglia, M., Brera, C., Onori, R. and Benelli, L. 1991, Contamination of human milk with ochratoxin A, pp. 105-108 In: Castegnaro, M., Plestina, R., Dirheimer, G., Chernozemsky, I.N. and Bartsch, H. (eds.), *Mycotoxins, Endemic Nephropathy and Urinary Tract Tumours.* WHO International Agency for Research on Cancer, Lyon.

Mortimer, P.H., di Menna, M.E. and White, E.P. 1978, Pithomycotoxicosis "Facial eczema" in cattle, In Mycotoxic Fungi, Mycotoxins, Mycotoxicoses, pp. 63-72 In: Wyllie, T.D. and Morehouse, L.G. (eds.) *Mycotoxic Fungi, Mycotoxins, Mycotoxicoses*, Marcel Dekker, New York.

Schneider, D.J., Miles, C.O., Wessels, J.C., Lategan, H.J. and Van Halderen, A. 1996, First report of field outbreaks of ergot alkaloid toxicity in the Republic of South Africa. *Onderstepoort J. Vet. Res.* 62: 97-108.

Shank, R.C. 1978, Mycotoxicoses in man: dietary and epidemiology considerations, pp. 1-19 In: Wyllie, T.D. and Morehouse, L.G. (eds.) *Mycotoxic Fungi, Mycotoxins, Mycotoxicoses*, Marcel Dekker, New York.

Shone, D.K., Philip, J.R. and Christie, G.J. 1959, Agalactia of sows caused by feeding the ergot of bulrush millet, *Pennisetum typhoides*. *Vet. Rec.* 71: 129-132.

Smith, J.F., di Menna, M.E. and McGowan, L.T. 1990, Reproductive performance of Coopworth ewes following oral doses of zearalenone before and after mating, *J. Reprod. Fertil.* 89: 99-106.

Speijers, G.J.A. and van Egmond, H.P. 1993, Worldwide ochratoxin A levels in food and feeds. In: Creppy, E.E., Castegnaro, M. and Dirheimer, G. (eds.) *Human ochratoxicosis and its pathologies* INSERM/ John Libbey Eurotext Ltd. 231: 85-100.

Spoendlin, M., Moch, H., Brunner, F., Brunner, W., Burger, H.-R., Kiss, D., Wegmann, W., Dalquen, P., Oberholzer, M., Thiel, G. and Mihatsch, M.J. 1995, Karyomegalic interstitial nephritis: further support for a distinct entity and evidence for a genetic defect. *Amer. J. Kidney Dis.* 25: 242-252.

Stoev, S. and Stoikov, D. 1993, Mycotoxic nephropathy (ochratoxicosis) with swine. *Vet. Sci.* 27: 57-61 (in Bulgarian).

Towers, N.R. and Sprosen J.M. 1992, Fusarium mycotoxins in pastoral farming: Zearalenone-induced infertility in sheep. In: Gopalakrishnakone, P. and Tan, K. (eds.) *Recent Advances in Toxinology Research*. 3: 272-284.

Wells, T.R., Kreutzer, W.A. and Lindsey, D.L. 1972, Colonisation of gnotobiotically grown peanuts by *Aspergillus flavus* and selected interacting fungi, *Phytopathology*. 62: 1238-1242.

Wolf, J.C. and Mirocha, C.J. 1977, Control of sexual reproduction in *Gibberella zeae* (*Fusarium roseum* "*Graminearum*"), *Appl. Environ. Microbiol.* 33: 546-550.

15

THE EFFECT OF LEAF AGE AND POSITION ON THE DYNAMICS OF MICROBIAL POPULATIONS ON AERIAL PLANT SURFACES

Marie-Agnès Jacques

INRA, Station de Pathologie Végétale
Domaine St Maurice
BP 94, 84143 Montfavet cedex
France

INTRODUCTION

Epiphytic microbial communities of terrestrial plants are non uniformly distributed in space and in time on leaf surfaces. The size and composition of microbial populations vary under the influence of biotic and abiotic factors related to the micro-organisms themselves (traits conferring epiphytic fitness, nutritional resource utilisation, abilities to compete for space, resistance or production of toxic compounds), to the host (its genotype, the age and the position of the leaves), and to the environmental conditions (micro- and macro-climate, activity of vectors and pathogens, application of pesticides and other chemicals). Moreover, the microbial population dynamics results from four processes -immigration, emigration, multiplication and death of cells as discussed in a previous chapter (Lindow, this volume) - which are under the influence of the previously mentioned factors.

Establishing the relative importance of the different factors and processes that affect microbial population dynamics is important not only for answering unresolved fundamental ecological questions but also to develop practical solutions for the control of plant pathogenic micro-organisms. The development of novel strategies aimed at controlling plant pathogenic epiphytic micro-organisms rely on gaining a deeper insight into the functioning of populations of phytopathogenic and beneficial members of the community. Therefore some basic questions have to be answered. For example, what factors affect the composition of microbial communities? Do these factors influence immigration/emigration or growth/death? To limit proliferation of plant pathogenic micro-organisms is it more effective to try to limit their immigration than their growth or is the relative importance of these factors and processes similar for a given population?

When considering approaches for the study of epiphytic microbial community structure, one may be tempted to adopt either a factor-oriented analysis or a process-oriented analysis. The diversity of the involved factors and the difficulty in discerning their influence on each process

Aerial Plant Surface Microbiology, edited by Morris et al.
Plenum Press, New York, 1996

233

usually dictates that the impact of each factor has to be studied separately. The choice of which factor to investigate normally depends on the model system available and the aims of the investigation. In our laboratory, we are interested in the mechanisms implicated in the colonisation of broad-leaved endive leaves during field-cultivation and storage by epiphytic bacteria implicated in post-harvest decay. The role of bacteria in the decay of these salads is presented in detail in a previous chapter (Morris and Nguyen-the, this volume). Broad-leaved endive forms loose heads similar to butterhead type lettuces; The outer leaves are spread out and the inner leaves are closed tightly together. The outer leaves are relatively well-exposed to airborne microbial contamination while inner leaves are covered by adjacent leaves and hence relatively more protected from aerial microbial contamination. As a consequence of this morphology and in an attempt to evaluate sorting methods based on predicted microbial load of minimally processed salads (pre-cut, washed and packaged), we were particularly interested in a factor-oriented approach. The factor of interest for our studies is the age and the position of the leaves.

This chapter focuses on the effect of leaf age and position on microbial population sizes and composition in the phyllosphere. A brief review of the literature concerning the effect of this factor on microbial populations (and particularly on bacterial populations) will be presented. This will be followed by a discussion of the hypothetical mechanisms by which leaf age has its effect and the implications for the experimental approach taken in studies of epiphytic microbial population structure. Specific examples will be provided from studies of the influence of leaf age and position on population dynamics of total mesophilic, fluorescent and pectolytic epiphytic bacteria for broad-leaved endive (Jacques and Morris, 1995a; Jacques et al., 1995). This work allows us to show how leaf age and position may interact with bacterial immigration and growth influencing the quality of leaves within the canopy as a habitat for bacteria.

MICROBIAL COMMUNITY STRUCTURE

Characterisation of microbial community structure requires the design of a sampling strategy which considers collection of samples in time and space, the choice of a sampling unit and methods for liberating micro-organisms from the leaf surface and recovering them for further analyses. These different issues were recently reviewed by Jacques and Morris (1995b) and earlier by Hirano and Upper (1991) and Kinkel (1991). Here we will focus only on the choice of the sampling unit which is fundamental to increase precision and soundness of the data on microbial community structure.

The choice of a sampling unit has received increasing attention since the analysis of the statistical implications of the use of individual samples vs. bulked samples for estimating population sizes of epiphytic bacteria (Hirano et al., 1982). These authors have shown that epiphytic bacterial populations are often lognormally distributed on individual leaves of various crops leading generally to over-estimations of the true means of bacterial population sizes if bulked samples are used (depending on the variance among population sizes and the number of leaves in the bulk). Moreover, it has also been shown that the use of bulked samples leads to under-estimations of variances if microbial populations follow skewed distributions such as the lognormal distribution (Hirano et al., 1982; Kinkel, 1992).

Individual leaves are often used as the sampling unit for various biological reasons. Incidence, severity and progress of foliar diseases of plants in the field or processed vegetable products in storage are often assessed on individual leaves (Rouse et al., 1985) or leaf pieces (Jacques and Morris, 1995a). This is the case, because inoculum thresholds for symptom expression are more meaningful for single plant organs than for a whole plant or several plants. Furthermore, interactions among micro-organisms are more likely to take place in restricted sites than across groups of leaves or entire plants. Hence, an understanding of the epidemiology of diseases or decay, and the ecology of microbial communities on aerial parts of plants may benefit from an analysis of individual samples rather than bulks of leaves.

Individual leaves also represent natural units for colonisation by micro-organisms. The relative abundance of the different components of the microbial community (Morris and Rouse, 1985; Hirano and Upper, 1991) vary substantially among leaves in the canopy. However, it is possible that the leaf, itself, is a bulked sample of multiple communities or numerous densely and sparsely colonised sites (Kinkel *et al.*, 1995). Additionally, it has recently been demonstrated that naturally-occurring epiphytic bacteria may be in aggregates composed of multiple microbial species (Morris *et al.*, 1994) that may be non homogeneously distributed among leaves on a plant.

The aerial parts of plants are colonised by various epiphytic microbial populations. The predominant colonists on leaves of terrestrial plants are bacteria, yeasts and filamentous fungi. The presence of protozoa and algae has also been recorded on some plants (Ruinen, 1961; Bernstein and Carroll, 1977). Extensive analyses of the literature concerning the composition of epiphytic microbial populations have already been published (Dickinson, 1976; Hirano and Upper, 1991; Jacques and Morris, 1995b) giving a large body of qualitative and quantitative data on the identity of micro-organisms and their presence on leaves through time.

The composition and the size of these communities are spatially and temporally dynamic. Successional trends in epiphytic microbial populations studied in a number of plants show bacteria as being the initial colonists on unfolded leaves while yeasts and filamentous fungi predominate during the middle of the growing season until leaf senescence (Blakeman, 1985). Temporal trends have been shown to occur at various scales, from daily to seasonal intervals. (Wildman and Parkinson, 1979; Mew and Kennedy, 1982; Hirano and Upper, 1989; Ercolani, 1991; Hirano *et al.*, 1994). Variations in microbial populations have been studied at different spatial scales ranging from their distribution on a leaf (Morris, 1985; Morris *et al.*, 1994; Kinkel *et al.*, 1995) to various positions in the canopy (Ercolani, 1991). The size and composition of the microbial populations vary spatially in the plant canopy reflecting aggregation in space and habitat heterogeneity within a plant canopy (Ercolani, 1979; Andrews *et al.*, 1980; Hirano *et al.*, 1982; Jacques *et al.*, 1995).

THE EFFECT OF LEAF AGE AND POSITION ON MICROBIAL POPULATIONS

What do we mean by leaf age? One could say that the answer to this question is obvious : it is the time since the leaf has emerged. But in practice the answer to this question depends on the sampling procedure adopted in each study. One can either specify the type of leaf, the sampling time or the time since the leaf emerged. If one is assessing the effect of leaf age on microbial populations for a given leaf type throughout the growing season, this is synonymous with assessing the effect of plant age (or growth stage) on epiphytic microbial populations residing on this particular type of leaf. At a given sampling time (relative to a given plant age or growth stage) one can analyse microbial populations on leaves of different ages on the plant. However, it can be difficult to separate the effect of leaf age from that of leaf position; The age of the leaves are often defined relative to the other leaves present on the plant but not as an absolute value expressed as the number of day after leaf emergence. The use of staggered sowing dates provides leaves of the same age (days after emergence) on plants at different growth stages for any given time of sampling. The underlying differences between the ways that leaf age is defined in practice are very important as they reflect differences in leaf exposure to biotic and abiotic environmental conditions (in quality and quantity of the different parameters) and the different physiological and phenological status of the plants and the leaves.

Effect on Means, Variances, and Composition of Naturally Occurring Populations

Leaf age and position has been shown to influence the sizes of the epiphytic microbial population colonising leaves of plants with various morphologies and physiological characteristics. These include evergreen trees (Ercolani, 1991; Périssol *et al.*, 1993), deciduous trees (Andrews *et al.*, 1980; Aylor, 1995), perennial and annual herbaceous plants (Dickinson *et al.*, 1975; Weller and Saetler, 1980; Jones *et al.*, 1985; Thompson *et al.*, 1993). Both the means and the variances of the population sizes have been shown to be influenced by leaf age and position (Jacques *et al.*, 1995).

Microbial population densities vary among leaves on a plant at a given sampling time. Inner/younger leaves of lettuce (*Lactuca sativa* L.), cabbage (*Brassica oleracea* L. var. *capitata*), whitloof chicory (*Cichorium intybus* L. var. *foliosum*), broad-leaved endive (*Cichorium endivia* L. var. *latifolia*) and sugar beet (*Beta vulgaris* L.) have been reported to carry smaller bacterial population densities than outer/older leaves at a given sampling time during field cultivation (Ercolani, 1976; Geeson, 1979; van Outryve *et al.*, 1989; Morris and Lucotte, 1993; Thompson *et al.*, 1993; Jacques *et al.*, 1995) and during storage of processed vegetables (Jacques and Morris, 1995a). Differences in population densities of total mesophilic bacteria between the youngest and the oldest leaves of broad-leaved endive could exceed 2 $\log_{10}$ units at harvest and these discrepancies persisted during storage after minimal processing for leaves separated by age group and stored in polypropylene packs (Jacques and Morris, 1995a; Jacques *et al.*, 1995). The differences recorded during storage were attributed mainly to physiological differences of leaves of various ages as all leaves were processed (after harvest leaves separated by age are cut, washed with chlorinated water, rinsed with tap water, spun dry, packaged in polypropylene packs) and stored in the same conditions (6°C, 7 days).

Microbial population sizes are affected by interactions among the age of the leaf they colonise, the age of the plant and environmental conditions. The size of the epiphytic bacterial populations in general and of *Pseudomonas syringae* pv. *savastanoi* in particular are influenced by interactions between the age of the leaves of olive (*Olea europaea* L.), the date of leaf emergence and the sampling time, but by none of these variables alone (Ercolani, 1979; Ercolani, 1991). It was noticed for 1 wk-old leaves of broad-leaved endive that the time of leaf emergence through the growing season had a significant influence on bacterial population size (Jacques *et al.*, 1995). The position of the emergent leaves on the head, the age of the plant, and the environmental conditions were different throughout the growing season. For a given type of leaf of broad-leaved endive, population densities of total mesophilic, fluorescent and pectolytic bacteria increased throughout the growing season (Jacques *et al.*, 1995) and during storage (Jacques and Morris, 1995a). Indeed bacterial numbers increased as leaves aged but plant age also contributed significantly to epiphytic bacterial densities. In a field of plants of different ages (staggered sowing dates) bacterial population sizes on leaves one week after emergence were generally significantly lower on older plants than on younger plants in two experiments out of three (Jacques *et al.*, 1995).

To summarise, the different effects of leaf age and position observed for densities of naturally occurring microbial populations were as follows: (1) microbial population sizes generally increase as leaves age through time for a given type of leaf; (2) at a given sampling date population sizes are greatest for the oldest leaves on the plant, (3) there are interactions among the age of the leaves and the age of the plant on which they emerge and the abiotic environmental conditions. In the examples given above the effect of age and/or position of the leaves in the canopy affect the population sizes of naturally occurring bacteria, filamentous fungi and yeasts.

Leaf-associated bacterial population sizes among leaves in a canopy follow highly skewed distributions that can be approximated either by a lognormal (Hirano *et al.*, 1982; Jacques *et al.*, 1995) or Weibull (Ishimaru *et al.*, 1991) distribution. We have found that the lognormal distribution provides a better fit for population sizes on leaves within individual age classes than for the entire collection of leaves of broad-leaved endive representing different age classes and positions (Jacques *et al.*, 1995). An example of a graphical evaluation of the distributions and confirmation with the Shapiro-Wilk statistic is given in Fig. 1. In this test, observations that follow a normal distribution approximate a straight line. Slopes of lines reflect the variability of the data. The variances of the population sizes were

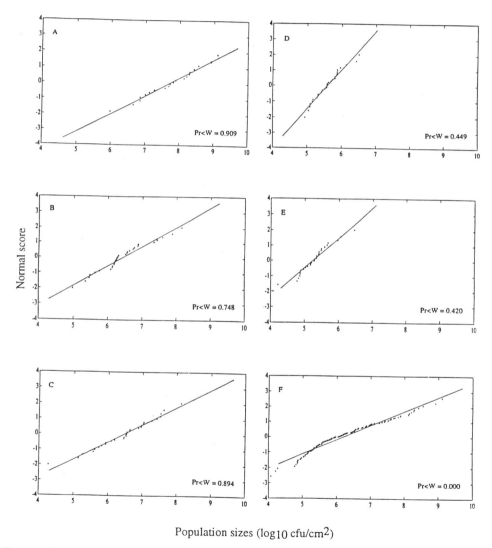

Population sizes (log₁₀ cfu/cm²)

Figure 1. Distributions of total bacterial population sizes at harvest in 1991 on leaves of 73-68 (A), 61-54 (B), 47-40 (C), 26-19 (D), 7-0 (E) day-old and on all leaves pooled (F). The probability of observing a smaller W value (Pr<W) is indicated on each graph. Values of Pr<W greater than 0.05 indicate a non significant deviation from the normal distribution. From Jacques *et al.* (1995) with permission.

different among the different age classes as indicated by the slopes of these lines. Population sizes are less variable on younger leaves than on older leaves therefore differentiation of leaves by age is one means to increase precision in determining population sizes. As detailed below it is likely that there are substantial differences among leaves of different ages and positions as habitats for bacteria which could explain the heterogeneity in variances.

In many of the previously cited works, the predominant members of the microbial communities were identified using either numerical taxonomic methods on phenetic data (Ercolani, 1991; Périssol et al., 1993) or by techniques grouping organisms by chemical composition (van Outryve et al., 1989; Thompson et al., 1993; Legard et al., 1994). These extensive studies identified large numbers of bacteria (sample sizes of 1701, 590, 1236, 324, 1780 bacteria identified respectively by Ercolani, 1991; van Outryve et al., 1989; Thompson et al., 1993; Périssol et al., 1993; Legard et al., 1994). However, no consistent general trend can be drawn from these surveys on the effect of leaf age and position on microbial community composition. It was noted that there was no consistent difference in the number of species isolated from each leaf type throughout the seasons on sugar beet and wheat (Triticum aestivum) (Thompson et al., 1993; Legard et al., 1994). However, Périssol et al. (1993) detected a lower species diversity on older leaves of evergreen oak compared to younger leaves and a greater species diversity in the litter than in the phyllosphere. From the data collected by Ercolani (1991) on olive leaves, it also seems that the older the leaves the lower the species diversity. It is generally agreed that a few microbial species are dominant on the leaf surface and diversity indexes are rarely available in studies analysing the effect of leaf age or position on epiphytic microbial community structure. The number of isolates required to show any significant influence of leaf age and position - at a given sampling time for several growing seasons - on the relative frequency of certain members of the community is likely to be too large to make such measures feasible. The differences in communities among individual leaves, the absence of species from some samples through time, and the superimposed seasonal patterns of occurrence of different species and groups add other difficulties to this type of investigation.

Effect on Introduced Populations

The effect of leaf age and position on the population dynamics of plant pathogenic bacteria and fungi under field or controlled conditions has been assessed after inoculation of the deleterious micro-organism. Such studies have been conducted as a means to understand disease development and how leaf age affects plant resistance. In these studies not only the population sizes of the phytopathogenic microbes were recorded but also the incidence and the severity of the disease. The general trends which can be drawn from these studies show that for plants of a given age, micro-organisms develop or multiply at a much faster rate in younger leaves than in older leaves. Hence, increased incidence and severity of diseases are associated with the youngest leaves. In other words the resistance of a plant to a pathogen increases as tissues age. This type of resistance has to be evaluated separately from plant-age resistance which is noticed in many cases (Suparyono and Pataky, 1989; Koch and Mew, 1991; Chandrashekar and Halloran, 1992). In most of the cases reported here the effect of leaf age cannot be separated from a possible effect of leaf position. This leaf-age related resistance was noticed for blast disease on rice (Oryza sativa L.) caused by Pyricularia oryzae Cavara (Roumen, 1992; Roumen et al., 1992) and bacterial blight (Xanthomonas campestris pv. oryzae) of the same crop (Koch and Mew, 1991), crown rust (Puccinia coronata var. lolii) on perennial ryegrass (Lollium perenne L.) (Plummer et al., 1992), blue mould (Peronospora tabacina) of tobacco (Nicotiana tabacum) (Wyatt et al., 1991), bacterial leaf spot (Pseudomonas cichorii) of chrysanthemum (Chrysanthemum morifolium) (Jones et al., 1985); aschochita (Aschochita fabae f. sp. lentis) blight of lentil (Lens culinaris

Medik.) (Pedersen and Morrall, 1994); pathogenic strains of *Pseudomonas syringae* pv. *lachrymans* colonising leaves of cucumber (*Cucumis sativus* L.) (Leben, 1988), Stewart's wilt (*Erwinia stewartii*) and Goss's wilt (*Clavibacter michiganense* subsp. *nebraskense*) of sweet corn (*Zea mays* L.) (Suparyono and Pataky, 1989), among others.

The increase in resistance with ageing of leaves differs among cultivars (Koch and Mew, 1991; Roumen *et al.*, 1992). A more rapid build-up of resistance in young leaves over time causes a reduction in the total number of successful infections resulting in sporulating lesions. Roumen *et al.* (1992) postulated that cultivar resistance characterized by a susceptible infection type and by a rapid increase of age-related resistance is less-likely to be quickly overcome by new strains of the fungus than cultivar resistance based on a resistant infection type.

At a given sampling time, *P. cichorii* is able to colonise and induce symptoms on mature coffee leaves but not on young leaves, whereas *P. syringae* pv. *garcae* behaves in an inverse manner on the same host (Oliveira *et al.*, 1991). For a given type of leaf the development of *Septoria nodorum* increased with the age of wheat (*Triticum aestivum*) leaves and plant age (use of staggered sowing dates), and for a given plant age, the younger leaves support a lesser development of the fungus than the older ones (Tiedemann and Firsching, 1993). For minimally processed leaves of broad-leaved endive stored in polypropylene packs at 6°C during 7 days the rate of soft rot and marginal necrosis recorded after inoculation of a pathogenic strain of *Pseudomonas fluorescens* was significantly higher for older leaves than for younger leaves stored in same conditions (Jacques and Morris, 1995a).

To summarise the major findings of the different studies cited above, under natural conditions of microbial contamination older leaves seem to support higher microbial densities than younger leaves, whereas the introduction of a micro-organism generally leads to the opposite conclusion i.e. younger leaves are more conducive to their development and eventually to a higher incidence and severity of the disease they may cause. However, there are several exceptions.

This apparent contradiction gives rise to two major points. Firstly, it is difficult to separate the effect of the phenological state of the leaves, their position and the age of the plant and the underlying differences of environment and physiological status of leaves. Secondly, although the time of exposure to inoculum varies both with the age and position of the leaves under conditions of natural contamination, it is similar for all leaves on inoculated plants. The differential susceptibility of leaves to decay may only reflect differences in the likelihood of reaching the threshold of population size under natural conditions of contamination. The leaf age-related resistance may also be dependent on the accumulation of defence responses elicited by the plant which are less in younger tissues than in older tissues. This plant reaction may however be ineffective for non pathogenic epiphytic micro-organisms.

This suggests that the influence of leaf age and position on microbial populations can be complex and may result from variations in (1) immigration and emigration rates of microbes to the leaf surface (depending on accessibility of leaves to airborne micro-organisms, availability of immigrants, activity of vectors and trapping efficiency of the leaf) and (2) factors influencing the growth and multiplication of the epiphytes on the leaf surface (variable with the physiological status of the leaf and its quality as a habitat for microbes, microclimate, and activity of leaf-associated microflora) (Andrews *et al.*, 1980; Jacques *et al.*, 1995). The data supporting these different hypotheses will be detailed in the subsequent sections to gain insight into the mechanisms of microbial population dynamics on leaves of different ages and positions. Epiphytic fungal and bacterial community dynamics and the relative importance of the different processes in relation to the microbial community dynamics have been discussed previously (Kinkel, 1991; Hirano and Upper, 1991) and in this book (see chapter by Lindow).

ALTERATION OF IMMIGRATION AND EMIGRATION LEVELS DUE TO AGE AND POSITIONAL VARIATIONS OF LEAVES WITHIN A PLANT

There is little substantial evidence in the literature that variations in the accessibility of leaves to airborne microflora due to differences in position or age can affect epiphytic microbial population sizes. One investigation was designed to ascertain which of numerous variables can affect the densities of epiphytic yeasts, filamentous fungi and bacteria colonising apple leaves (Andrews *et al.*, 1980). Proximity of leaves to the periphery of the canopy and height of leaves in the canopy were shown to have some influence on microbial densities whereas the directional orientation of the samples had hardly any impact on microbial counts. A consistent pattern of higher population sizes of bacteria, yeasts and filamentous fungi was observed at lower elevations. However most of the time this effect was not significant for the different microbial populations. Indeed, it must be noted that the difference between filamentous fungal population sizes at low versus high elevation never reach one $\log_{10}$ unit. The lateral position of the leaves had a significant effect most of the time for yeasts. This effect was seldom significant for bacteria or fungi. Densities of bacteria were slightly higher on interior leaves in mid season and lower in late season. No consistent pattern was apparent for fungi or yeasts. If accessibility of leaves to airborne microflora was the main cause of the differences among the densities of the micro-organisms colonising leaves at different heights or lateral position in the tree, variations in the age of sampled leaves and in microclimate (light impact, UV radiation, temperature and humidity) may also have had an impact.

Another investigation providing some insight into the alteration of immigration rates of bacteria onto leaves of different positions has been conducted in our laboratory (Jacques *et al.*, 1995). Bacterial densities were quantified simultaneously on 1 wk-old leaves emerging on plants of three different ages. For two experiments out of three, total mesophilic, fluorescent and pectolytic bacterial population sizes were significantly higher on leaves emerging on younger plants than leaves emerging on older plants (Fig. 2). As a consequence of broad-leaved endive morphology, it is likely that leaves emerging on younger plants are exposed to airborne bacteria while leaves emerging on older plants are covered by adjacent leaves and hence more protected from this inoculum source. Therefore we attempted to determine whether or not bacterial immigration rates were altered on newly emerged leaves due to the presence of adjacent leaves. In one of our experiments total and pectolytic bacterial densities remained unaffected by the removal of adjacent leaves, but fluorescent pseudomonad densities substantially increased on emerging leaves (Fig. 3) indicating that the adjacent leaves could act as a barrier for contamination of emerging leaves. These experiments provide further evidence that differences in bacterial population sizes on leaves of different positions can be attributed in part to variations in accessibility of leaves to airborne bacteria. Nevertheless, as it was observed here and previously (Hirano and Upper, 1992; Thompson *et al.*, 1995a) the total bacterial community and the different sub-components may behave differently on the same leaves, indicating clearly that the dynamics of different populations are governed differently by the processes and factors which regulate them.

In a subsequent study, we were interested in the effect of leaf age on post-harvest decay of minimally processed broad-leaved endive (Jacques and Morris, 1995a). Naturally occurring decay was recorded for leaf pieces separated by age and stored in polypropylene packs. At the end of storage, younger leaves appear less susceptible to marginal necrosis and soft rot than older leaves. These differences probably reflect differences in the likelihood of a given population sizes *i.e.*, the threshold population size for decay. Indeed population sizes of fluorescent and pectolytic bacteria were significantly lower on younger leaves than on

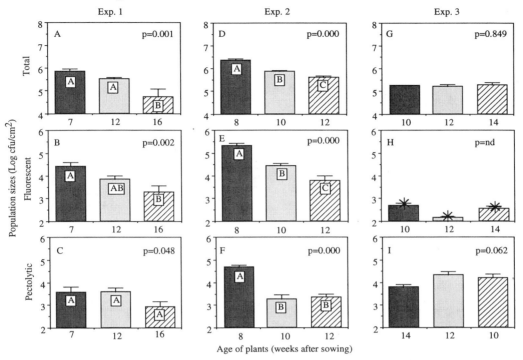

Figure 2. Mean densities of total (A, D and G), fluorescent (B, E and H) and pectolytic (C, F and I) bacteria on leaves emerging on plants of different ages of broad-leaved endive interplanted in the same field. Points covered by a star represent the mean level of sensitivity of the assay at that date. Error bars represent the standard error of the mean, and mean densities associated with different letters are significantly ($p<0.05$) different on the basis of Tukey's HSD test. From Jacques et al. (1995) with permission.

older leaves at the end of storage and quite different from the threshold determined for a strain of P. fluorescens after inoculation. However, leaf age had no notable effect on the threshold for the development of decay, and the effect of leaf age on decay after inoculation of leaf pieces of different ages with a pathogenic strain of P. fluorescens was not consistent among experiments, showing clearly the complexity of the phenomenon.

In an analysis of the effect of the canopy growth on spread of late leaf spot of groundnut caused by Cercosporidium personatum Savary and van Santen (1992) found that primary gradients of the disease are modified by canopy structure. A shift in the prevailing direction of disease progress in older crops (staggered sowing dates) was attributed to a modified turbulence pattern above and within the crop at high leaf area indices. The disease spread and spore dispersal gradients varied with the age of crops reflecting a decreasing accessibility of the foliage to the pathogen corresponding to the closing of the canopy and the increase in its leaf area index.

A model was presented describing the vertical variation of the aerial concentration of Venturia inequalis ascospores released from a ground level source in an orchard (Aylor, 1995). The aerial concentration of ascospores decreased with height as a consequence of a rapid increase of wind speed and turbulent eddy diffusivity with height above the ground leading to a dilution of the ascospores in the air.

The studies by Andrews et al. (1980) and by our laboratory (Jacques et al., 1995) raise the question of the origin of immigrants, their availability through time and the vectors

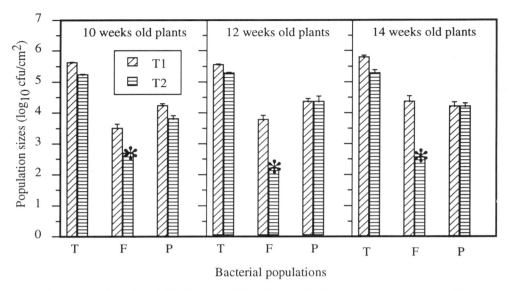

Figure 3. Mean densities of total (T), fluorescent (F), and pectolytic (P) bacteria on newly emerged leaves on plants of different ages of broad-leaved endive interplanted in the same field. In treatment T1, inner leaves were excised 1 week before sampling to expose leaves that would emerge during the following week. Treatment T2 plants were not groomed. Points covered by a star represent the mean level of sensitivity of the assay at that date. Error bars represent the standard error of the mean. From Jacques *et al.* (1995) with permission.

of microbial immigration and emigration. Inoculum sources of micro-organisms include plant foliage, soil and seeds. Various studies have noted particularities of leaf-associated bacteria that are thought to be adaptations to the peculiar conditions prevalent at the leaf surface, such as colony pigmentation, for example, as a means of protection against UV light. Earlier studies indicated that predominant leaf-associated bacteria are usually different from the predominant soil bacteria leading to the conclusion that the soil does not appear to serve as an important source of inoculum for epiphytes (Stout, 1960; Pieczarka and Lorbeer, 1975). Some studies have shown that seed-associated bacteria (Leben, 1961; Fryda and Otta, 1978) or bacteria introduced as seed dressing (de Leij *et al.*, 1995; Thompson *et al.*, 1995b) can colonise both the rhizosphere and the phyllosphere of the hosts. Nevertheless, the ability of these inocula to survive and migrate in the phyllosphere was dependent on the maintenance of high relative humidity (Leben, 1961; Fryda and Otta, 1978). However, the genetically modified fluorescent pseudomonad introduced as a seed dressing of sugar beet and wheat originally isolated from the phyllosphere of sugar beet successfully developed high population densities in field experiments both in the phyllosphere and in the rhizosphere of the inoculated hosts (de Leij *et al.*, 1995; Thompson *et al.*, 1995b).

Lindemann *et al.* (1982) show that in agricultural areas, plant foliage was the main source of airborne bacteria. Higher concentrations of airborne bacteria were measured above cropped fields than above bare soil and the concentration of airborne micro-organisms increased as crops matured. A net upward flux of bacteria was positively correlated with wind speed but not with canopy-level air temperatures or relative humidity (Lindemann and Upper, 1985). Several plant pathogenic micro-organisms were reported to be removed and dispersed by water splash due to rain or sprinkler irrigation (Graham and Harrison, 1975; Venette and Kennedy, 1975; Fitt *et al.*, 1989; McCartney and Butterworth, 1992). Insects and arthropods, as well as pollen and human activities, have been reported to be occasionally efficient vectors of epiphytic micro-organisms (Johnson *et al.*, 1993; Bailey *et al.*, unpublished results).

In the case of apple leaves (Andrews *et al.*, 1980), if accessibility of leaves to airborne inoculum was the main cause of higher microbial densities at lower elevation than at higher elevation in the trees, then it seems likely that the major inoculum source was either the ground cover plants, if any, or the nearby soil rather than micro-organisms dispersed from the apple tree canopy or from distant sources. In the case of broad-leaved endive, plant foliage seems to be the main source of airborne bacteria as population sizes increase on newly emerging leaves (1 wk-old leaves) throughout the growing season while the crop developed (Jacques *et al.*, 1995). Lindow (1994) noted that epiphytic bacterial population sizes and composition on pear flowers and fruit and citrus leaves were strongly influenced by ground cover crop species in the vicinity of the trees. The removal of ground cover crop plants resulted in reductions of epiphytic bacterial populations on pear and citrus of about 30-fold. All these studies suggest that plant foliage seems to be the major source of immigrants for plants moved either by wind or water splash; other sources and vectors being only of occasional importance.

Depending on the morphology of the plant and the major source of immigrants, the ability of leaves at different heights to efficiently trap micro-organisms is a function of physical, anatomical and chemical characteristics which evolve with age. The trapping efficiency of a leaf depends also on the size and the surface characteristics of the particles to be retained (Forster, 1977). Anatomical features and topography of leaf surfaces of a given age are incredibly diverse (Juniper, 1991) leading to a very difficult generalisation regarding their efficacy as particle traps. Wax projections increase the contact angle between a water drop and the leaf surface leading to a low wetability of the leaf surface. Hence, micro-organisms arriving in aerosols will be less likely to be retained on such surfaces. The production and the regeneration of wax is very rapid on young leaves but slows considerably once a leaf is fully expanded (Hallam and Juniper, 1971). Hence the different leaves of a plant (juvenile, mature and senescent) may differ markedly in their waxiness. Degradation of wax through ageing of leaves (Forster, 1977, Blaker and Greyson, 1987) often combined with increased injury of leaves will enhance its wetability and may increase its particle-trapping efficiency (Juniper and Jeffree, 1983). However, efficiency of spore capture was attributed to the presence of undegraded wax structure in the case of Sitka spruce needles as the removal of wax greatly decreased the capture of spores as did an increasing age of needles (Forster, 1977). Both leaf age and position are known to affect the choice of leaves by insects due to their different nutrient richness and composition (Fiala *et al.*, 1990). Leaf surface features act as repellents for insects and hence could reduce the deposition of particles carried by these vectors (Allen *et al.*, 1991). This mechanism may be more efficient on young leaves where the density of leaf features is higher.

The amount of wax on a leaf is also known to be a function of light, being greater at high light intensity (Juniper and Jeffree, 1983). Hence, whether a leaf is under direct solar radiation or in shade will be critical for its trapping efficiency and its quality as a habitat for microbes, as the amount and the quality of wax govern the wetability of the leaf surface and indirectly affects the loss of nutrients from the leaf.

FACTORS INFLUENCING GROWTH, MULTIPLICATION, AND SURVIVAL OF EPIPHYTES ON THE SURFACE OF LEAVES OF DIFFERENT AGES AND POSITIONS

Evidence of a predominant effect of leaf physiology on bacterial densities among leaves of different ages was provided after inoculation of broad-leaved endive leaves with a strain of *P. fluorescens* (T53) (Jacques and Morris, 1995a). Whereas population sizes of T53 were equivalent on leaf pieces of both ages after inoculation, the mean population sizes of T53 were significantly lower on younger leaf pieces than on older leaf pieces after 7 days of storage at 6°C. Abiotic environmental conditions were found to be no more different among packs of different ages than among packs of a given age based on measurements of the atmosphere composition (Jacques M.-A., unpublished data).

Variations in leaching of nutrients, anatomical features and microclimate may be part of the physiological nature of the effect of leaf-age and position on epiphytic microbial growth. Leachate composition and quantity vary among leaves of the same plant depending on the physiological age of the leaf and its position (Tukey, 1971; Fiala *et al.*, 1985; Fiala *et al.*, 1990). Loss of nutrients increases as the leaf matures concomitantly with the wetting properties of the leaf surface which are determined by the fine structure of the wax (Juniper, 1991). The increase in leaching concerns mineral nutrients (Tukey, 1966), carbohydrates (Tukey, 1966; Fiala *et al.*, 1990), amino and organic acids (Tukey, 1966).

The amount and the composition of the leachates are influenced by environmental factors (Tukey, 1971; Schönherr and Baur, this volume). The pattern of carbohydrate leached depends on the photosynthetic activity of the leaf (Fiala *et al.*, 1985) and hence varies with the position of leaves in the canopy. The leaching of carbohydrates parallels the light intensity received by the plant. Leaching is higher during periods of high light intensity and lower during periods of darkness. High temperatures have a similar effect on carbohydrate leaching than high light intensities. The amount of wax is also a function of the light intensity received by the leaf. Hence the availability of nutrients at the leaf surface is largely dependent on the age and the position of the leaf in the canopy and on their interactions. It is also dependent on the nature of the micro-organism itself and its capacity to produce surfactants. This was shown for epiphytic bacteria (Bunster *et al.*, 1989; Akit *et al.*, 1981) and it was shown that this production was a key factor in the pathogenicity of some strains of *P. fluorescens* (Hildebrand, 1989). Schönherr and Baur (this volume) also suggest that these surfactants may increase cuticle permeability thus enabling micro-organisms to increase their local nutritional resources.

Solar radiation has a detrimental effect on the survival of conidia on the top of the canopy. Conidia of *Beauveria bassiana* survived substantially longer within the canopy of wheatgrass (*Agropyron cristatum* L.) and alfalfa (*Medicago sativa* L.) than those at the top of the canopy (Inglis *et al.*, 1993). However, temperature and relative humidity were only slightly more favourable for conidial survival at the middle than at the top of the canopy and feeding by spore-consuming insects was not limited to the top of the canopy.

The other factors that can account for the effect of leaf age and position on microbial growth or survival on leaf surfaces include differential production of toxic compounds and variation of microclimate parameters. As a consequence of canopy development with plant age it has been suggested that there is a reduction of sunlight penetration and air movement to lower foliage in parallel with an increase of humidity and leaf wetness period (Agrios, 1988). For two growing seasons of alfalfa in Missouri, relative humidity and leaf wetness tended to be higher within the canopy than above it whereas mean daily temperatures above and within the canopy generally differed by only 1°C (Emery and English, 1994). Considerable variations of leaf surface temperature were found with leaf height in a canopy of ryegrass (Burrage, 1971). The upper levels of the canopy did not reach the highest surface temperature due to lower density of the crop and a resulting higher degree of air movement and thus greater air transfer. The pattern of radiation received by a leaf is dependent on its height and its angle in the canopy, its anatomical features and more generally on its physical and chemical composition (Burrage, 1971). As a consequence of the variations of all these parameters, we can speculate upon the existence of different carrying capacities for micro-organisms among leaves of different ages and positions.

CONCLUSIONS: THE IMPLICATIONS OF HETEROGENEITY OF MICROBIAL POPULATIONS DUE TO POSITION AND AGE OF LEAVES

The differences observed in the means and the variances of the microbial population sizes among leaves of different ages and positions have important implications for sampling strategies aimed at quantifying epiphytic microbial populations and subsequently on determining ability of plant material to be processed or stored. The separation of leaves by their age and position allow a more precise determination of the mean of microbial population sizes and hence would increase the sensitivity of statistical tests aimed at comparing the effect of treatments on epiphytic bacteria. Undoubtedly, this separation will have important consequences for the design of experiments of tests for biological control of foliar diseases and studies of the ecology of the interactions between host plants and micro-organisms. Hence, we propose that sampling procedures should not be totally random, but should take leaf age into account.

The precise determination of microbial population dynamics as influenced by host factors is required for successful prediction and enhancement of biological, chemical and integrated control methods. The food industry and particularly the production of minimally-processed vegetables could benefit from taking into account the effect of leaf age and position on the means and variances of microbial population sizes. The design of sampling plans to determine microbiological quality of the raw material before processing will be more meaningful for leaves separated by age classes. In order to maintain an acceptable quality of the final product sorting methods and processing could be improved by taking into account leaf age and positions as illustrated for broad-leaved endive (Jacques and Morris, 1995a).

The testing of material for plant breeding should take account of the physiological state of the plant and also of the different leaves of the plant as it may affect the rate of the microbial multiplication and disease severity. Heritable components of such resistances may be of great value for plant breeders. Also," a rational and cost-effective approach toward fungicide use must take into account both cultivar resistance and age-related changes" (Pelletier and Fry, 1990).

The variability in community structure results from the differential impacts of the population processes on each population. The leaf-to-leaf variability which results from differences in their age and position affects these processes. Our understanding of the mechanism of community dynamics will benefit from an understanding of the leaf-to-leaf variability within sub-samples of the leaves differentiated by their age and/or their position on the plant as they represent substantial different habitats for micro-organisms.

ACKNOWLEDGMENTS

Some of the work reported in this chapter was financially supported by Le Conseil Régional Provence-Alpes-Côte d'Azur. I thank Mark J. Bailey and Cindy E. Morris for helpful comments on the manuscript.

REFERENCES

Agrios, G.N. 1988, Plant pathology, 3rd edition, Academic Press, San Diego, CA.
Akit, J., Cooper, D.G., Mannien, K.I. and Zajic, J.E. 1981, Investigation of potential biosurfactant production among phytopathogenic corynebacteria and related microbes, *Curr. Microbiol.* 6:145-150.

Allen, E.A., Hoch, H.C., Steadman, J.R. and Stavely, R.J. 1991, Influence of leaf features on spore deposition and the epiphytic growth of phytopathogenic fungi, pp. 87-110 *In* : Andrews, J.H. and Hirano, S.S. (eds.) *Microbial Ecology of Leaves*, Springer Verlag New York Inc.

Andrews, J.H., Kenerley, C.M. and Nordheim, E.V. 1980, Positional variation in phylloplane microbial populations within an apple tree canopy, *Microb. Ecol.* 7:71-84.

Aylor, D.E. 1995, Vertical variation of aerial concentration of *Venturia inequalis* ascospores in an apple orchard, *Phytopathology* 85:175-181.

Blakeman, J.P. 1985, Ecological succession of leaf surface microorganisms in relation to biological control, pp. 6-30 *In* : Windels, C.E. and Lindow, S.E. (eds.) *Biological Control on the Phylloplane*, American Phytopathological Society, St Paul MN.

Bernstein, M.E. and Carroll, G.E. 1977, Microbial populations on Douglas fir needle surfaces, *Microb. Ecol.* 4:41-52.

Blaker, T.W. and Greyson, R.I. 1987, Developmental variation of leaf surface wax of maize, *Zea mays*, *Can. J. Bot.* 66:839-846.

Bunster, L., Fokkema, N. J. and Schippers, B. 1989, Effect of surface active *Pseudomonas* spp. on leaf wetability, *Appl. Environ. Microbiol.* 55:1340-1345.

Burrage, S.W. 1971, The micro-climate at the leaf surface, pp. 91-101 *In* : Preece, T.F. and Dickinson, C.H. (eds.) *Ecology of Leaf Surface Micro-organisms*, Academic Press London.

Chandrashekar, M. and Halloran, G.M. 1992, Leaf scorch (*Kabatiella caulivora* (Kirch.) Kirk.) of subterranean clover (*Trifolium subterraneum* L.) : influence of host cultivar, growth stage and pathogen isolates on the disease severity, *Euphytica* 61:181-186.

de Leij, F.A.A.M., Sutton, E., Whipps, J.M., Fenlon, J.S. and Lynch, J.M. 1995, Field release of a genetically modified *Pseudomonas fluorescens* on wheat : establishment, survival and dissemination, *Biotechnology* 13:1488-1492.

Dickinson, C.H. 1976, Fungi on the aerial surfaces of higher plants, pp. 293-324 *In* : Dickinson, C.H. and Preece, T.F. (eds.) *Microbiology of Aerial Plant Surfaces*, Academic Press London.

Dickinson, C.H., Austin, B. and Goodfellow, M. 1975, Quantitative and qualitative studies of phylloplane bacteria from *Lollium perenne*, *J. Gen. Microbiol.* 91:157-166.

Emery, K.M. and English, J.T. 1994, Development of foliar diseases of alfalfa in relation to microclimate, host growth and fertility, *Phytopathology* 84:1263-1269.

Ercolani, G.L. 1976, Bacteriological quality assessment of fresh marketed lettuce and fennel, *Appl. Environ. Microbiol.* 31:847-852.

Ercolani, G.L. 1979, Distribuzione di *Pseudomonas savastanoi* sulle foglie dell'olivo, *Phytopath. Medit.* 18:85-88.

Ercolani, G.L. 1991, Distribution of epiphytic bacteria on olive leaves and the influence of leaf age and sampling time, *Microb. Ecol.* 21:35-48.

Fiala, V., Derridj, S. and Jolivet, E. 1985, Influence de la teneur en glucides solubles des feuilles de *Zea mays* L. sur le choix du site de ponte de la pyrale, *Ostrinia nubilalis* Hbn. (*Lepid. Pyralidae*), *Agronomie* 5:927-932.

Fiala, V., Glad, C., Martin, M., Jolivet, E. and Derridj, S. 1990, Occurence of soluble carbohydrates on the phylloplanes of maize (*Zea mays* L.) : variations in relation to leaf heterogeneity and position on the plant, *New Phytol.* 115:609-615.

Fitt, B.D.L., McCartney, H.A. and Walkate, P.J. 1989, The role of rain in dispersal of pathogen inoculum, *Annu. Rev. Phytopathol.* 27:241-270.

Forster, G.F. 1977, Effect of leaf surface wax on the deposition of airborne propagules, *Trans. Brit. Mycol. Soc.* 68:245-250.

Fryda, S.J. and Otta, J.D. 1978, Epiphytic movment and survival of *Pseudomonas syringae* on spring wheat, *Phytopathology* 98:1064-1067.

Geeson, J.D. 1979, The fungal and bacterial flora of stored white cabbage, *J. Appl. Bacteriol.* 46:189-193.

Graham, D.C. and Harrison, M.D. 1975, Potential spread of *Erwinia* spp. in aerosols, *Phytopathology* 65:739-741.

Hallam, N.D. and Juniper, B.E. 1971, The anatomy of the leaf surface, pp. 3-37 *In* : Preece, T.F. and Dickinson, C.H. (eds) *Ecology of Leaf Surface Micro-organisms*, Academic Press London.

Hildebrand, P.D. 1989, Surfactant-like characteristics and identity of bacteria associated with broccoli rot in atlantic Canada, *Can. J. Plant Pathol.* 11:205-214.

Hirano, S.S. and Upper, C.D. 1989, Diel variation in population size and ice nucleation activity of *Pseudomonas syringae* on snap bean leaflets, *Appl. Environ. Microbiol.* 55:623-630.

Hirano, S.S. and Upper, C.D. 1991, Bacterial community dynamics, pp. 271-294 *In* : Andrews, J.H. and Hirano, S.S. (eds.) *Microbial Ecology of Leaves*, Springer Verlag New York Inc.

Hirano, S.S. and Upper, C.D. 1992, Population dynamics of *Pseudomonas syringae* in the phyllosphere, pp. 21-29 *In* : Galli, E., Silver S. and Witholdt, B., (eds.) Pseudomonas : *Molecular Biology and Biotechnology*, ASM Washington, DC.

Hirano, S.S., Nordheim, E.V., Arny, D.C. and Upper, C.D. 1982, Lognormal distribution of epiphytic bacteria populations on leaf surfaces, *Appl. Environ. Microbiol.* 44:695-700.

Hirano, S.S., Clayton, M.K. and Upper, C.D. 1994, Estimation of and temporal changes in means and variances of populations of *Pseudomonas syringae* on snap bean leaflets, *Phytopathology* 84:934-940.

Inglis, G.D., Goettel, M.S. and Johnson, D.L. 1993, Persistence of the entomopathogenic fungus, *Beauveria bassiana*, on phylloplanes of crested wheatgrass and alfalfa, *Biol. Control* 3:258-270.

Ishimaru, C., Eskridge, K.M. and Vidaver, A.K. 1991, Distribution analysis of naturally occurring epiphytic population of *Xanthomonas campestris* pv. *phaseoli* on dry beans, *Phytopathology* 81:262-268.

Jacques, M.-A. and Morris, C.E. 1995a, Bacterial population dynamics and decay on leaves of different ages of ready-to-use broad-leaved endive, *Int. J. Food Sci. Technol.* 30:221-236.

Jacques, M.-A. and Morris, C.E. 1995b, MiniReview : issues related to the quantification of bacteria from the phyllosphere, *FEMS Microbiol. Ecol.* 18:1-14.

Jacques, M.-A., Kinkel, L.L. and Morris, C.E. 1995, Population sizes, immigration and growth of epiphytic bacteria on leaves of different ages and positions of field-grown endive (*Cichorium endivia* var. *latifolia*), *Appl. Environ. Microbiol.* 61:899-906.

Johnson, K.B., Stockwell, V.O., Burgett, D.M., Sugar, D. and Loper, J.E. 1993, Dispersal of *Erwinia amylovora* and *Pseudomonas fluorescens* by honey bees from hives to apple and pear blossoms, *Phytopathology* 83:473-484.

Jones, J.B., Chase, A.R., Harbaugh, B.K. and Raju, B.C. 1985, Effect of leaf wetness, fertilizer rate, leaf age and light intensity before inoculation on bacterial leaf spot of chrysanthemum, *Plant Dis.* 69:782-784.

Juniper, B.E 1991, The leaf from the inside and the outside : a microbe's point of view, pp. 21-42 *In* : Andrews, J.H. and Hirano, S.S. (eds.) *Microbial Ecology of Leaves*, Springer Verlag New York Inc.

Juniper, B.E., and Jeffree, C.E. 1983, Plant surfaces, Edward Arnold, London. 93 p.

Kinkel, L.L. 1991, Fungal community dynamics, pp. 253-570 *In* : Andrews, J.H. and Hirano, S.S. (eds.) *Microbial Ecology of Leaves*, Springer Verlag New York Inc.

Kinkel, L.L. 1992, Statistical consequences of combining population samples, *Phytopathology* 82:1168.

Kinkel, L.L., Wilson, M. and Lindow, S.E. 1995, Effects of scale on estimates of epiphytic bacterial populations. *Microbial Ecol.* 29:283-297.

Koch, M.F. and Mew, T.W. 1991, Effect of plant age and leaf maturity on the quantitative resistance of rice cultivars to *Xanthomonas campestris* pv. *oryzae*, *Plant Dis.* 75:901-904.

Leben, C. 1961, Microorganisms of cucumber seedlings, *Phytopathology* 51:553-557.

Leben, C. 1988, Relative humidity and the survival of epiphytic bacteria with buds and leaves of cucumber plants, *Phytopathology* 78:179-185.

Legard, D.E., McQuilken, M.P., Whipps, J.M., Fenlon, J.S., Fermor, T.R., Thompson, I.P., Bailey, M.J. and Lynch, J.M. 1994, Studies of seasonal changes in the microbial populations on the phyllosphere of spring wheat as a prelude to the release of a genetically modified microorganisms, *Agric. Ecosys. Environ.* 50:87-101.

Lindemann, J., Constantinidou, W.R., Barchet, W.R. and Upper, C.D. 1982, Plants as sources of airborne bacteria, including ice nucleation-active bacteria, *Appl. Environ. Microbiol.* 44:1059-1063.

Lindemann, J. and Upper, C.D. 1985, Aerial dispersal of epiphytic bacteria over bean plants, *Appl. Environ. Microbiol.* 50:1229-1232.

Lindow, S.E. 1994, The role of immigration in establishment of epiphytic bacterial populations and practical implications), *Mole. Ecol.* 3:614.

McCartney, H.A. and Butterworth, J. 1992, Effects of humidity on the dispersal of *Pseudomonas syringae* from leaves by water splash, *Microb. Releases* 1:187-190.

Mew, T.W. and Kennedy, B.W. 1982, Seasonal variation in populations of pathogenic pseudomonads on soyabean leaves, *Phytopathology* 72:103-105.

Morris, C.E. 1985, Diversity of epiphytic bacteria on snap bean leaflets based on nutrient utilization abilities : biological and statistical considerations, *Ph D Thesis*, University of Madison Wisconsin.

Morris, C.E. and Lucotte, T. 1993, Dynamics and variability of bacterial population density on leaves of field-grown endive destined for ready-to-use processing, *Int. J. Food Sci. Technol.* 28:201-209.

Morris, C.E. and Rouse, D.I. 1985, Role of nutrients in regulating epiphytic bacterial populations, pp. 63-82 *In* : Windels, C.E. and Lindow, S.E. (eds.) *Biological Control on the Phylloplane*, American Phytopathological Society, St Paul, MN.

Morris, C.E., Jacques, M.-A. and Nicot, P.C. 1994, Microbial aggregates on leaf surfaces : characterization and implications for the ecology of epiphytic bacteria, *Mole. Ecol.* 6:613.

Pedersen, E.A. and Morrall, R.A.A. 1994, Effects of cultivar, leaf wetness duration, temperature and growth stage on infection and development of ascochyta blight of lentil, *Phytopathology* 84:1024-1030.

Pelletier, J.R. and Fry, W.E. 1990, Characterization of resistance to early blight in three potato cultivars : receptivity, *Phytopathology* 80:361-366.

Périssol, C., Roux, M. and Le Petit, J. 1993, Succession of bacteria attached to evergreen oak leaf surfaces, *Eur. J. Soil Biol.* 29:167-176.

Oliveira, J.R., Romeiro, R.S. and Muchovej, J.J. 1991, Population tendencies of *Pseudomonas cichorii* and *P. syringae* pv. *garcae* in young and mature coffee leaves, *J. Phytopathol.* 131:210-214.

Pieczarka, D.J. and Lorbeer, J.W. 1975, Micro-organisms associated with bottom rot of lettuce grown in organic soil in new York state, *Phytopathology* 65:16-21.

Plummer, R.M., Hall, R.L. and Watt, T.A. 1992, Effect of leaf age and nitrogen fertilisation on sporulation of crown rust (*Puccinia coronata* var. *lolii*) on perennial ryegrass (*Lollium perenne* L.), *Ann. Appl. Biol.* 121:51-56.

Roumen, E.C. 1992, Effect of leaf age on components of partial resistance in rice to leaf blast, *Euphytica* 63:271-279.

Roumen, E.C., Bonman, J.M. and Parlevliet, J.E. 1992, Leaf age related partial resistance to *Pyricularia oryzae* in tropical lowland rice cultivars as measured by the number of sporulating lesions, *Phytopathology* 82:1414-1417.

Rouse, D.I., Nordheim, E.V., Hirano, S.S. and Upper, C.D. 1985, A model relating the probability of foliar disease incidence to the population frequencies of bacterial plant pathogens, *Phytopathology* 75:505-509

Ruinen, J. 1961, The phyllosphere. I. An ecologically neglected milieu, *Plant Soil* 15:81-109.

Savary, S. and van Santen, G. 1992, Effect of crop age on primary gradients of late leaf spot (*Cercosporidium personatum*) on groundnut, *Plant Pathol.* 41:265-273.

Stout, J.D. 1960, Bacteria of soil and pasture leaves at Claudlands showgrounds, *New Zeal. J. Agric. Res.* 3:413-430.

Suparyono and Pataky, J.K. 1989, Influence of host resistance and growth stage at the time of inoculation on Stewart's wilt and Goss's wilt development and sweet corn hybrid yield, *Plant Dis.* 73:339-345.

Tiedemann, A.V. and Firsching, K.H. 1993, Effects of ozone exposure and leaf age of wheat on infection processes of *Septoria nodorum* Berk., *Plant Pathol.* 42:287-293.

Thompson, I.P., Bailey, M.J., Fenlon, J.S., Fermor, T.R., Lilley, A.K., Lynch, J.M., McCormack, P.J., McQuilken, M.P., Purdy, K.J., Rainey, P.B. and Whipps, J.M. 1993, Quantitative and qualitative seasonal changes in the microbial community from the phyllosphere of sugar beet (*Beta vulgaris*), *Plant Soil* 150:177-191.

Thompson, I.P., Ellis, R.J. and Bailey, M.J. 1995a, Autecology of a genetically modified fluorescent pseudomonad on sugar beet, *FEMS Microbiol. Ecol.* 17:1-14.

Thompson, I.P., Lilley, A.K., Ellis, R.J., Bramwell, P.A. and Bailey, M.J. 1995b, Survival, colonisation and dispersal of genetically modified *Pseudomonas fluorescens* SBW25 in the phytosphere of field grown sugar beet, *Biotechnology* 13:1493-1497.

Tukey, H.B., Jr. 1966, Leaching of metabolites from above-ground plant parts and its implication, *Bull. Torrey Bot. Club* 93:385-401.

Tukey, H.B., Jr. 1971, Leaching of substances from plants, pp. 67-80 *In* : Preece, T.F. and Dickinson, C.H. (eds.) *Ecology of Leaf Surface Micro-organisms*, Academic Press London.

van Outryve, M.F., Gosselé, F. and Swings, J. 1989, The bacterial microflora of witloof cichory (*Cichorium intybus* L. var. *foliosum* Hegi) leaves, *Microb. Ecol.* 18:175-186.

Venette, J.R. and Kenedy, B.W. 1975, Naturally produced aerosols of *Pseudomonas glycinea*, *Phytopathology* 65:737-738.

Weller, D.M. and Saetler, A.W. 1980, Colonization and distribution of *Xanthomonas phaseoli* and *Xanthomonas phaseoli* var. *fuscans* in field-grown navy beans, *Phytopathology* 70:500-506.

Wildman, H.G. and Parkinson, D. 1979, Microfungal succession on living leaves of *Populus tremoides*, *Can. J. Bot.* 57:2800-2811.

Wyatt, S.E., Pan, S.Q. and Kúc, J. 1991, β-1,3-glucanase, chitinase and peroxidase activities in tobacco tissues resistant and susceptible to blue mould as related to flowering, age and sucker development, *Physiol. Mole. Plant Pathol.* 39:433-440.

16

SPATIAL AND TEMPORAL VARIATIONS IN SIZE AND PHENOTYPIC STRUCTURE OF POPULATIONS OF *PSEUDOMONAS SYRINGAE* ON FRUIT TREES

Jacques L. Luisetti[*]

Station de Pathologie Végétale
Institut National de la Recherche Agronomique, B.P. 57
49071 Beaucouzé cedex
France

INTRODUCTION

Most pathovars of *Pseudomonas syringae* (*P. syr.*) are known to be epiphytes on their host plants. According to Leben (1965), this means that these phytopathogenic bacteria are able to survive on the aerial parts of the plants when the conditions are not favourable (mainly high temperature and low humidity), and to multiply significantly when they become favourable again (moisture after rain or dew). Quantitative variations have been reported for the epiphytic populations of *P. syr.* pathovars occurring on fruit trees (Crosse, 1959; English and Davis, 1960; Panagopoulos and Crosse, 1964; Panagopoulos, 1966; Ercolani, 1969; Gardan *et al.*, 1972; Luisetti and Paulin, 1972; Latorre and Jones, 1979; Burr and Katz, 1984; Roos and Hatting, 1986; Bordjiba and Prunier, 1991) depending mainly on the stage of vegetative development but also on the plant genotype or on the climatic conditions (Latorre *et al.*, 1985). As an example, Figure 1 shows the variations of the epiphytic populations of *P. syr.* pv. *persicae*, the causal agent of a severe die-back of peaches: a high level in spring followed by a strong decrease during summer and a significant increase during leaf fall.

These epiphytic populations of *P. syr.* pathovars were considered to be of importance for disease development (Crosse, 1966). Symptoms of pear blight during spring are frequently observed after a period of negative temperatures which allows the phytopathogenic bacteria to gain entry into the plant. The infection also requires the occurrence of sufficient quantities of epiphytic *P. syr.* pv. *syringae* (Ercolani, 1969; Luisetti and Paulin, 1972; Luisetti, 1978). The incidence of bacterial canker of cherry is related to the epiphytic inoculum of *P. syr.* pv. *mors-prunorum* (Crosse, 1959). The incidence of the bacterial

[*] present address: INRA / CIRAD, BP 180, 97455 Saint Pierre. La Réunion

Aerial Plant Surface Microbiology, edited by Morris et al.
Plenum Press, New York, 1996

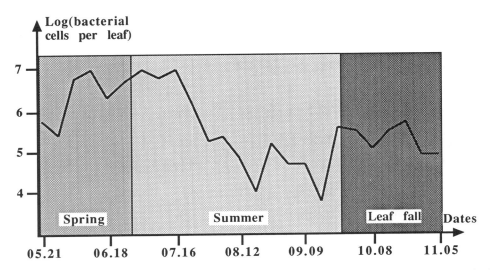

Figure 1. Dynamics of epiphytic populations of *P. syringae* pv. *persicae* on peach leaves from spring to leaf fall.

die-back of peaches has been clearly demonstrated to depend on the level of the epiphytic populations of *P. syr.* pv. *persicae* during leaf fall (unpublished data, Figure 2).

Moreover, the epiphytic populations on fruit trees are responsible for the contamination of the forming buds, which results in the overwintering of the bacterium. On fruit trees, the epiphytic stage of the phytopathogenic bacterium can be mostly considered as the link between two successive pathogenic stages. This linkage is well illustrated by the life cycle of *P. syr.* pv. *persicae* on peaches (Figure 3).

Even if the role of the epiphytic populations of phytopathogenic pseudomonads in disease development is well established, some points remain to be elucidated. They mainly

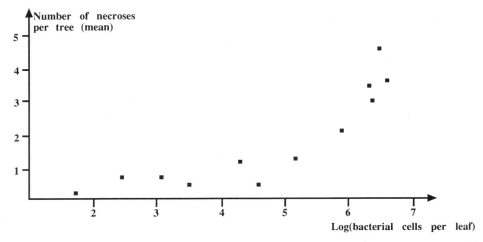

Figure 2. Relationship between the epiphytic populations of *P. syringae* pv. *persicae* on peach leaves during leaf fall (weighted average of 10 estimations) and the disease incidence observed during the following winter.

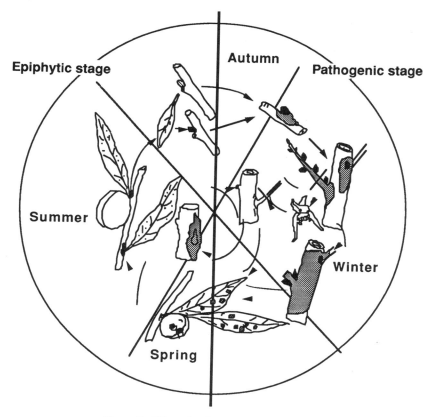

Figure 3. Life cycle of *P. syringae* pv. *persicae*.

concern the nature of the interactions between the epiphytic bacteria and the plants they can be recovered from.

The contribution of our laboratory to a better understanding of the role of the epiphytic populations includes two aspects: their spatial distribution at the level of the orchard, the tree or the leaf, and their phenotypic variability.

SPATIAL DISTRIBUTION OF *PSEUDOMONAS SYRINGAE* PV. PERSICAE EPIPHYTIC ON PEACH TREES

Investigations on the spatial distribution of epiphytic populations were performed on peaches infected by *P. syr.* pv. *persicae* (Prunier *et al.*, 1972). Trees grown out of the area where the disease occurred are naturally free of any *P. syr.* pv. *persicae*. Therefore, epiphytic populations of this pathogen could only be recovered from diseased trees or from healthy trees within a diseased orchard (Luisetti *et al.*, 1976). Experiments were performed in order to determine the distribution of the epiphytic *P. syr.* pv. *persicae* populations within a diseased orchard, or within a tree and their location on the leaf. This information appears to be useful to understand how the epiphytic bacteria spread within an orchard, or to determine the sample size or the sampling procedure best suited to estimate the quantity of epiphytic bacteria in an orchard or on a tree. This can also give indications on the nature of the interaction between epiphytic bacteria and the host on which they live.

Spatial Distribution at the Scale of an Orchard

A one-hectare orchard (576 trees of the susceptible variety "Redwing") was established in the area where the die-back occurred, but far (more than 3 kilometres) from any diseased peach orchard. Classical cultural practices were applied until the trees were three years old. During the autumn of the third year, wound inoculations were performed twice (in October and in November) on one-year old shoots with an aggressive strain (M24S) of *P. syr.* pv. *persicae* resistant to streptomycin on only exactly 16 trees located in the middle of the square orchard. Leaves were randomly picked from each non inoculated tree (10 per tree) at different times (the following May, August and October) to estimate the epiphytic population levels (Luisetti *et al.*, 1982).

In early spring all the 560 non inoculated trees of the experimental orchard were individually observed and appeared completely healthy while the 16 inoculated trees were more or less damaged. No dead trees were observed at this time.

During May, epiphytic populations were recovered from 26 out of the 40 concentric quarter rings and the estimated concentrations varied widely from 10^2 to 10^8cfu/leaf (Figure 4). A significant (p = 0.01) centripetal gradient of bacterial concentrations was observed.

The summer sampling indicated that there was no longer a significant gradient of concentrations. Moreover, the mean bacterial concentration on the leaves was much lower than in the spring although the proportion of contaminated samples was greater (30/40). The bacterial concentration within the contaminated concentric quarter rings ranged from 10 to 10^4cfu/leaf (Figure 5).

In autumn, all but one of the concentric quarter rings were found to harbour epiphytic populations of *P. syr.* pv. *persicae*. Moreover, for 34 of the 40 concentric quarter rings, more than 50% of the trees were epiphytically infested.

It appeared that some trees had no detectable epiphytic populations of the pathogen until leaf fall. Moreover, two years after the inoculation within the central focus, more than 40% of the non inoculated trees remained completely healthy and without any detectable epiphytic population of the pathogen, while trees close by were dead or showed symptoms of die-back (Figure 6).

Spatial Distribution at the Scale of a Tree

All the foliar bunches of two weakly diseased young trees were collected during early spring and analysed individually to estimate their epiphytic population level. Additionally, later in spring or in summer, all the leaves of some shoots of other diseased trees were removed and analysed individually. In both cases the same method as above was used for the population estimation.

The population concentration within the foliar bunches ranged from undetectable to 10^6cfu/leaf; about 10 to 20% were without detectable epiphytic population. Nevertheless, more than 60 - 70% showed bacterial concentrations over 10^4cfu/leaf (Figure 7).

The analysis of 9 young shoots with a variable number of leaves indicated that the proportion of polluted leaves varied greatly, but that in most cases, some leaves had no detectable epiphytic populations of the pathogen (Table 1). Furthermore, the analysis of all the individual leaves carried by the four shoots of a branch, revealed that 10 to 60% of the leaves (depending on the shoot) were free of epiphytic populations of the pathogen. The proportion of bacteria-free leaves decreased from the tip to the base of the branch. The probability for a leaf to be bacteria-free appeared to be highest for young leaves (Figure 8).

NW   NE

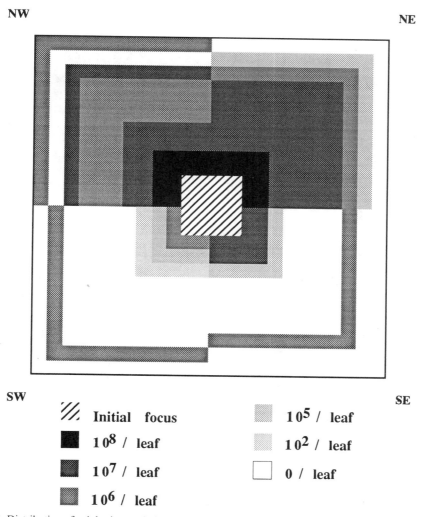

SW   SE

Figure 4. Distribution of epiphytic populations of *P. syringae* pv. *persicae* on healthy peach trees in May from symptoms developed in a central focus.

Spatial Distribution at the Scale of a Leaf

From a young and weakly diseased tree, 62 leaves were randomly sampled. Each individual leaf was first pressed on the surface of a general medium (yeast extract, peptone glucose agar, "LPGA") plus 100 ppm streptomycin to obtain the imprint of bacterial cells present on both faces, then transferred to a flask containing 20 ml sterile distilled water and washed for 30 minutes on a rotary shaker. Finally, after sampling a small quantity to estimate the concentration of *P. syr.* pv. *persicae*, the leaf was ground within the washing suspension. Both washing and grinding suspensions were sampled, serially diluted and plated on LPGA medium amended with streptomycin. All plates were incubated at 23°C for 4 days. The population was expressed either as the number of colony forming units (cfu) per leaf (imprinting method) or as the logarithm of cfu per leaf (washing or grinding method). Low

NW   **NE**

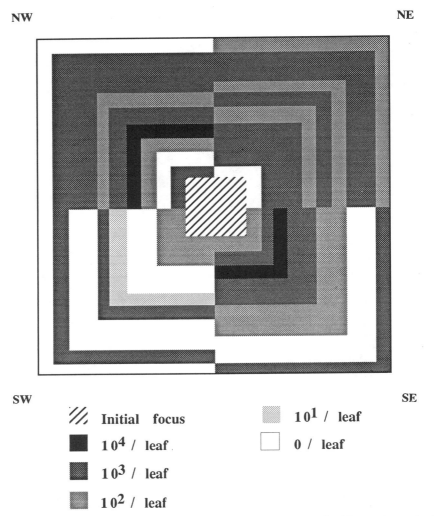

SW   **SE**

▨ **Initial focus**   ▨ 10^1 / **leaf**

■ 10^4 / **leaf**   ☐ 0 / **leaf**

■ 10^3 / **leaf**

▨ 10^2 / **leaf**

Figure 5. Distribution of epiphytic populations of *P. syringae* pv. *persicae* on healthy peach trees in August from symptoms developed in a central focus.

numbers of colonies developed from leaf prints on agar medium, both for the upper and lower sides of the leaves. Developing colonies were rarely confluent even if the epiphytic population estimated by leaf grinding was greater than 10^6 cfu/leaf.

The estimation of the epiphytic population of 62 individual leaves provided three different types of results (Table 2). Some leaves (about a third) showed a restricted number of colonies on printing plates (1 to 7) and the washing and grinding methods provided estimations that did not differ significantly. For about 50% of the leaves, if a few (sometimes no) colonies were observed after printing, no bacteria were detected in the washing suspension while significant populations were recovered after grinding. For the remaining 18% of the leaves, only the printing method revealed colonies while washing and grinding suspensions appeared free of the pathogen.

NW NE

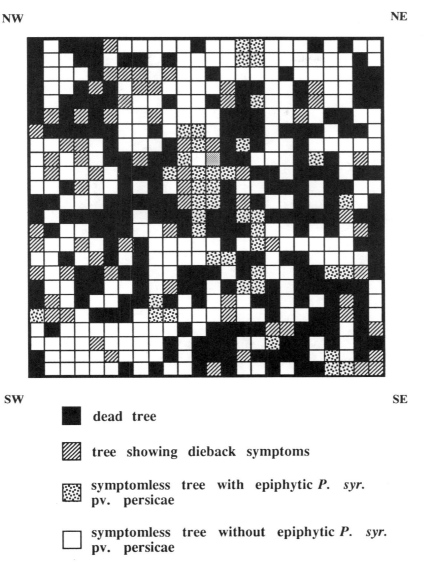

SW SE

■ dead tree

▨ tree showing dieback symptoms

▦ symptomless tree with epiphytic *P. syr.* pv. persicae

☐ symptomless tree without epiphytic *P. syr.* pv. persicae

Figure 6. Distribution of bacterial die-back on peach trees in an orchard two years after inoculation of *P. syringae* pv. *persicae* in a central focus.

To explain these results, several hypotheses may be proposed: bacterial cells could be either present as isolated individuals or grouped in microcolonies, and these cells may be located either on the leaf surface or within the tissue, for example in the substomatal chamber or inside superficial wounds. The first type of leaves listed in Table 2 would support isolated cells and also superficial microcolonies, while for the second type, the microcolonies would only be internal. For the third type of leaves, bacteria would be present only as individual cells located on the surface (Figure 9).

The distribution of the epiphytic populations of *P. syr.* pv. *persicae* on peach appears to be rather heterogeneous at any scale. The spatial distribution is also subject to rapid

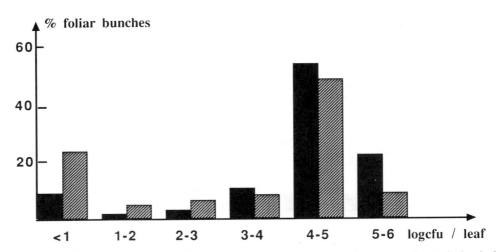

Figure 7. Distribution of the total foliar bunches from two peach trees (68 and 174) according to the level of infestation by epiphytic populations of *P. syringae* pv. *persicae*.

changes with time. Similar results were also obtained for *P. syr.* pv. *syringae* on pear and on peach (unpublished data). Little has been published on the spatial distribution of the epiphytic populations of plant pathogenic bacteria. Hirano and Upper (1989) observed daily variations in population size for *P. syr.* pv. *syringae* on bean leaflets. Kinkel *et al.* (1995) observed large variations in the size of bacterial populations on potato plants and snap bean plants at three different biological scales (including leaflet disks, entire leaflets and whole plants). Furthermore, Hirano and Upper (1985) considered each leaf as an independent biological unit with its own dynamics. All these authors reported a high level of variability whatever the scale, crop, whole plant or plant part they investigated. Therefore, the sampling design and sample size have to be adjusted according to the scale of observation in order to obtain an appropriate estimation of the size of epiphytic populations. The heterogeneous distribution of epiphytic bacteria on the leaf seems to indicate that colonisation sites play an important role in their dynamics in providing nutrients for multiplication and protection against unfavourable conditions such as drought, ultraviolet radiation, etc. Since the host ranges for epiphytic multiplication and for pathogenicity are identical for most *P. syr.*

Table 1. Epiphytic populations of *P. syringae* pv. *persicae* on the leaves of all the shoots of one peach branch (sampling in July)

Shoot number	Number of leaves	% contaminated leaves	Range of population (log cfu / leaf)
1	26	69	2,40 - 4,66
2	2	100	2,88 - 3,81
3	25	12	2,70 - 4,14
4	2	0	0
5	12	33	2,40 - 4,24
6	23	39	2,40 - 3,18
7	1	100	2,88
8	4	50	2,88 - 3,00
9	22	64	2,40 - 4,73

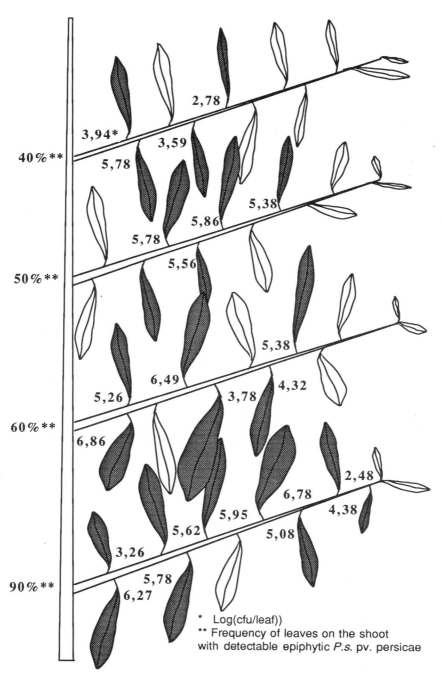

Figure 8. Epiphytic populations (log) of *P. syringae* pv. *persicae* on leaves of a peach branch.

Table 2. Estimation of epiphytic populations of *P. syringae* pv. *persicae* for individual leaves, successively by imprinting, washing, and grinding

Estimation of bacterial population sizes		
Leaf print (No. of colonies)	Leaf washing (Log (cfu/leaf))	Leaf grinding (Log (cfu/leaf))
20*		
5	3,48	3,34
5	3,00	2,88
1	2,70	2,70
2	3,18	2,88
7	4,08	3,20
7	3,40	4,34
31*		
5	0	2,88
6	0	3,11
5	0	2,88
6	0	3,26
0	0	2,40
0	0	2,88
0	0	4,65
5	0	4,73
1	0	4,51
11*		
5	0	0
2	0	0
6	0	0
7	0	0

* number of leaves (out of 62) showing similar results

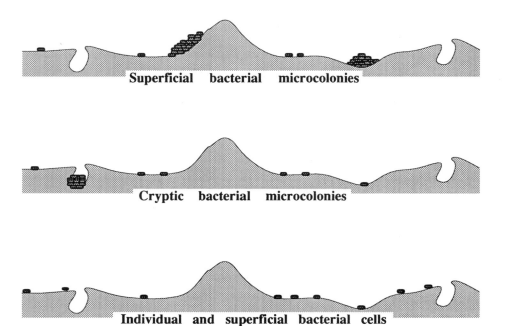

Figure 9. Hypothetical location of the bacterial epiphytic populations on the leaf.

pathovars, it could be supposed that both properties are more or less genetically linked. Some recent results seem to confirm this hypothesis (Yessad, 1992). Nevertheless, the nature of the interaction between epiphytic bacteria and the host plant remains to be elucidated.

PHENOTYPIC VARIABILITY, PATHOGENICITY, AND HOST RANGE

Though phenotypic variations do exist within any pathovar of *P. syr.*, they are greater within the pathovar *syringae,* which is considered to be pathogenic to a very broad host range. Our objective was to investigate the phenotypic variations within the populations of pathovar *syringae* epiphytic on pears. Two aspects were considered: the utilisation of different carbon sources and the host range.

Leaves (throughout one vegetative period) and buds (during the following winter) were sampled regularly from pears in an experimental orchard. The epiphytic bacteria were recovered from these samples by washing and the issued suspension and its dilutions were plated on King's medium B. For each sample a large number (between 10 and 40) of colonies resembling *P. syr.* were collected for further investigation, providing then a collection of 438 isolates. A quick characterisation based on the L.O.P.A.T. test (Lelliott *et al.*, 1966) confirmed that all belonged to *P. syr.* and induced a hypersensitive reaction on tobacco leaves. A first step was to analyse the variability in the utilisation of several carbon sources (esculin, sorbitol, inositol, erythritol, DL-lactate, D(-)-tartrate and L(+)-tartrate) and to determine the pathogenicity on three host-plants (pear, lilac and bean). A high level of variability was observed both in the utilisation of 7 carbon sources and in the host range (one third of the 438 isolates were not pathogenic on any of the three plants). A group of 57 isolates, representative of the phenotypic and pathogenic variations observed within the whole ·population, was selected for more complete characterisation including assimilation profiles on a wider range of carbon sources, serotyping, *in vitro* toxin production, ice nucleation activity and pathogenicity on a broader host range.

The assimilation profiles of the 57 selected isolates were determined using API 100 microplates (99 different carbon sources). A clustering was achieved using the Ascending Hierarchic Classification (AHC) method. In addition, DNA/DNA hybridisations using the *P. syr.* pv. *syringae* type strain (CFBP 1392 = NCPPB 281 = ATCC 19310) were performed on a subsample of 15 isolates to confirm that they belonged to this genomic species.

All the 15 isolates used for the DNA/DNA hybridisation showed a similarity coefficient to the type strain above 70%, and were considered to belong to the same genomic species.

Variations were observed only for 31 out of the 99 carbon sources tested. The AHC method separated ($p = 0.05$) the 57 isolates into 5 classes (Table 3).

Class 1 represented 14 isolates utilising propionate but not glutarate and transaconitate. Class 2 (6 isolates) was characterised by the use of benzoate and the non utilisation of β-hydroxybutyrate as sole carbon source. Most of the 21 isolates of class 3 did not metabolise DL-lactate and D-glucosamine. Class 4 (15 isolates) was characterised by the utilisation of ethanolamine, dl-α−amino-n-valerate, l-tyrosine and putrescine by most isolates. A lone isolate defined the last class: it was able to catabolise 5-keto-D-gluconate, D(-)-tartrate, and l tyrosine but not malonate, glutarate, β-hydroxybutyrate and mesotartrate, D(+)-malate, D-galacturonate, protocatechuate, p-hydroxybenzoate and quinate.

Looking at the distribution of the classes according to the dates of sampling, we observed that isolates of class 1 were recovered from blossoms twice in early spring where they represented 100% of the strains present in the samples and once from leaves in late

Table 3. Differential utilisation of 31 carbon sources by 57 isolates of *P. syr.* pv. *syringae*, epiphytic on pears

Carbone source	Number of isolates in the cluster	Cluster 1 14	Cluster 2 6	Cluster 3 21	Cluster 4 15	Cluster 5 1
				Total number of positive isolates		
tryptamine	1	-	-(5*)	-	-	-
D(+) cellobiose	1	-(13)	-	-	-	-
5-keto-D-gluconate	2	-	-	-(20)	-	+
D(-) tartrate	3	-	-	-	-(13)	+
raffinose	4	-	-	-	-(11)	-
benzoate	7	-	+	-	-(14)	-
trehalose	11	-(9)	-	-	-(9)	-
ethanolamine	12	-(12)	-(4)	-	+(8)	-
DL-α–amino-valerate	13	-	-	-	+(13)	-
propionate	17	+	-	-	-(12)	-
L-tyrosine	20	-	-	-(12)	+(10)	+
putrescine	22	-	-	-(13)	+(14)	-
DL-lactate	27	+(7)	+	-(19)	+(12)	-
D-glucosamine	31	-(8)	+(5)	-(14)	+(12)	+
glutarate	42	-(11)	+(5)	+(19)	+	-
malonate	46	+	+	+(12)	+	-
D-lyxose	46	+(13)	+	+(11)	+	+
trans-aconitate	46	-(11)	+	+	+	+
DL-β-hydroxybutyrate	50	+	-(5)	+(20)	+	-
α-keto-glutarate	55	+(13)	+	+(20)	+	+
caprate **	55	+(13)	+	+	+	-
D-sorbitol ***	56	+(13)	+	+	+	+

* number of isolates showing this response** plus caprylate*** plus L-serine

spring but mixed with isolates of other classes. Isolates of class 2 were recovered once from very young fruits and once from leaves. Nevertheless, in the 15 other samples, isolates from class 3 and class 4 were the most frequent.

Pathogenicity and Host Range

In addition to the three hosts previously used (pear, lilac and bean), 8 other plant species were inoculated with the 57 isolates plus two *P. syr.* pv. *syringae* reference strains isolated from lesions on pear (CFBP 1392 and 2027.37) and confirmed to be pathogenic on pear, lilac and bean. For pear, lilac, bean, maize, pea, fodder beet and cabbage, young plants were used. For cherry laurel (*Prunus lauro-cerasus*), Tahiti lime (*Citrus latifolia*), tangelo (*Citrus reticulata* blanco x *C. paradisi*) and pomelo (*Citrus paradisi*), young and immature leaves were picked from growing trees and deposited, after rapid disinfection in a solution of sodium hypochlorite, on a Petri dish containing 0,8% agar.

Pin prick inoculations were performed on all plants except on beet and cabbage for which the bacterial suspensions calibrated at 10^7 and 10^5 cfu/ml were applied by spraying. Symptoms were recorded after an incubation of seven days either in a controlled environment chamber (20°C, 16 hr/8 hr light/dark period and 95% relative humidity) for whole plants or in a growth cabinet (20°C, 16 hr/8 hr light/dark period) for the detached leaves.

Only two of the 57 strains tested were not pathogenic to any of the test plants. They originated from samples collected at different times of the year. Two isolates were pathogenic on only one (lime) and two (lime and cabbage) hosts. On the other hand, none were pathogenic on all 11 plants, and the broadest observed host range included 9 plant species

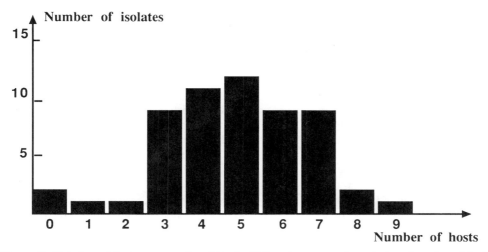

Figure 10. Variations of the hosts range size within the 57 *P. syringae* pv. *syringae* isolates from the epiphytic microflora of pear trees (11 hosts inoculated).

(one isolate). The majority of isolates (50 out of 57) were pathogenic to at least 3 - 7 different plant species (Figure 10).

Pear, bean and cherry laurel were the most susceptible plants (45 isolates) and there was an absolute conformity between susceptibility to pear and to bean, even among all the 438 isolates of the initial collection. More than 50% of the isolates were able to develop expanding symptoms on pea, cabbage, beet, lilac and maize. The three *Citrus* species were the least susceptible hosts since only 30 - 35% of the isolates induced spots on leaves of these plants after inoculation (Figure 11).

Not more than 3 isolates showed the same host range. A significant proportion of isolates pathogenic on pear (one third) did not induce any lesions on lilac.

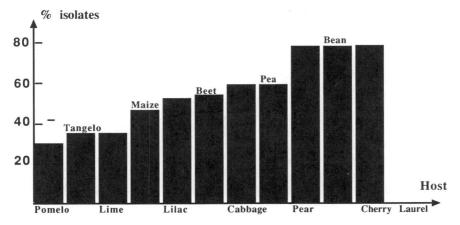

Figure 11. Pathogenicity to 11 different hosts of 57 *P. syringae* pv. *syringae* isolates from the epiphytic microflora of pear trees.

The 15 isolates selected for DNA/DNA hybridisation clearly belonged to the same genomic species still named up to now "*P. syr.* pv. *syringae*". The other 42 isolates which are not phenotypically different from them also belonged to this species. The extension of this name to the entire collection of 438 isolates recovered from the epiphytic microflora of pears appeared safe to assume. We could then consider that all the isolates from the epiphytic microflora roughly characterised as "*P. syr.* pv. *syringae*" were not mis-identified despite the variations observed between them.

The capability of the 57 isolates to utilise 99 carbon sources was as variable as reported for the pathovar *syringae* whatever the host (Gardan *et al.*, 1991). Based on the profiles of assimilation of these 99 substrates, five clusters could have been defined which could be considered as five phenons, each well characterised. These phenons seemed to coexist all along a vegetative period within the same orchard, perhaps within the same tree. Some clear-cut changes with time were observed, but the low number of isolates for each sampling time did not allow any definitive conclusion.

Based on the inoculation of 12 different plants, less than 4% of the strains were unable to cause any symptom. Perhaps most, if not all isolates would have been found pathogenic to at least one plant species if the host range study had included other fruit trees such as peach, plum, cherry, apricot and kiwi, or some herbaceous plants which are clearly considered as hosts for *P. syr.* pv. *syringae*. The conclusion would then be that the epiphytic populations of *P. syr.* pv. *syringae* are truly pathogenic to plants as suggested by the occurrence of the hypersensitive reaction when infiltrated into tobacco leaves. Nevertheless, a large fraction (80%) of these populations was pathogenic only to the host they were recovered from. The remaining bacteria were pathogenic to plant species which may have been present near the plant they were recovered from, and one may hypothesise that migration may have occurred (airborne?) from the host to the plant sampled. These results would suggest that the epiphytic populations of *P. syr.* pv. *syringae* are of epidemiological importance not only for the host which carries them but also for other plants.

The non pathogenicity to lilac of some isolates (one third) which was confirmed for all the 438 isolates collection would suggest that the definition of the pathovar *syringae* (Young, 1991) has to be revised. Many isolates clearly pathogenic to pear failed to develop symptoms on lilac, although, in their nutrient utilisation profiles, they did not differ from those pathogenic to both hosts. However, Yessad-Carreau *et al.* (1994) reported that strains isolated from pear lesions were pathogenic to both pear and lilac. Lilac may not be the best reference host for pathovar *syringae*.

CONCLUSION

The structure of the epiphytic populations of *P. syr.* is rather complicated as illustrated by the above results. The epiphytic populations on fruits trees are heterogeneously distributed whatever the scale used in the investigation. Within an orchard, the epiphytic populations harboured at the same moment by individual trees may widely vary and some trees may remain free of epiphytic bacteria. Similarly, on an individual tree the epiphytic populations may be quantitatively different between individual branches or shoots depending upon their position within the tree. Moreover, although leaves are the plant organ most frequently harbouring epiphytic populations, a high proportion of them, including both the youngest and some old ones, may remain free of epiphytic bacteria. At scale of a tree, a large size of epiphytic populations is generally correlated with a high proportion of leaves carrying epiphytic populations. Since similar observations have been reported for *P. syr.* on several woody plants (grapevine, pear and peach) and also on some herbaceous plants (bean, pea), one may conclude that a heterogeneous distribution is one of the main characteristics of the epiphytic populations of *P. syr.* This conclusion could be extended to some pathovars of *Xanthomonas campestris* known to have an epiphytic phase. Even at the scale of a leaf, the distribution of epiphytic bacteria appears heterogeneous, but only little information is available on their precise location. However, such information is essential for developing research to investigate the interactions between epiphytic bacteria and the plants that support them.

Another aspect of the complexity within the epiphytic populations has been illustrated by the phenotypic variability of epiphytic *P. syr.* pv. *syringae* recovered from pear trees. It concerns the capability of this pathovar, not only to metabolise different carbon sources, but also to induce symptoms on some hosts. These results are of importance for the epidemiology of the diseases incited by this ubiquitous pathogen.

In this chapter, only two aspects concerning epiphytic populations were developed. Since epiphytic populations are important, indeed essential for disease development, more investigations are required to provide a better understanding of the interaction between epiphytic bacteria and host plants. They would have to deal with the bacteria, with the plant and with the role of the environmental conditions.

ACKNOWLEDGMENTS

I thank Martine Devaux, J. L. Gaignard, A. Drouhard, and J. P. Lafuste for participating in the research on peach dieback and pear blight reported here. I also thank L. Gardan for having performed DNA–DNA hybridisations.

REFERENCES

Bordjiba, O. and Prunier, J.P. 1991, Establishment of an epiphytic phase by three species of *Pseudomonas* on apricot trees, *Acta Hort.* 293: 487-494.

Burr, T.J. and Katz, B.H. 1984, Overwintering and distribution pattern of *Pseudomonas syringae* pv. *papulans* and pv. *syringae* in apple buds, *Plant Dis.* 68: 383-385.

Crosse, J.E. 1959, Bacterial canker of stone-fruits. 4. Investigation of a method for measuring the inoculum potential of cherry trees, *Ann. appl. Biol.* 47: 306-317.

Crosse, J.E. 1966, Epidemiological relations of the Pseudomonad pathogens of deciduous fruit trees, *Annu. rev. Phytopathol.* 4: 291-310.

English, H. and Davis, J.R., 1960, The source of inoculum for bacterial canker and blast of stone fruit trees, *Phytopathology* 50: 634.

Ercolani, G.L. 1969, Sopravvivenza epifitica di popolazioni di *Pseudomonas mors-prunorum* Wormald da Ciliego e di *P. syringae* Van Hall da Pero sulla pianta ospite di provenizza e sull'altra pianta, *Phytopathol. Mediterr.* 8, 3: 197-206.

Ercolani, G.L. 1969, Sopravvivenza di *Pseudomonas syringae* Van Hall sul Pero in rapporto all'epoca della contaminazione, in Emilia, *Phytopathol. Mediterr.* 8, 3: 207-216.

Gardan, L., Luisetti, J. and Prunier, J.P., 1972, Variations in inoculum level of *Pseudomonas morsprunorum persicae* on the leaf surface of peach trees, *Proc. 3rd Int. Conf. Plant Path. Bact.*, Wageningen: 87-94.

Gardan, L., Cottin, S., Bollet, C. and Hunault, G. 1991, Phenotypic heterogeneity of *Pseudomonas syringae* Van Hall, *Res. Microbiol.* 142: 995-1004.

Hirano, S.S. and Upper, C.D. 1983, Ecology and epidemiology of *Pseudomonas syringae*, *Biotechnology* 3: 1073-1078.

Hirano, S.S. and Upper, C.D. 1989, Diel variation in population size and ice nucleation activity of *Pseudomonas syringae* on snap bean leaflets, *Appl. Environ. Microbiol.* 55: 623-630.

Latorre, B.A. and Jones, A.L. 1979, Evaluation of weeds and plant refuse as potential sources of inoculum of *Pseudomonas syringae* in bacterial canker of Cherry, *Phytopathology* 69: 1122-1125.

Latorre, B.A., Gonzales, J.A., Cox, J.E. and Vial, F. 1985, Isolation of *Pseudomonas syringae* pv. *syringae* from cankers and effect of free moisture on its epiphytic populations on sweet cherry trees, *Plant Dis.* 69: 409-412.

Leben, C. 1965, Epiphytic microorganisms in relation to plant disease, *Annu. Rev. Phytopathol.* 3: 209-230.

Lelliott, R.A., Billing, E. and Hayward, A.C. 1966, A determinative scheme for the fluorescent plant pathogenic Pseudomonads, *J. Appl. Bacteriol.* 29: 470-489.

Luisetti, J. 1978, L'influence du gel sur le développement des phytobactérioses. Lutte contre les gelées, pp. 89-98 In: *Journées nationales d'information INVUFLEC*, Angers.

Luisetti J. and Drouhard A. 1982. Dépérissement Bactérien du Pêcher: quelques caractéristiques de la dissémination de l'agent responsable, *Fruits*, 37, 337-343.

Luisetti, J. and Paulin, J.P. 1972, Etudes sur les bactérioses des arbres fruitiers. 3. Recherche de *Pseudomonas syringae* à la surface des organes aériens du Poirier et étude de ses variations quantitatives, *Ann. Phytopathol.* 4: 215-227.

Luisetti J., Prunier J.P., Gardan L., Gaignard J.L. and Vigouroux A. 1976, Le Dépérissement Bactérien du Pêcher. *INVUFLEC*, Paris.

Panagopoulos, C.G. 1966, Studies on the source of inoculum for blast and black pit of *Citrus*, pp. 340-345 In: *Proc.1rst Congr. Mediterr. Phytopathol. Union.*

Panagopoulos, C.G. and Crosse, J.E. 1964, Frost injury as a predisposing factor in blossom blight of pear caused by *Pseudomonas syringae* Van Hall, *Nature* 202: 1352.

Prunier, J.P., Luisetti, J. and Gardan, L. 1972, Etudes sur les bactérioses des arbres fruitiers. II. Caractérisation d'un *Pseudomonas* non-fluorescent, agent d'une bactériose nouvelle du Pêcher, *Ann. Phytopathol.* 2: 181-197.

Roos, I.M.M. and Hatting, M.J. 1986, Resident populations of *Pseudomonas syringae* on stone fruit tree leaves in South Africa, *Phytophylactica* 18: 55-58.

Yessad, S. 1992, *Etude de la relation entre le pouvoir pathogène et l'aptitude épiphyte de Pseudomonas syringae pv. syringae, agent causal du dessèchement bactérien du Poirier*, Ph.D. Thesis, University of Paris XI, Orsay, France.

Yessad-Carreau, S., Manceau, C. and Luisetti, J. 1994, Occurence of specific reactions induced by *Pseudomonas syringae* pv. *syringae* on bean pods, lilac and pear plants, *Plant Pathol.* 43: 528-536.

Young, J.M. 1991, Pathogenicity and identification of the lilac pathogen *Pseudomonas syringae* pv. *syringae* Van Hall 1902, *Ann. Appl. Biol.* 118: 283-298.

17

QUANTIFYING MICROBIAL COMPETITION ON LEAVES

Linda L. Kinkel,[1] Miriam R. Newton,[1] and Kurt J. Leonard[1,2]

[1] Department of Plant Pathology
University of Minnesota
Saint Paul, Minnesota 55108
[2] USDA ARS Cereal Rust Laboratory
Saint Paul, Minnesota 55108

INTRODUCTION AND DEFINING TERMS

The concept of competition occupies a central place in theories of ecology and evolution. Over the last three decades, extensive studies of the role of competition in regulating the population dynamics of various organisms have been accompanied by the development of a rich body of theory relating to competition (Diamond, 1978; Grime, 1979; Tilman, 1982; Roughgarden, 1983). However, not all ecologists have accepted the supremacy of competition as a mechanism for regulating natural populations (Roughgarden, 1985; Connell, 1983; Connor and Simberloff, 1986; Goldberg and Barton, 1992). In response to the intense focus on competition by some ecologists, a vociferous debate has erupted about the significance of interspecific competition in natural communities (Lewin, 1983 a, b). In recent years, researchers have provided strong evidence for the importance of disturbance, the physical environment, and extra-population movement (immigration and emigration) in determining the dynamics of specific populations (Dayton, 1971; Roughgarden, 1986). Today this debate continues, and the tension generated by the competing hypotheses has provided a fertile ground for both theory and experimentation.

Among phyllosphere ecologists there has been comparatively little corresponding debate about the relative significance of interspecific competition, disturbance, the physical environment, or extra-population movement in determining epiphytic population dynamics. Despite the publication of many studies documenting the influence of the physical environment on microbial population growth and survival on leaves and recent work that suggests that immigration can have a significant effect on epiphytic community size (Kinkel et al., 1989), we have little understanding of the relative quantitative significance of these factors in regulating phyllosphere populations. Similarly, though there have been a large number of studies reporting the existence of microbial competition on leaves (Blakeman and Brodie, 1977; Fokkema et al., 1983; Lindemann and Suslow, 1987; Adee et al., 1990; Kinkel and Lindow, 1993; Wilson and Lindow, 1994 a, b), we have only very limited insight into the

Aerial Plant Surface Microbiology, edited by Morris et al.
Plenum Press, New York, 1996

importance of competition to epiphytic populations, and on the potential selective advantage of being a good competitor. Consequently, we have little understanding of the role of microbial competition in such phenomena as pathogen race shifts and the evolution of increased aggressiveness in pathogen populations. Additionally, this lack of insight strongly limits our abilities to design effective competition-based biological control strategies.

The development of a general understanding of the significance of competition to microbial population dynamics on leaves will require at least two factors: a precise distinction between microbial competitive ability and microbial fitness; and a generalisable and density independent strategy for quantifying both the impact of competition on fitness, and the competitive abilities of coexisting micro-organisms. Fitness, formally defined, is the contribution of an individual to the next generation (Roughgarden, 1979; Hamilton, 1967; Rieger et al., 1991). For microbes on leaves, this may be reflected in population size or number of reproductive propagules produced (e.g. spores), depending upon the ecology of the organism. Fitness is influenced by many different factors, only one of which is competition. Ultimately it is the relative fitness of coexisting organisms within a particular environment that will determine which will predominate. To illustrate the importance of the distinction between competitive ability and fitness, consider the case in which two populations are inoculated onto individual leaves in a 1:1 mixture. Following incubation, populations of the two microbes are present on the leaves in a 4:1 mixture. These data indicate that the two strains differ in relative fitness. However, the strains may or may not be competitors, and the relative success of one population over the other does not prove the competitive superiority of that strain. In fact, although on the same leaf, the two distinct populations may not interact at all. Thus, the predominance of one strain over the other indicates only that strain's superior relative fitness which may or may not reflect competitive interactions between the two strains. This view of competitive ability contrasts with that held by some researchers who hold that the superior competitor is that strain that ultimately "wins" the contest (Tilman, 1987; Thompson, 1987). Such a perspective may be adequate for organisms existing in stable habitats characterised by one or a few dominant limiting resources, but is inadequate for microbes existing in the spatially and temporally heterogeneous phyllosphere environment. In fact, a synonymy between competitive ability and predominance may fundamentally obscure our ability to determine those factors that regulate epiphytic populations, for example in settings where immigration rates may determine predominance.

So how is competitive ability defined? Conceptually, competitive ability consists of two distinct components. First, competitive ability describes the ability of a particular organism to reduce the relative fitness of coexisting organisms. Second, competitive ability refers to the competence of an organism in resisting reduction in its own fitness as a function of a coexisting organism. In fact, many different types of interactions between individuals or populations can affect relative fitnesses, including predation, antibiosis, parasitism, or competition (Kinkel and Lindow, 1996). The process of resource competition occurs when the interaction between the two organisms is mediated via the demand for one or more limiting resources (Peters, 1991). Because the process of competition cannot generally be measured directly, its intensity is measured by its effect on competing organisms. Characteristically, the intensity of competition is quantified as the reduction in reproductive output, biomass, or vigour of competing organisms. One factor that can significantly confound the measurement of both the intensity of competition between populations and competitive ability is the density dependence of competition (Firbank and Watkinson, 1985; Rouse et al., 1984; Kardin and Groth, 1989; Snaydon, 1991; Holmer and Stenlid, 1993). Specifically, the effects of competition on a population are a function of the population densities of the target population and its competitor. At higher population densities and with constant competitive ability, a superior competitor will have a greater inhibitory effect on the population of an inferior competitor than at a lower population density. Because the intensity of competition, or the extent to which the relative fitness of the poorer competitor is reduced by the superior competitor, is a function of population density, a general understanding of the significance of competition to coexisting populations cannot be achieved by experimental investigation at only one population density (Firbank and Watkinson, 1985). Additionally, though the intensity of competition's effect on reproductive output for an organism is greater at a higher than at a lower population density, it does not follow that the competitive ability of the interacting organisms has itself changed as a function of density. Therefore, a density-independent measure of microbial competitive ability is needed for the development of a general model to quantify the significance of competition to microbial fitness on leaves.

In this chapter, we present a model for quantifying microbial competitive abilities independent of microbial densities, and for distinguishing differences among microbes in relative fitness from differences in competitive ability. We illustrate the use of the model for quantifying competitive interactions between *Puccinia graminis tritici* strains, and between *Pseudomonas syringae* and *Stenotrophomonas maltophilia* strains, on wheat leaves. Finally, we discuss how the model can be used to enhance strategies for selecting biocontrol organisms, to quantify and predict the significance of competition to coexisting populations having different ecological traits, and to expand our understanding of the role of competition in phyllosphere population dynamics in the field.

A MODEL FOR QUANTIFYING COMPETITIVE INTERACTIONS AMONG MICROBES ON LEAVES

The model is based upon the monomolecular equation (Campbell and Madden, 1990), and is derived in detail in Newton *et al.* (in press). We provide a brief summary here. Let reproductive output refer to the contribution of an individual to the next generation, so that this describes spore production by a fungus or final population size of a bacterium after a single or specified number of generations. The model makes two assumptions: 1. Leaves have limited resources for the support of microbial growth and thus there is a limit to reproductive output on a leaf; and 2. As the density of reproductive individuals increases, fewer resources are available to each individual and thus the reproductive output per individual decreases. Stated mathematically,

$$dQ/dU = b (M - Q) \tag{1}$$

where Q = reproductive output per leaf (spores produced or final bacterial population per leaf)

- U = number of reproductive individuals (e.g. number of sporulating lesions or initial bacterial population size)
- M = maximum reproductive output on a leaf (maximum number of spores that can be produced or maximum bacterial population size that can be achieved)
- b = reproductive efficiency (quantifies the proportional rate of approach of reproductive output per leaf to the maximum as a function of increasing density of reproductive units; specifically, b determines the proportion of Q/M achieved at a given U such that a greater b describes a higher proportion of the "carrying capacity", M, attained at a given number of reproductive individuals, U).

Solving the differential equation, $Q = M - Ce^{-bU}$, where C is a constant of integration. Since Q = 0 when U = 0, C = M. Therefore, for a microbial strain or species existing alone on leaves, $Q = M(1-e^{-bU})$ (Fig. 1). Figure 1, showing the general form of the curve described

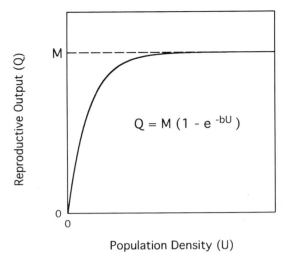

Figure 1. Reproductive output as a function of the number of reproductive individuals (population density) for a single microbial population on a leaf. The curve illustrates the finite limit (M) to the total reproductive output from any leaf, and the decreasing reproductive output per individual as total reproductive output approaches the maximum. The parameters Q, U, M, and b are as described in the text.

by this equation, thus illustrates the influence of intrastrain competition on the reproductive output per individual.

In the case where two microbial strains coexist on individual leaves, we quantify the ability of strain 2 to compete for resources limiting to reproductive output of strain 1 using the interaction coefficient β_2. The coefficient β may be either positive or negative, and its value describes the effect of a single individual of strain 2 on the reproductive output of strain 1 (interstrain interaction or interstrain competition) relative to the effect of an individual of strain 1 on its own reproductive output (intrastrain interaction or intrastrain competition). Values of β that are negative ($\beta < 0$) describe the situation where strain 2 enhances the reproductive output of strain 1. A β-value of zero indicates that strain 2 has no effect on the reproductive output of strain 1. When $0 < \beta < 1$, strain 2 has a negative effect on the reproductive output of strain 1, but the interstrain competitive effects of strain 2 on strain 1 are less than intrastrain competition in strain 1. A β-value of 1 describes the case where intra- and inter-strain competitive effects on reproductive output per individual of strain 1 are equal. Finally, values of β that are greater than 1 indicate that the per individual effect of strain 2 on the reproductive output of strain 1 are larger than the per individual effect of strain 1 on its own reproductive output, or that interstrain competition > intrastrain competition. Mathematically, the value of β_2 quantifies the number of reproductive individuals of strain 1 on a leaf which would reduce the reproductive output of strain 1 per reproductive individual on that leaf by the same amount as a single reproductive individual of strain 2. Therefore, the quantity $\beta_2 U_2$ is equal to the number of strain 1 reproductive individuals that would cause the same effect on the per individual reproductive output of strain 1 as do U_2 reproductive individuals of strain 2. This provides a means to convert U_2, or the number of reproductive individuals of strain 2, into an equivalent number of strain 1 reproductive individuals. Thus, the apparent "total" number of reproductive individuals accessing resources limiting to strain 1 is $U_1 + \beta_2 U_2$. Consequently, the proportion of strain 1 individuals as a percent of the apparent total reproductive individuals in the mixed population is $U_1/(U_1+ß_2U_2)$. Finally, the maximum reproductive output by strain 1 that can be attained on a leaf colonised by both strains will equal the maximum reproductive output for strain 1 when alone on leaves multiplied by the proportion of strain 1 individuals out of the apparent total: $M_1U_1/(U_1+ß_2U_2)$. This describes the reproductive output of strain 1 when leaves are colonised by both strains 1 and 2.

Substitution of the mixed strain terms into the single strain equation above (equation 1) provides a means for analysing interstrain competition. Specifically, $M_1U_1/(U_1+ß_2U_2)$ replaces M in the single strain equation as the maximum reproductive output for strain 1 when coexisting with strain 2. Likewise, U in equation 1 must be replaced by $U_1+ß_2U_2$ to describe the apparent total number of reproductive individuals which competitively affect reproductive output of strain 1. Thus, the equation that describes reproductive output for strain 1 as a function of the number of reproductive individuals of both strains 1 and 2 becomes:

$$Q_1 = (U_1/\{U_1+ß_2U_2\})\ M_1\ [1-e^{(-b_1(U_1+ß_2U_2))}] \tag{2}$$

An analogous equation can be developed for strain 2.

APPLICATION OF THE MODEL TO TWO EXPERIMENTAL SYSTEMS

Methods

Interactions between strains of *Puccinia graminis tritici* (Pgt) and between *Pseudomonas syringae* (Ps) and *Stenotrophomonas maltophilia* (Sm) on individual wheat leaves were quantified using the model. Urediniospores of Pgt colour mutant strains 22 (orange) and 41 (grey) were inoculated singly and in combination (1:1 ratio) onto 6-day old wheat plants (variety McNair 701). Fresh urediniospores were suspended in a light mineral oil and sprayed onto wheat leaves at 4 different total concentrations (urediniospores/ml). Following inoculation, plants were kept in a dew chamber at room temperature for 15 hours to promote infection. Plants were then placed into a controlled environment chamber (16 hour days, 22 C daytime and 18C night-time; 80% RH) for 24 days. Uredinia (lesions) were quantified on individual leaves 9 days after inoculation. In terms of the model, the number of lesions per leaf represents U, or the number of reproductive individuals. Urediniospores were collected from each individual leaf starting 10 days after inoculation and spores were collected every 2 days for 14 days or until the leaf died. Spores from serial collections from each individual leaf were bulked and quantified at the conclusion of the experiment. These data were used to calculate average spore production per leaf per day for every leaf. Average spore production per leaf per day is used to represent Q, or reproductive output per leaf in the model.

Bacterial isolates (Ps and Sm) were inoculated singly and in combination (1:1 ratio) onto 10-day old wheat leaves (McNair 701) at 4 different total inoculum concentrations (cfu/ml). Bacterial suspensions were sprayed onto leaves until runoff. Immediately following inoculation, bacterial populations on individual leaves were determined using leaf wash-dilution plating procedures. The time zero population size estimates are represented by U, or the number of reproductive individuals per leaf. Following inoculation, plants were placed individually into plastic bags to maintain high relative humidity. Bagged plants were kept in a controlled environment chamber at 22 C daytime and 18 C night-time for 48 hours. After 48 hours, population sizes on individual leaves were assessed as described above. Log-transformed population sizes (cfu) per leaf represent Q, or the reproductive output per leaf in the model.

For both the fungal and the bacterial systems, non-linear regression was used to estimate the model parameters M (maximum reproductive output or carrying capacity per leaf) and b (reproductive efficiency) for each strain when present alone on leaves. Following the estimation of M and b, non-linear regression was used on the data from the leaves on which populations were co-inoculated to determine ß-values for each strain.

Results: Pgt

Competition occurred between Pgt strains 22 and 41 on wheat leaves, and influenced relative fitnesses of the two strains. Strain 41 had a significantly greater competitive effect on strain 22 than strain 22 had on strain 41 ($\beta_{22} = 0.70$: β_{22} significantly less than 1, t = 3.72, p = 0.0002; $\beta_{41} = 1.32$: β_{41} significantly greater than 1, t = 2.96, p = 0.0034). Thus, a single pustule of strain 22 had a weaker inhibitory effect on the reproductive output (spore production) of strain 41 than one pustule of strain 41 had on itself (for strain 41, interstrain competition < intrastrain competition). In contrast, a single pustule of strain 41 had a greater effect on the reproductive output of strain 22 than one pustule of strain 22 had on itself (for strain 22, interstrain competition > intrastrain competition). Based upon these interaction coefficients, we expect that at the same total pustule density (pustules/leaf), strain 41 will produce fewer spores per pustule when alone on leaves than when co-inoculated with strain 22. Likewise, we expect that strain 22 will produce more spores per pustule when alone on leaves than when in combination with strain 41 at the same total pustule density. The data (Figs. 2 & 3) illustrate this relationship, and also show the value of the model for quantifying competitive effects. Specifically, consideration of the spore production data alone for strain 22 would lead to the conclusion that competitive interactions with strain 41 negatively influence reproductive output (Fig. 2). However, for strain 41 we see an apparent increase in spore production per pustule in the presence of strain 22 relative to spore production when strain 41 is alone on leaves when controlled for total pustule density. In the absence of the model, one may conclude that competition from strain 22 does not negatively influence strain 41. However, the model clarifies that competition from strain 22 does negatively influence the reproductive output of strain 41 ($\beta_{22} > 0$), but that the interstrain competitive effect of strain 22 is less than the intrastrain competitive effect of strain 41 on itself.

The interaction coefficients for a given strain pair characterise the overall interaction by providing a measure of the relative intensity of inter- and intra-strain competition for each individual strain. The data show that strain 41 has stronger competitive ability than strain 22 both as a function of its strong competitive effect on strain 22 ($\beta_{41} > 1$) and its relative resistance to inhibition by strain 22 ($\beta_{22} < 1$). Because a single interaction coefficient simultaneously measures both the effect of strain 41 on strain 22 and the response of strain 22 to strain 41, it does not provide a means of distinguishing the competitive effect of strain 41 from the competitive response of strain 22. However, for any strain interaction, the strain having the greater interaction coefficient has the greatest competitive ability, or the greatest effect on the relative fitness of the coexisting strain and better resistance to reductions in relative fitness in the presence of the coexisting strain. Additionally, though competitive ability for a strain as measured by the interaction coefficient is specific to a particular competitor (the interaction coefficients do not provide a strain-independent measure of competitive ability), comparison of interaction coefficients for a particular microbial strain in combination with a variety of other strains can provide a means for quantifying and comparing the differences in competitive ability among those other strains relative to the target strain.

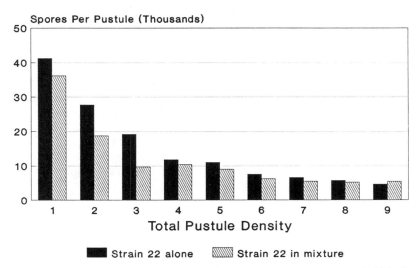

Figure 2. Mean number of spores produced per pustule for *Puccinia graminis tritici* strain 22 in the presence and in the absence of strain 41. Pustule density classes are 1: 50-100; 2:101-150; 3:151-200; 4:201-250; 5:251-300; 6:301-350; 7:351-400; 7:401-450; 8:451-500; 9: > 500.

Though strain 41 has greater competitive ability, strain 22 had superior relative fitness (greater reproductive output) in the mixture at all population densities (pustule densities) investigated (data not shown). Thus, the strain that predominates in the mixture is not the strain that has the greater competitive effect or the greater resistance to competition. This raises a fundamental question about the significance of competitive ability in conferring a selective advantage among these Pgt strains, and therefore about the importance of compe-

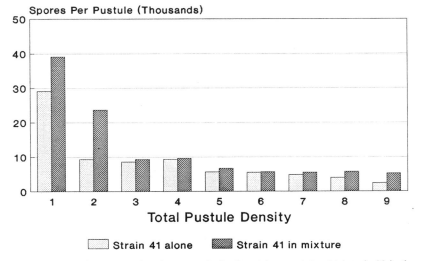

Figure 3. Mean number of spores produced per pustule for *Puccinia graminis tritici* strain 22 in the presence and in the absence of strain 41. Pustule density classes are as described for figure 2.

tition to these microbes. Though the data show that competition influences relative fitness even at low pustule densities (Figs. 2 & 3), in this case the differences in M between the two strains (2.50×10^5 for strain 22 vs. 1.82×10^5 for strain 41) are more important than the differences in competitive ability in determining which strain predominates in the presence of a competitor. In other words, the greater M compensates for the poorer competitive ability of strain 22 in influencing relative fitness in the presence of a competitor. Thus, though microbial competition occurs and influences reproductive output at pustule densities commonly observed in the field, other factors that influence reproductive output, especially host resistance (which can influence M), may be more significant in relation to relative fitness for these strains than competitive ability. The extent to which this may be generally true among a diverse collection of Pgt strains is important to determining those factors that place the greatest selection pressure on this pathogen and thus that are most important in its evolution.

Results: Bacteria

There were no significant differences between Ps and Sm in either carrying capacity (M) or reproductive efficiency (b) (data not shown). However, based upon the interaction coefficients we conclude that Ps had greater competitive ability than Sm ($\beta_P = 1.03$, $\beta_s = 0.14$; $\beta_P > \beta_s$). These ß-values describe the situation where the effect of a reproductive individual of Ps on the per-individual reproductive output of Sm is approximately the same as the effect of a single reproductive individual of Sm on itself (β_P not significantly different from 1). At 7 out of 7 different initial total densities (reproductive individual densities), the reproductive output of Sm when co-inoculated onto leaves in a 1:1 mixture with Ps was not significantly different from its per-individual reproductive output when inoculated singly onto leaves. In contrast, the effect of Sm on the reproductive output of Ps is substantially less than the effect of Ps on its own per-individual reproductive output. In terms of competitive influence, Sm is nearly invisible to Ps, as quantified by the fact that β_s is not significantly different from 0. Ps populations when co-inoculated with Sm were always greater than one-half the population density of Ps when alone on leaves at 7 of 7 total inoculum densities. Thus, the data and the interaction coefficients illustrate that, on a per individual basis, interstrain competition has a smaller influence on reproductive output for Ps than does intrastrain competition, while for Sm interstrain and intrastrain competition have virtually the same influence on reproductive output.

Though Ps and Sm differ in competitive ability, there are no distinct differences in relative fitness between the strains when co-inoculated onto leaves. Specifically, population sizes of the two strains are approximately equal over a range of initial inoculum densities (data not shown). As in the fungal system, experimental data from a diverse collection of strain combinations is necessary to determine the relationships among competitive ability (ß), carrying capacity (M), reproductive efficiency (b), and relative fitness for bacteria on leaves.

FURTHER APPLICATIONS OF THE MODEL

The model described here has the potential to significantly enhance our understanding of phyllosphere microbial interactions and their role in microbial fitness on leaves. At the most basic level, the model clarifies the distinction between microbial competitive ability and relative fitness. Furthermore, as shown for the Pgt strains on wheat leaves, the model illustrates that microbes that have the strongest competitive ability may not have the greatest relative fitness. Having clarified this distinction, the model offers the opportunity to investigate the relationships between b, M, ß, and relative fitness in both experimental and simulation settings. For example, we have used simulation modelling to determine the relative fitness benefits of a 20% advantage in either b, M, or ß while holding the other parameters constant. Using the base values of b = 0.02, M = 2 x 10^5, or ß = 1.0, a 20% advantage in competitive ability translates to only a 10% advantage in relative fitness at reproductive population densities near the maximum. In contrast, a 20% advantage in M confers a 20% advantage in relative fitness independent of population density. Finally a 20% advantage in b confers the greatest benefit in relative fitness at very low population densities, and this advantage is quickly lost with increasing population density. Specifically, a 20% advantage in b confers no net advantage in relative fitness by the time the population has reached 0.5 of the potential reproductive maximum. Experimental data summarising the frequency with which population densities at which the presence of a competitor reduces relative fitness (or at which differences in M or b confer a fitness advantage) are achieved in the field will help to elucidate those factors that place the greatest selection pressure on specific epiphytic populations. Corresponding data on the effects of the physical environment on competitive ability independent of its effect on microbial carrying capacities are also needed.

A second, practical application of the model is that it provides us with a useful criterion for the evaluation and selection of biocontrol organisms. The question of what to optimise when selecting biological control agents for use in the phyllosphere has generally not been answered very satisfactorily. When resource competition is the primary mechanism by which the pathogen population is inhibited, researchers often focus on selecting the "best competitor" for use in biocontrol. However, to be effective, the biocontrol agent must also be a good colonist. Relative to biocontrol potential, carrying capacity alone is an insufficient measure of colonisation ability. More specifically, a good colonist in this context is an organism that is capable of reaching the population densities necessary for competitive or pre-emptive exclusion of the target organism.

How can competitive and colonisation abilities be measured in a manner that appropriately quantifies biocontrol potential? Based upon our model parameters, we suggest that this be done in the following manner. First, we propose the concept of *effective density* (ED), defined as the number of individuals of a biocontrol organism necessary for complete competitive exclusion of the target (pathogen) population. Mathematically, the effective density for microbial strain 2 to completely exclude microbial strain 1 is quantified as:

$$ED_2 = M_1/ß_2 \tag{3}$$

where M and ß are as defined previously. The effective density predicts the population size necessary for "success" for a particular antagonist. Based upon this model, the population density of Sm necessary for complete competitive exclusion of Ps is 3.6 x 10^6 (Table 1), while the effective density of Ps for excluding Sm is 4.74 x 10^5. As a second step to evaluating the potential of an organism in biocontrol, the effective density for an organism must be indexed relative to the carrying capacity for that organism on leaves. Specifically, we hypothesise that the most efficient biocontrol organism will be that organism for which the ratio of the effective density to the carrying capacity is minimised. This *biocontrol index* (BI) can be stated mathematically as:

$$\overline{BI}_2 = ED_2/M_2 \tag{4}$$

and describes the proportion of its carrying capacity at which strain 2 will competitively exclude strain 1. We predict that the organism that reaches its effective density at the smallest proportion of its potential carrying capacity will be most efficient and effective in the

Table 1. Competition model parameters, effective population densities, and biocontrol
index values for *Stenotrophomonas maltophilia* and *Pseudomonas syringae* when
co-inoculated onto wheat leaves

Bacterial strain	M[1]	ß[2]	Effective density[3]	Biocontrol index[4]
Sm[5]	5×10^5	0.14	3.60×10^6	7.20
Ps[6]	5×10^5	1.03	4.74×10^5	0.95

[1] Maximum reproductive output (bacterial 'carrying capacity') on a leaf
[2] The competition coefficient, defined as the effect of a single individual of the strain on the
reproductive output of the coexisting strain relative to the effect of an individual of the coexisting
strain on its own reproductive output
[3]The effective density (ED) for a strain is the population size necessary for complete competitive
exclusion of the coinoculated strain. Calculated for strain 1 as $M_2/ß_1$
[4]The biocontrol index measures the proportion of the carrying capacity for a strain at which it will
competitively exclude the coinoculated strain. Calculated for strain 1 as ED_1/M_1
[5]*Stenotrophomonas maltophilia;* [6]*Pseudomonas syringae*

competitive inhibition of that target population. In the example of Sm and Ps (Table 1), it is
clear that Sm will never exclude Ps, and Ps will exclude Sm only when present at very near
its own carrying capacity. Thus, in this case our model would suggest that neither organism
represents a strong prospective biocontrol agent for the other. However, the model parameters provide an important new means for quantifying and comparing the potential efficiency
of biocontrol organisms that incorporates differences in both competitive and colonisation
abilities among prospective biological control agents.

SUMMARY

Is resource competition important to microbes in the phyllosphere? Though evidence
indicates that competition occurs and that resources are sometimes limiting for microbial
growth on leaf surfaces (Blakeman and Brodie, 1977; Fokkema *et al.*, 1983; Adee *et al.*,
1990; Kinkel and Lindow, 1993; Wilson and Lindow 1994 a, b), we currently have little
understanding of the general significance of resource competition to microbial fitness on
leaves. The model described here will help in determining the relationships between
competitive ability and relative fitness for microbes existing in mixed populations on leaf
surfaces at a range of total population densities. Specifically, having determined the competitive abilities of co-inoculated microbes on leaves, the model provides both a means for
determining the population densities at which competition significantly reduces the relative
fitness of those microbes, and a quantification of the magnitude of that reduction as a function
of population density. Such information is fundamental to understanding the role of competitive interactions in determining microbial population dynamics on leaves, and to predicting the potential for resource competition-based biological control strategies to achieve
success.

ACKNOWLEDGMENTS

Portions of this work were funded by a University of Minnesota Graduate School
Grant-in-Aid to Linda Kinkel, a University of Minnesota Department of Plant Pathology
Flor Fellowship to Miriam Newton, Minnesota Agricultural Experiment Station Project
22-18H, and USDA research funds. We thank Dr. William R. Bushnell, USDA-ARS Cereal
Rust Laboratory, University of Minnesota, St. Paul for supplying Pgt culture SR22 and Dr.
James D. Miller, USDA-ARS Northern Crop Science Laboratory, North Dakota State

University, Fargo for culture SR41. We thank Dr. Roger K. Jones, University of Minnesota Department of Plant Pathology for supplying the bacterial strains used in this work. We appreciate the thoughtful review of Dr. Mark Wilson, Auburn University, whose comments contributed significantly to the final manuscript.

REFERENCES

Adee, S. R., Pfender, W. F., and Hartnett, D. C. 1990, Competition between *Pyrenophora tritici-repentis* and *Septoria nodorum* in the wheat leaf as measured with de Wit replacement series. *Phytopathology* 80:1177-1182.

Blakeman, J. P and Brodie, I. D. S. 1977, Competition for nutrients between epiphytic micro-organisms and germination of spores of plant pathogens on beetroot leaves. *Phys. Pl. Pathol.* 10:29-42.

Campbell, C. L. and Madden, L. V. 1990, *Introduction to Plant Disease Epidemiology.* John Wiley and Sons, Inc., New York.

Connell, J. H. 1983, On the prevalence and relative importance of interspecific competition: evidence from field experiments. *Am. Nat.* 122:661-696.

Connor, E. F. and Simberloff, D. 1986, Competition, scientific method, and null models in ecology. *Am. Sci.* 74:155-162.

Dayton, P. K. 1971. Competition, disturbance, and community organization: the provision and subsequent utilization of space in a rocky intertidal community. *Ecol. Monogr.* 41:351-389

Diamond, J. M. 1978, Niche shifts and the rediscovery of interspecific copmetition. *Am. Sci.* 66:322-331.

Firbank, L. G. and Watkinson, A. R. 1985, On the analysis of competition within two-species mixtures of plants. *J. Appl. Ecol.* 22:503-517.

Fokkema, N. J., Riphagen, I., Poot, R. J., and deJong, C. 1983, Aphid honeydew, a potential stimulant of *Cochliobolus sativus* and *Septoria nodorum* and the competitive role of saprophytic mycoflora. *Trans. Br. mycol. Soc.* 81:355-363.

Goldberg, D. E., and Barton, A. M. 1992, Patterns and consequences of interspecific competition in natural communities: A review of field experiments with plants. *Am. Nat.* 139:771-801.

Hamilton, T. H. 1967, *Process and Pattern in Evolution.* Macmillan, New York.

Holmer, L. and Stenlid, J. 1993, The importance of inoculum size for the competitive ability of wood composing fungi. *FEMS Microbiol. Ecol.* 12:169-176.

Kardin, M. K., and Groth, J. V. 1989, Density-dependent fitness interactions in the bean rust fungus. *Phytopathology* 79:409-412.

Kinkel, L. L., Andrews, J. H., and Nordeheim, E. V. 1989, Fungal immigration dynamics and community development on apple leaves. *Microb. Ecol.* 18:45-58.

Kinkel, L. L. and Lindow, S. E. 1993, Invasion and exclusion among coexisting *Pseudomonas syringae* strains on leaves. *Appl. Environ. Microbiol.* 59:3447-3454.

Kinkel, L. L. and Lindow, S. E. 1996, Microbial competition and plant disease biocontrol. In: Andow, D., Ragsdale, D. and Nyvall, R. (eds.), *Ecological Interactions and Biological Control.* Westview Press, Boulder, Colorado (in press).

Lewin, R. 1983, Santa Rosalia was a goat. *Science* 221:636-639.

Lewin, R. 1983, Predators and hurricanes change ecology. *Science* 221:737-740.

Lindemann, J. and Suslow, T. V. 1987, Competition between ice nucleation active wild-type and ice nucleation deficient deletion mutant strains of *Pseudomonas syringae* and *P. fluorescens* biovar I and biological control of frost injury on strawberry blossoms. *Phytopathology* 77:882-886.

Lindow, S. E. 1987, Competitive exclusion of epiphytic bacateria by Ice- mutants of *Pseudomonas syringae.* *Appl. Environ. Microbiol.* 53:2520-2527.

Rieger, R., Michaelis, A., and Green, M. M. 1991, *Glossary of Genetics,* 5th Edition. Springer-Verlag, Berlin.

Roughgarden, J. 1979, *Theory of Population Genetics and Evolutionary Ecology: An Introduction.* Macmillan, New York.

Roughgarden, J. 1986, A comparison of food-limited and space-limited communities, pp. 492-516 In: Diamond, J. and Case, T.J. (eds.). *Community Ecology.* Harper and Row, New York.

Roughgarden, J., Iwasa, Y. and Baxter, C. 1985, Demographic theory for an open marine population with space-limited recruitment. *Ecology* 66:54-67.

Rouse, D. I., MacKenzie, D. R. and Nelson, R. R. 1984, Density dependent sporulation of *Erysiphe graminis* f. sp. *tritici. Phytopathology* 74:1176-1180.

Snaydon, R. W. 1991, Replacement or additive designs for competition studies? *J. Appl Ecol.* 28:930-946.

Thompson, K. 1987, The resource ratio hypothesis and the meaning of competition. *Functional Ecol.* 1:297-303.

Tilman, D. 1987, On the meaning of competition and the mechanisms of competitive superiority. *Functional Ecol.* 1:304-315.

Wilson, M. and Lindow, S. E. 1994a, Ecological similarity and coexistence of epiphytic ice-nucleating (Ice+) *Pseudomonas syringae* strains and a non-ice-nucleating (Ice-) biological control agent. *Appl. Environ. Microbiol.* 60:3128-3137.

Wilson, M. and Lindow, S. E. 1994b, Coexistence among epiphytic bacterial populations mediated through nutritional resource partitioning. *Appl. Environ. Microbiol. 60:4468-4477.*

18

PREDICTING BEHAVIOR OF PHYLLOSPHERE BACTERIA IN THE GROWTH CHAMBER FROM FIELD STUDIES

Christen D. Upper[1,2] and Susan S. Hirano[2]

[1] U.S. Department of Agriculture
   Agricultural Research Service
[2] Department of Plant Pathology
   University of Wisconsin
   Madison, Wisconsin 53706

INTRODUCTION

Greenhouse and growth chamber have served the plant sciences well. Use of a controlled environment facilitates experimental manipulation, diminishes environmental variability and provides the means to separate variables and test hypotheses. Such fundamental studies as those that led to the discovery and/or elucidation of mechanisms of phenomena such as photoperiodism, phototropism, photosynthesis, phytohormones and many others were greatly facilitated by use of controlled environmental facilities. The eminent success of such an approach for many aspects of plant-microbe interactions is well documented. For example, screening for resistance to a number of diseases has been successfully conducted in controlled environments for decades.

One common assumption that is intrinsic to use of such facilities is that results of experiments in controlled environments should be relevant to the phenomenon of interest in the field. In the examples listed in the preceding paragraph, and in many other cases, this has turned out to be an appropriate assumption. Indeed, results from studies in controlled experimental facilities have been so useful in so many cases, that this assumption has been more or less accepted as fact. However, there are some kinds of studies for which controlled environments may not serve particularly well as surrogates for the field. One need only search the literature of biological control of plant diseases to find ample examples of micro-organisms that provide excellent disease control in controlled environments but behave quite differently under field conditions (*cf.* Andrews, 1990).

In the remainder of this chapter we will defend the thesis that in the case of plant-microbe interactions, particularly in the phyllosphere, it is probably wise to assure oneself that the behavior of a system in the greenhouse or growth chamber does mimic behavior of the same system in the field before engaging in extensive laboratory studies. To

Aerial Plant Surface Microbiology, edited by Morris et al.
Plenum Press, New York, 1996

support this thesis we will provide several examples from our own research in which experiments in the laboratory and field do not yield similar results.

EXAMPLES OF PLANT–BACTERIAL INTERACTIONS THAT ARE DIFFERENT IN THE FIELD AND LABORATORY SETTINGS

Disease Symptoms in the Field Differ from Those in the Growth Chamber

The phytopathogen *Pseudomonas syringae* pv. *syringae* (Pss) is the causal agent of bacterial brown spot of snap beans (*Phaseolus vulgaris* L.). Although symptoms of the disease are expressed on the foliage of the susceptible host, the small necrotic lesions that form on developing bean pods are the economically important phase of the disease. Blemished pods may cause negative consumer reaction, and those that are malformed do not perform properly in processing machinery. Thus, fields with more than some low level of diseased pods may be bypassed, with the result equivalent to a total crop loss for the grower. Typical field symptoms of the disease on pods are shown in Fig. 1A. When growth chamber or green house grown pods are inoculated with the brown spot pathogen, the symptoms expressed are seen as dark-green, water soaked and sunken lesions (Fig. 1B). The symptoms of the disease are clearly different in the two environments. Or does Pss cause two distinct diseases, one in the growth chamber and a different one in the field?

There has been substantial interest in studies of mechanisms of plant bacterial interactions at the molecular level with Pss, including studies of brown spot disease (cf. Niepold *et al.*, 1985; Mukhopadhyay *et al.*, 1988; Willis *et al.*, 1990; Willis *et al.*, 1994; Collmer and Bauer, 1994). However, the disease that has been studied is the one in the controlled environment. If what we are ultimately interested in is mechanisms of disease

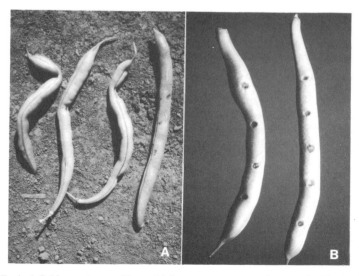

Figure 1. (A) Typical field symptoms of bacterial brown spot on snap bean pods. (B) Typical laboratory symptoms on inoculated pods from growth chamber grown bean plants. The pods were inoculated by placing 10 μl droplets of a Pss suspension (10^8 to 10^6 cfu/ml) on the pod surface, then pricking through the droplets with a 25 ga needle.

causation in the field, might we be led astray when we study the disease illustrated in Fig. 1B to try to understand the mechanism of causation of the disease illustrated in Fig. 1A? Although there is no compelling evidence to date that the laboratory pod assay is not a useful pathogenicity screen for Pss, it is intriguing that the same "molecular machinery" in the bacterium should induce such different reactions in the host. We suggest that frequent tests to show that experiments will provide the same results regardless of whether the symptoms are those shown in Fig. 1A or 1B would provide assurance that the "right" disease is being studied.

Phyllosphere Population Sizes of Pss Differ on Bean Cultivars Grown in the Field, but Not in the Growth Chamber

Population sizes of Pss on bean leaves in the field are strongly influenced by the cultivar of snap bean upon which they grow (Upper *et al.*, 1987). For example, population sizes on cultivar (cv.) Eagle are normally about two orders of magnitude higher than those on cv. Bush Blue Lake 94 (BBL94) (Fig. 2A). In the field, cv. Eagle is much more susceptible

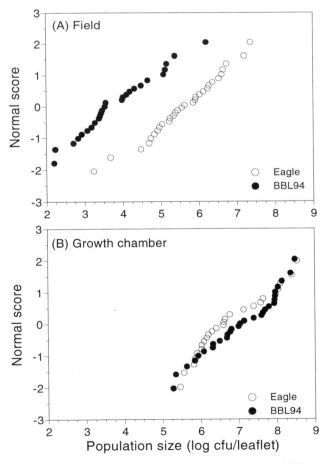

Figure 2. Population sizes of Pss on individual leaflets from (A) field-grown and (B) growth chamber-grown plants of snap bean cultivars Eagle and Bush Blue Lake 94.

to brown spot disease than is BBL94. Although population sizes of the pathogen are strongly influenced by weather conditions, with growth following intense rains and population decline occurring during dry, hot weather, population sizes of Pss on the two cultivars generally change in parallel, and the cultivar differences persist all season. Total numbers of bacteria recoverable from the two cultivars, however, are quite similar. Thus, the differences apply only to specific components of that community, and are not reflected in the size of that community (at least as can be determined by dilution plating).

When we mist-inoculated plants of the two cultivars in the growth chamber with Pss, the bacterium achieved approximately equal population sizes on both cultivars (Fig. 2B). Thus, different factors must affect regulation of Pss populations by the host in the growth chamber as compared to the field. To study the impact of host genes on such populations, it will be necessary to either conduct the experiments in the field or determine conditions that will mimic field conditions in the laboratory. We have used the former method to assist plant breeders to select bean germplasm that minimises numbers of Pss on leaves (Upper et al., 1987).

Conditions Suitable for Growth of Pss on Bean Leaves in the Field Differ from Those in the Growth Chamber

Daily changes in sizes of naturally occurring populations of Pss on bean leaves in Wisconsin are typically quite small (less than five fold) (Hirano et al., 1994a) (Fig. 3A). However, on occasion, changes may exceed 100- or even 1,000-fold (Fig. 3A). These large changes are very frequently associated with intense rains (cf. Hirano et al., 1995, 1996). Large and rapid decreases measured immediately after intense rains are likely due to rain-mediated washoff of bacteria from leaf surfaces (Lindemann and Upper, 1985). Large increases are due to periods of growth following rain (Hirano and Upper, 1989; Hirano et al., 1995, 1996). Under conditions in the field in Wisconsin, bean leaves are wet with dew most nights. Yet bursts of growth of Pss do not occur in the absence of intense rains. Although the mechanism by which intense rains trigger the onset of rapid growth of Pss remains unknown, such growth can be prevented by protecting plants from rain with plastic shelters, or by absorbing the energy of falling rain on inert window screens (Hirano et al., 1995, 1996). Thus, the momentum of falling rain is required for the growth-triggering phenomenon in the field.

In the growth chamber or greenhouse, however, all that is required for growth of Pss on a bean leaf is high humidity. When bean plants are mist-inoculated with a suspension of Pss and maintained under moist conditions or at saturated humidity, the bacterium is able to achieve population sizes approximately equivalent to those on bean leaves in the field (Fig. 3B). Even when leaves were maintained at humidities below saturation, Pss grew to relatively large numbers (Peterson and Upper, unpublished). In the laboratory setting, population sizes of Pss on leaves subjected to simulated rain of sufficient momentum to trigger growth in the field were essentially the same as those on control leaves (i.e., plants not subjected to the simulated rain). Hence, because growth of Pss on bean leaves in the growth chamber does not require raindrops with sufficient momentum, the mechanism of rain-triggered growth of the bacterium must be studied in the field. Further, because the growth patterns of Pss differ in the laboratory and field, meaningful studies of the regulation of bacterial population dynamics and sizes must also be done in the field. We do not have a suitable way to reproduce in the growth chamber the patterns of growth that occur in the field.

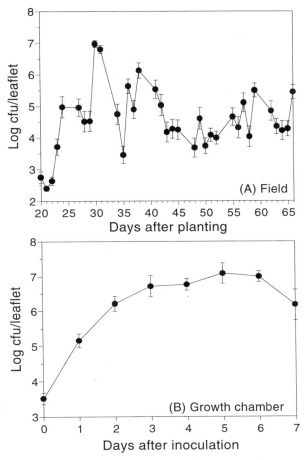

Figure 3. Population dynamics of Pss on (A) field-grown and (B) growth chamber-grown bean plants.

Relative Fitness of Mutant Strains of Pss

Lindow has devised a system of alternating hot, dry conditions with moist conditions in the laboratory to study traits associated with epiphytic fitness of Pss (Lindow, 1991, 1993; Lindow *et al.*, 1993; Beattie and Lindow, 1994a, 1994b, 1994c). With this system, Tn5-mutants were identified that grew and/or survived less well than the wild type strain (Lindow, 1993; Lindow *et al.*, 1993). These mutants, designated epiphytic fitness mutants, were considered less fit than the parental strain because they achieved smaller population sizes on bean leaves than the parent. When four of the mutants were examined under field conditions, the reduction in population size of the mutants relative to the wild type strain was greater in the field than in the laboratory (Beattie and Lindow, 1994b). Nonetheless, the general behavior of the mutants relative to each other was similar to those observed under laboratory conditions. Thus, although the laboratory experiments were qualitatively reproducible in the field setting, the results were quantitatively different. However, under the field conditions used, the applied strains of Pss grew on leaves quite well, even in the absence of rain. Thus, experiments in plots near Berkeley seem to more nearly resemble those in the greenhouse or growth chamber than those conducted in the field in Wisconsin. These results probably are indicative that the regulation of bacterial population sizes on [bean] leaves are very sensitive to environmental conditions, and that careful attention to such conditions is probably necessary in studies designed to determine how such populations are regulated.

In the past decade there has been quite a lot of activity directed toward use of molecular genetics to elucidate mechanisms of pathogenesis in bacteria. Both for reasons of experimental necessity and government regulations, these studies have been conducted under laboratory conditions. As more has become known about the genes that may affect pathogenesis, however, it has become appropriate to determine if the phenotypes determined in the laboratory are predictive of those that the same genotype produces in the field.

Willis *et al.* (1990) have isolated a gene, designated *lemA*, that is required for brown spot lesion formation in Pss. DNA sequence analysis of *lemA* demonstrated that the predicted protein sequence bears significant similarity to bacterial two-component regulatory proteins (Hrabak and Willis, 1992; Willis *et al.*, 1994). Under controlled conditions, the *lemA* mutant NPS3136 (*lemA1*::Tn*5*) was found to be indistinguishable from its wild type parent in its ability to grow on or within leaves after spray inoculation or infiltration, respectively (Willis *et al.*, 1990). When the mutant was sprayed onto bean leaves in the field, it proved substantially less fit than its wild type parent (Hirano *et al.*, 1994b). After two weeks, population sizes of the mutant were approximately 50-fold smaller than those of the parent. When the experiment was repeated, this time with inoculated seed, the population sizes of the mutant and parent strains remained approximately equal from planting to seedling emergence. As soon as the leaf habitats were fully emerged, population sizes of the mutant fell below those of the parent. Thus, on leaves in the field, the mutant was less fit than the parent. In the laboratory or on germinating seeds, however, the two strains achieved approximately the same population sizes, indicating comparable fitness in these two very different environments, and one more case where we have not successfully modelled field behavior in the growth chamber.

The most intriguing difference between behavior of mutant and parent strains in the laboratory and field was observed when the two strains were applied simultaneously to the same leaves (or seeds) (Hirano *et al.*, 1994b). On the same leaf with its wild type parent, strain NPS3136 behaved exactly as it had when applied alone. However, the parent, when in the presence of the mutant in the field, mimicked the behavior of the mutant. In the laboratory this suppression of the parent by the nonpathogenic mutant did not occur, or was so small as to be difficult to measure.

On the basis of these experiments, one wonders if the phenotypes of other bacterial mutants that affect plant-bacterial interactions will be the same in the field as in the laboratory. Such studies of mutants are the basis of our current and developing knowledge of the mechanisms of plant-bacterial interactions. As noted below, logistics and regulations preclude doing the early stages of screening and mechanism evaluation in the field. Thus, should it not be worth the effort to assure that the behavior of these interactions in the greenhouse or growth chamber is adequately predicted by behavior of the same interactions in the field? By taking what actually happens in the field to the greenhouse or growth chamber, should we not have greater confidence that what we learn in the controlled environment reflects what happens in the field?

CONCLUDING REMARKS

What is the general lesson to take home from all of this? We have proposed that it is probably worth the effort to use the behavior of interest in a plant-microbe interaction in the field as the criterion for developing laboratory systems for studying that interaction. In those cases where this can be readily achieved, it should make the eventual transition from the laboratory to the field much smoother. However, we have also provided several examples that suggest that we are not very good — or at least, not very successful — at following our own advice. In all of the cases illustrated here we have completely failed to find conditions under which our field data will predict behavior of the Ps-bean system in the growth chamber. For the most part, we have coped with the situation by avoiding the problem. We are interested in the ecology of bacteria on bean leaves, and feel that there is no better place to study such a system than in the field. We face serious limitations, however, when we seek to elucidate the mechanism of a phenomenon such as the effect of the momentum of rain on growth of Pss, a problem that would be much easier to study in the laboratory, if only we knew how. For many scientists who work on plant-microbe interactions in the phyllosphere, performing their experiments in the field is simply impractical or even impossible. For example, the logistics of screening thousands of mutants to search for a particular phenotype would be unfathomable under the intrinsic variability encountered in the field. Such a venture may be highly impractical. Further, in most cases, such mutants would probably have been made in ways that would render them subject to regulation as engineered organisms. It may be some years before the responsible regulatory agencies in most countries will permit such a screen to be conducted in the field. Thus, the impractical experiment is also legally impossible as well. Then what can be done? At the very least, when better ways to assure that laboratory experiments mimic the field are not available, frequent experiments should be performed to compare results from experiments done in the two different environments. Although it is neither practical nor possible to conduct screens of Tn5 mutants in the field, it is both possible and practical to test the phenotype of the mutant that is chosen from a laboratory screen to make sure that the phenotype is the one desired in the environment that is of ultimate importance — the field.

REFERENCES

Andrews, J.H., 1990, Biological control in the phyllosphere: Realistic goal or false hope, *Can. J. Plant Pathol.* 12:300-307.

Beattie, G.A., and Lindow, S.E., 1994a, Survival, growth, and localization of epiphytic fitness mutants of *Pseudomonas syringae* on leaves, *Appl. Environ. Microbiol.* 60:3790-3798.

Beattie, G.A., and Lindow, S.E., 1994b, Comparison of the behavior of epiphytic fitness mutants of *Pseudomonas syringae* under controlled and field conditions, *Appl. Environ. Microbiol.* 60:3799-3808.

Beattie, G.A., and Lindow, S.E., 1994c, Epiphytic fitness of phytopathogenic bacteria: Physiological adaptations for growth and survival, pp. 1-27 In : Dangl, J.L. (ed.) *Bacterial Pathogenesis of Plants and Animals: Molecular and Cellular Mechanisms*, vol. 192. Springer-Verlag Heidelberg.

Collmer, A., and Bauer, D.W., 1994, *Erwinia chrysanthemi* and *Pseudomonas syringae*: Plant pathogens trafficking in extracellular virulence proteins, pp. 43-78 In : Dangle, J.L. (ed.) *Bacterial Pathogenesis of Plants and Animals: Molecular and Cellular Mechanisms*, vol. 192. Springer-Verlag Heidelberg.

Hirano, S. S., Baker, L.S., and Upper, C.D., 1996, Raindrop momentum trigger growth of leaf-associated populations of *Pseudomonas syringae* on field-grown snap bean plants, *Appl. Environ. Microbiol.* 62: 2560-2566.

Hirano, S.S., Clayton, M.K., and Upper, C.D., 1994a, Estimation of and temporal changes in means and variances of populations of *Pseudomonas syringae* on snap bean leaflets, *Phytopathology* 84:934-940.

Hirano, S.S., Ostertag, E.M., Savage, S.A., Willis, D.K., and Upper, C.D., 1994b, Contribution of the regulatory gene *lemA* to fitness of *Pseudomonas syringae* pv *syringae* in the phyllosphere and spermosphere under field conditions, *Molec. Ecol.* 3:607.

Hirano, S.S., Rouse, D.I., Clayton, M.K., and Upper, C.D., 1995, *Pseudomonas syringae* pv. *syringae* and bacterial brown spot of snap bean: A study of epiphytic phytopathogenic bacteria and associated disease, *Plant Dis.* 79:1085-1093.

Hirano, S.S., and Upper, C.D., 1989, Diel variation in population size and ice nucleation activity of *Pseudomonas syringae* on snap bean leaflets, *Appl. Environ. Microbiol.* 55:623-630.

Hrabak, E.M., and Willis, D.K., 1992, The *lemA* gene required for pathogenicity of *Pseudomonas syringae* pv. *syringae* on bean is a member of a family of two-component regulators, *J. Bacteriol.* 174:3011-3020.

Lindemann, J., and Upper, C.D., 1985, Aerial dispersal of epiphytic bacteria over bean plants, *Appl. Environ. Microbiol.* 50:1229-1232.

Lindow, S.E., 1991, Determinants of epiphytic fitness in bacteria,pp. 295-314 In : Andrews, J.H. and Hirano, S.S. (eds.) *Microbial Ecology of Leaves*, Springer Verlag New York.

Lindow, S.E., 1993, Novel method for identifying bacterial mutants with reduced epiphytic fitness, *Appl. Environ. Microbiol.* 59:1586-1592.

Lindow, S.E., Andersen, G., and Beattie, G.A., 1993, Characteristics of insertional mutants of *Pseudomonas syringae* with reduced epiphytic fitness, *Appl. Environ. Microbiol.* 59:1593-1601.

Mukhopadhyay, P., Williams, J., and Mills, D., 1988, Molecular analysis of a pathogenicity locus in *Pseudomonas syringae* pv. *syringae*, *J. Bacteriol.* 170:5479-5488.

Niepold, F., Anderson, D., and Mills, D., 1985, Cloning determinants of pathogenesis from *Pseudomonas syringae* pathovar *syringae*, *Proc. Natl. Acad. Sci. USA* 82:406-410.

Upper, C.D., Hirano, S.S., Rouse, D.I., Kmiecik, K.A., and Bliss, F.A., 1987, Host avoidance of ice nucleation active *Pseudomonas syringae* on bean, pp. 1027 In : Civerolo, E.L., Collmer, A., Davis, R.E., and Gillaspie, A.G. (eds) *Plant Pathogenic Bacteria*, Martinus Nijhoff Publishers Dordrecht, Netherlands.

Willis, D.K., Hrabak, E.M., Rich, J.J., Barta, T.M., Lindow, S.E., and Panopoulos, N.J., 1990, Isolation and characterization of a *Pseudomonas syringae* pv. *syringae* mutant deficient in lesion formation on bean, *Mol. Plant-Microbe Interact.* 3:149-156.

Willis, D.K., Rich, J.J., Kinscherf, T.G., and Kitten, T., 1994, Genetic regulation in plant pathogenic Pseudomonads, pp. 167-193 In : Setlow, J.K. (ed.) *Genetic Engineering: Principles and Methods*, vol. 16, Plenum Press, New York.

19

PHYLLOSPHERE ECOLOGY

Past, Present, and Future

John H. Andrews

Department of Plant Pathology
University of Wisconsin-Madison
1630 Linden Drive
Madison, Wisconsin 53706-1598

INTRODUCTION

In a session at the 6th International Symposium on the Microbiology of Aerial Plant Parts (from which this book issues), all participants were asked to consider three things: 1) the most important successes in phyllosphere research to date (that is, facts or subject matter considered to be known or largely known); 2) the most important areas for research over the next five years; and 3) creative visions for the future. Discussions were held in multiple small groups, each under the direction of a moderator, and subsequently in an assembled body.

For the most part, the following chapter attempts to present a consolidated and somewhat reorganised compilation of the ideas developed in that process. To provide some unifying themes, the comments and suggestions have been assigned to one of the following four categories: i) habitat - including the leaf (or aerial plant part) and atmosphere; ii) microbial population dynamics - including adhesion, grown, death, immigration, and emigration; iii) community structure - species composition, biological control, energy flow, and nutrient cycling; and iv) experimental approaches related to sampling, measurement, identification, and detection. I consider first those aspects that were recognized as being relatively well known or unknown future priorities, and conclude with the speculations for the future. Since this is a synopsis and perspective - not a review article - only a few key references, or those not within the mainstream of phyllosphere microbiology, are cited. For referenced specific information, readers are referred to the earlier chapters in this volume; to the five existing phyllosphere volumes (Preece and Dickinson, 1971; Dickinson and Preece, 1976; Blakeman, 1981; Fokkema and van den Heuvel, 1986; Andrews and Hirano, 1991); and to the original literature.

Known and Unknown

The Aerial Habitat. At least from the time of the first volume in the phyllosphere microbiology series (see above), and particularly from the contributions in micrometeorol-

Aerial Plant Surface Microbiology, edited by Morris et al.
Plenum Press, New York, 1996

285

ogy, it has been recognized that the phyllosphere is a highly dynamic habitat subject to pronounced cyclic and noncyclic variation. A thin shell or boundary layer of air surrounds leaves and other aerial plant parts. This attenuates wind turbulence and the rate of change of some physical factors such as humidity in that narrow zone between the leaf surface and the bulk atmosphere. Even on a microscale, however, the aerial parts of plants are subject to appreciable change in time and space with respect to such factors as temperature, wind speed, radiation, humidity, dew, rainfall, and chemical milieu. What needs to be determined soon is the extent to which these factors individually and in concert shape the epiphytic community and the importance of tolerance to fluctuations in determining success in phyllosphere colonisation.

Because of the large body of literature in plant anatomy, morphology, and physiology, we know quite well the water and gas relations of leaves, and their structure, particularly as this relates to the surface layers (Juniper and Jeffree, 1983). We also know that plants can respond to stress, and even to such seemingly minor stimuli as touch (Braam and Davis, 1990). Undoubtedly, these responses often carry over to influence the microbial community, though to what extent is unknown. There is a carrying capacity (K) set by the plant habitat for epiphytic micro-organisms and this varies spatially and through time. But, what specific factors set K? To what extent is it influenced by nutrients and antimicrobial compounds that diffuse onto the leaf surface *vs.* those that are deposited in the form of pollen, insect honeydew, debris, and pollutants? How does K vary by position upon or among leaves, by season, plant species, level of soil fertility, etc.? What role does insect feeding or wounding play in exudate chemistry? The questions and details here seem endless; we are just beginning to get some answers (Schönherr and Baur, this volume; Derridj, this volume) that follow on from the pioneering work by Tukey and others in the 1960's. This needs to be pursued actively.

Finally, we know that the phyllosphere habitat can be colonised, internally and externally, by diverse microbes. Some of these colonists are strictly epiphytes; some are solely endophytes (emerging only to sporulate or if the host dies); and some, most notably the bacteria, are both to varying degrees (see Community Structure and Processes). The plant sets the stage in the colonisation process. It both affects, and is affected by, its associated micro-organisms. The best evidence for the former is the general host property of resistance to most micro-organisms; the most striking example of the latter is enhanced fitness imparted to plants by certain endophytes. There is also the long-standing observation that epiphytic communities influence plant susceptibility to disease. Research on the molecular basis of disease resistance and, related to this, of the virulence of pathogens, is at the cutting edge of plant pathology (Kombrink and Somssich, 1995). What isn't known is how and to what extent the plant habitat can be manipulated to alter endophytism or promote biological control.

Population Dynamics. A lot is known in a general way about the dynamics of phylloplane microbial populations. The nutritional physiology of bacteria and fungi is known in considerable detail based on laboratory culture. Their metabolic processes and versatility in nature are considerably less well understood. We have known for a long time that epiphytic population sizes vary enormously, even among adjacent, visually "identical" leaves and within sectors of a leaf canopy. The range in size of at least some bacterial populations is log normally distributed while certain fungal populations have been described as normally distributed. Undoubtedly, such variation reflects primarily the marked heterogeneity in habitat noted above. But to what extent? We also know that in agroecosystems there is a complicated interplay among chemicals (such as fungicides and insecticides), pathogens, insects, and epiphytic microbes. Modelling these interactions and predicting outcomes from various starting conditions will be immensely complex.

What are the factors that determine epiphytic fitness or the colonising ability of phyllosphere microbes? Will these traits be fundamentally the same or different for different taxa? What do phylloplane residents have that transients or soil inhabitants lack? These are some of the most interesting and important questions in leaf microbiology. There is now some evidence that, at least for certain bacteria, motility and osmotolerance are important (Lindow 1991; Beattie and Lindow, 1995). The ability to nucleate ice, among species capable of doing so, does not seem to be important in contributing to the epiphytic phase. Evidence for the importance of antibiotic or extracellular polysaccharide production and tolerance to UV light is equivocal. By using an inactivation mutagenesis approach it may be possible to implicate other traits that are not intuitive as colonisation factors (Lindow, 1991; Beattie and Lindow, 1995).

Something of the demographic factors accounting for population change is also known. Populations must change by some function of birth, death, immigration, and emigration (see Lindow, this volume). In some situations birth and probably death seem to be the overriding factors (Hirano and Upper, 1989), while in others immigration (and consequently emigration from somewhere) is important (Kinkel et al., 1989). Much is known from the large body of literature in plant pathology about the adhesion, preinfectional growth, and transport of pathogens. How relevant is this to the biology of non-pathogens? At this point little more has been done than to quantify populations, describe general trends, and identify some of the likely underlying conditions. Why, for instance, do bean cultivars grown in the field vary by up to 1,000-fold in population sizes of *Pseudomonas syringae* pv. *syringae*, yet show no appreciable differences in the growth chamber (Upper and Hirano, this volume)? Why do intense rains trigger bursts of rapid growth of *P. syringae* pv. *syringae* but not of other bacteria associated with the same leaves? To these and many other questions, the answers, for the moment, remain elusive.

As alluded to above, plant pathologists have studied for decades the adhesion of fungal pathogens as the first stage in the series of steps in preinfectional growth. In a previous chapter, Romantschuk (this volume) has presented the current status of knowledge about adhesion of phytopathogenic bacteria to plant surfaces. However, there is virtually no information on how saprophytic bacteria and yeasts adhere to leaves. Adhesion is clearly a major attribute in microbial ecology because the ability of micro-organisms to adhere is generally - if not universally - required for growth on surfaces, infection, morphogenesis, and vector-associated dissemination. Correlates of adhesiveness typically include invasiveness, virulence, competitive or colonising ability, persistence and, in medical ecosystems, resistance to antibiotic chemotherapy. Adhesion is thus a general consideration for growth at any surface, whether this be on plants; on teeth, skin, or mucosa; and on inanimate objects such as rocks, pipelines, ship hulls, and medical implants. Among the unanswered questions are to what extent extracellular polymers contribute to adhesion on leaves and whether or not specific lock-and-key type mechanisms occur between microbe and cuticle.

Finally, we know that phylloplane populations must change in gene frequency as well as in size, though until recently little was known about this other than for populations of plant pathogens. Research at the NERC Institute of Virology and Environmental Microbiology in England (Bailey et al., this volume) has shown that the fluorescent pseudomonads colonising sugarbeet leaves are phenotypically closely related though comprising more than 100 distinct ribotype groups. Only a small subset, perhaps one-fifth of the ribotypes, occurs at two or more sampling times and one-tenth at two or more seasons. Peaks in abundance among the latter tend to occur at the same time each season. This apparent turnover in clones through time is perhaps conceptually not unlike the turnover in *E. coli* clones in the intestine (Linton et al., 1978). Why does this occur and what are the implications? Related work (Bailey et al., this volume) shows that naturally occurring mercury-resistance plasmids can move by conjugative transfer between donor and recipient *Pseudomonas fluorescens* cells on leaf surfaces. Thus the genetic structure of epiphytic populations changes through time by the colonisation of new genotypes, and gene transfer among receptive cells, as well as by mutation. How significant the changes may be and whether they can be manipulated, as in the case of biological control, remain to be determined.

Community Structure and Processes

From many studies dating back to the early work of Ruinen, we have a good knowledge of the fraction of the phylloplane community that can be detected and identified by plating techniques. Application of molecular methods to studies of the microbial ecology of leaves is anticipated and will facilitate enumeration and identification of microbes that can not be cultured. At this point it is anyone's guess as to what the eventual estimate of the biodiversity of phylloplane communities will be. The microbes represented to date by conventional approaches are genetically and physiologically diverse. As is the case typically also for plant and animal communities, there are a few abundant microbe species and a great preponderance of relatively rare species.

Beyond identity, we also know that there is a general successional sequence of micro-organisms on the phylloplane. This is characterized by a increase in abundance and diversity over time and an initial preponderance of bacteria with an eventual increase in yeasts and filamentous fungi. However, this is only a very broad trend that tends to be disrupted by local events related, for example, to weather and vegetation patterns. Colonists can originate from the unfolding buds themselves, or from soil, seed, or the air. The importance of each as a source pool remains unknown. Similarly, we still do not know how important the role of "founder" strains or species may be in determining the outcome of colonisation. Species composition of bacteria and fungi in the phyllosphere microbial community does not reflect that of the soil or rhizosphere, though soilborne microbes can be isolated from leaves sporadically. Host-specific biotypes may exist, but there is no strong evidence yet for host species specificity among phylloplane microbes other than the phytopathogens.

The phylloplane community extends within the leaf as well as residing on its surface. That bacteria colonise substomatal cavities has long been known; it is now clear that aerial plant organs as well as the roots are frequently inhabited by bacterial and fungal endophytes. Colonisation of cavities offers plant pathogens a bridgehead in invasion and, similarly, may provide a protected repository of inoculum for growth on the phylloplane under favourable conditions. Through their production of toxic secondary metabolites which deter herbivores and pathogens, fungal endophytes appear to be mutualists with plants. How common are bacterial endophytes and what is their role? We are just now beginning to recognise some of the implications of the enlarged concept of a phylloplane community.

For bacteria in some microbe-plant systems, especially involving the rhizosphere, there are considerable data implicating particular competitive and survival mechanisms such as production of siderophores or antibiotics. The evidence varies from largely circumstantial to elegant work involving deletion mutants and detection of the putative chemical at physiologically active concentrations *in situ* (Weller and Thomashow, 1990; Hamdan *et al.*, 1991). Little of this sort of work has been published to date for the phyllosphere. Competition has been demonstrated routinely but the mechanisms usually are obscure. Dating back to the work of Blakeman and colleagues in the 1970's (see Preece and Dickinson, 1971; Dickinson and Preece, 1976) nutrient competition has been implicated, though experimentally it is much more difficult to demonstrate conclusively than antibiosis. Progress has been made recently by replacement series analysis of competitive interactions between near isogenic bacterial strains (Wilson and Lindow, 1994) and application of selective nutrients such as salicylate, which are available to one but not the other competing strain (Wilson and Lindow,

1995). There is almost unlimited scope for mechanistic studies of microbial interactions in the aerial habitat.

Dozens if not hundreds of papers in the past three decades testify to the fact that biological control of pathogens occurs naturally or can be induced on leaf surfaces. Antagonists have been manipulated more-or-less successfully in the aerial habitat (leaves, flowers, fruit) and for the control of post-harvest diseases of fruit (Droby et al., this volume). The successes typically have been inconsistent, however, and without sufficient economic impact to have been adopted on a significant scale by growers. Thus we know considerably more about what doesn't succeed than what does. We have learned much, mostly from molecular biology and soil microbiology systems, about how biological control works and that several mechanisms generally operate concurrently. Microbial antagonism on fruit surfaces is a rapidly expanding realm of biocontrol, driven by the need to find alternatives to fungicides. The continuing challenge here will be to develop agents that can exert their activity consistently and effectively in the environment of the commodity without presenting significant hazards to human health.

Overall, the major question in biocontrol is whether effective, dependable strategies will be forthcoming. If these are to be developed, they will likely result from a wise choice of a host-pathogen system that is amenable to biological control; integrating microbial antagonism with other approaches to control; attacking the disease cycle at its most vulnerable point for biocontrol (which may mean, for example, in the leaf litter rather than on living leaves); and improving colonisation of antagonists. These are the directions for future research and they point out that we need a much better understanding of the basic microbial ecology of plant-microbe interactions, especially of the factors that determine competitive success in this habitat.

Broadly speaking, improvements in biological control could come from manipulating either the host plant or the biocontrol agent, and participants suggested several possible approaches for the future. Instead of- or in addition to - incorporating resistance genes, the most obvious options involve changing the plant environment to promote colonisation by antagonists. While there are numerous ways this could be accomplished, probably the most successful will be to genetically alter the host by conventional breeding or recombinant DNA methods. As an example, work is already underway to transform plants to produce specific substrates that can be catabolised exclusively or preferentially by the desired microbial strain (Savka and Farrand, 1992; Wilson et al., 1995). Thus, developing plant mutants altered in various chemical or physical properties may prove just as worthwhile as the more conventional approach of developing microbial mutants. Other possibilities include improving formulation of antagonist inoculum, strain improvement, applying multiple strains or species of antagonists complementary in mode of action, and enhancing endophytes as biocontrol agents. There has been little study of organisms other than fungi, bacteria, and nematodes as biocontrol agents. What role do viruses, myxomycetes, and soil animals play in biocontrol in leaf litter?

The role of phyllosphere micro-organisms in energy flow and nutrient cycling, identified in the previous volume as an important and virtually unexplored area (Andrews, 1991), remains so today. Among the most interesting aspects is how tightly linked are soil, plant, and epiphytic community. The continuum starts with nutrients and inhibitory compounds that are washed from leaves, and ultimately the fallen leaves themselves, that locally influence soil chemistry and biology. Since soil nutrition affects the nutrient status of the plant, this impact quite possibly carries over to the composition of the leaf exudates. These, in turn, affect the epiphytic (and endophytic?) community quantitatively and probably also qualitatively. So the research issues will involve testing the extent to which these probable associations are real: How do changes at one level in the system affect other levels? What role do epiphytes play in energy flow and nutrient cycling in natural ecosystems and in agroecosystems (particularly in the tropics where epiphytic biomass is relatively high and turnover rapid)? Do diseased or "stressed" plants have a significantly different microflora from healthy plants? To what extent does the microbial community associated with living leaves set the stage for decomposition of litter?

Experimental Approaches

Participants at the symposium on aerial plant surface microbiology generally felt that the behavior of plant-microbe systems in the greenhouse or growth chamber was relatively well known and predictable, unlike that in the field. This situation is merely a specific example of the advantages and limitations of controlled *vs.* natural experiments in biology. The laboratory, growth chamber, or greenhouse offer the highest possible regulation of independent variables and replication and randomisation of control and experimental treatments. This is at the cost of restrictions in time and space, realism, and generality (Diamond, 1986). The field offers essentially the opposite permutation: no regulation of independent variables and the lowest ability to match conditions or sites. The advantages, however, include a maximum assessment over time and space, high realism, and high generality.

It is by now a standard tenet of biological control that successful control under defined conditions is generally not highly correlated with control in nature. Likewise, in efforts to complete Koch's Postulates, plant pathologists have long recognized that disease symptoms frequently are not reproduced when plants are raised and inoculated under controlled conditions. Upper and Hirano (this volume) have pointed out numerous instances of the striking disparity in plant-microbe behavior in the two environments. As Diamond's (1986) assessment clearly shows, however, these constraints do *not* mean that growth chamber studies are pointless, only that we need to be sure that they are pertinent to the question being asked.

It is interesting to review the five previous phyllosphere volumes in the context of how approaches to the study of leaf surface microbiology have changed since 1970. As our discipline has evolved there has been a general trend away from descriptive accounts to an hypothesis-testing approach. Research now tends to have a more quantitative basis as reflected in the questions asked in the experimental design, and in validation and data analysis.

Of course, there have also been changes in instrumentation, related mainly to microscopy and molecular biology. Major advances in specimen preparation for microscopy, type of label for detection, and mode of examination are now available to researchers in microbial ecology. For example, cryofixation and other preparative procedures designed to minimise artifacts are now routine; moreover tissue printing methods have been developed to localise on a replica of the surface of interest molecules such as proteins, nucleic acids, and carbohydrate moieties (Reid *et al.*, 1991). Nonisotopic labels based on bioluminescence, chemiluminescence, fluorescence or colorimetry can be substituted for isotopes such as ^{32}P or ^{125}I. Cell analytical procedures have advanced from microspectrophotometry through autoradiography to flow cytometry (fluorescence activated cell sorting) (Javois, 1994).

Probably the most exciting advance in microscopy potentially relevant to the phyllosphere is confocal laser microscopy coupled with digital image analysis (Shotton, 1993). In this approach, the specimen is scanned by a spot of laser light focused through an epifluorescence microscope. Emitted light from only one plane in the specimen is admitted by the confocal aperture, hence the photomultiplier tube receives only focused light which is then displayed point-by-point on a charged couple device or other high resolution monitor. Consequently, interference from out-of-focus objects is eliminated and 3-dimensional images can be reconstructed by computer by optically sectioning through specimens. The cooled charge couple device, a solid-state silicon matrix of light-sensitive pixels, is the most sensitive camera system available to date. Light that is only marginally visible by eye can be readily resolved. Digital imaging allows images recorded electronically to be refined, pseudocolours to be imposed to enhance clarity, optical sectioning to be performed, and contrast and intensity to be varied for quantification. It has been used in diverse areas of microbial ecology, including the analysis of population growth, cell motility, bacterial attachment kinetics, viability, and biofilm architecture and chemistry (Caldwell *et al.*, 1992). Confocal microscopy and image analysis should be ideal for coping with the contoured topography of leaf surfaces.

Molecular methods offer precision and high sensitivity to the characterization of microbial populations and have revolutionised the capability to detect, identify, and quantify microbes. The applications that seem to have the most potential for use in quantifying populations are based either on immunochemistry (e.g., ELISA) or nucleic acid hybridisation (e.g., dot blots, PCR, probes). For example, nucleic acid probes bound to homologous target sequences can be visualised by radioactive or nonradioactive detection systems. The latter include fluorescence *in situ* hybridisation or FISH, used in cytology and to monitor bacteria in various habitats. Probes, including FISH, have been used in phylloplane studies (Li *et al.*, 1994), along with reporter molecules, such as the ß-glucuronidase and *ice* gene systems. These and related tools provide the means for increasingly detailed and accurate studies of microbe behavior on leaves (see Mechaber *et al.*, 1996).

THE FUTURE (5–10 YEARS)

Obviously, the ideas expressed in this section are speculative and some are wilder than others. Whatever else happens in the coming decade it seems clear that the rapidly breaking fields of technology, biotechnology, and molecular biology will drive the advances in biology, including those in phyllosphere microbial ecology. Not only will the new techniques and information be incorporated into different areas of biology, but the development will spur advances in the more "traditional" disciplines such as plant physiology and plant population biology.

Major developments in the computational sciences can be expected. Computer simulation models will exist for population dynamics of epiphytes and release of biocontrol agents. Advances in electronics will enable environmental conditions on the phylloplane at the microbial scale to be monitored by biosensors. Maybe we will even be able to see and recover microbes by non-destructive methods. In part because of these developments it will be possible to predict major changes in epiphytic communities, such as successional trends in the dominant microflora, and the impact of drought or fungicides. Through advances in image analysis, confocal microscopy, and related methods, researchers will be able to assess microbial colonisation patterns on leaves relatively quickly and accurately. Some participants felt that there might even be virtual reality videos of a plant surface voyage.

We can expect major advances in understanding of communication within microbial communities and between microbe and host plant. As noted earlier, in this active research area information is accumulating rapidly for host-pathogen (Kombrink and Somssich, 1995) and host-mutualist (Fisher and Long, 1992) systems. What remains to be seen is to what extent and how epiphytic microbes that appear to be free-living communicate with their hosts. Probably we shall find that cell-to-cell signalling and the expression of key regulatory genes are important in determining epiphytic fitness. Can phylloplane bacteria regulate gene expression as a function of their cell numbers, to mutual benefit, the way they do by autoinduction in certain highly evolved symbioses, such as between marine animals and luminous bacteria (Meighen, 1991; see also Hwang *et al.*, 1994) ? Or have there simply not been the selection pressures in the aerial habitat for these associations to evolve? Will there be evidence for reciprocal or unidirectional DNA transfer between epiphytic microbe and host, or is this phenomenon restricted to biological oddities such as the *Agrobacterium*-plant relationship (Winans, 1992).

Given the new array of molecular tools, it is easy to imagine that within the decade we will know much more about epiphytic community structure in taxonomic, physiological, and genetic terms. The taxonomic information, however, is likely to be restricted to a few microbe species or groups of experimental interest, such as the pseudomonads and possibly some of the yeasts, for which probes will be developed. We will probably know the extent of genetic variability for these populations through time or across sites, based on analysis by RAPD's, RFLP's, or other methods as appropriate, depending on the level at which comparisons are to be made. We should know, at least for the major bacteria, how genetic structure of epiphytic communities changes temporally and spatially and, from assessment based on reporter genes, the significance of gene transfer on the phylloplane.

There will be some commercially successful examples of foliar biological control within the decade. As noted earlier, these will most likely come by capitalising on particular disease cycles and cropping situations conducive to control and by integrating cultural, chemical, and biological methods. One can imagine a microbe engineered either for colonising ability or antagonistic potential. Possibly a strain of *Bacillus* incorporating both disease and insect control potential will be available, heralding an era of multidisciplinary efforts in biocontrol. Plant varieties may be released that will have been selected for their ability to support antagonistic microflora.

Biocontrol, broadly construed, will also be mediated by the plant itself. Within the past decade plants have been found to be versatile and adaptable factories for foreign gene expression (Weising *et al.*, 1988). They have been altered genetically to impart resistance to certain herbicides, insects, and plant viruses, and to produce compounds of industrial or pharmaceutical value such as plastics and proteins. It is now a routine matter to insert genes into plants and achieve expression in the transformants. The coming decade will be marked by similar advances, including the control of other plant pathogens. While the most obvious examples will involve expression of resistance genes, perhaps control will be accomplished also by production *in planta* of antibodies or antibiotics, or by immunising plants against pathogens.

CONCLUSION

The 6-volume series on phyllosphere microbiology provides a good historical record of the evolution of the discipline. These texts report conference proceedings held at 5-year intervals from 1970-1995. The general picture that emerges conveys what might be described as three general eras: 1) early work characterized largely by observational and descriptive studies on the microbiology of aerial plant parts ("natural history era"); 2) efforts to put the discipline on a solid mathematical basis and to define the extent of inherent variation ("quantitative era"); and, most recently, 3) experimental, hypothesis-driven studies on phylloplane communities, populations, and processes ("mechanistic era"). Far from being unique to phyllosphere microbiology, this progression could characterise most scientific endeavours, especially those in biology.

This chapter is different from its predecessor (Andrews, 1991) because it represents a collective opinion rather than that of one individual. Nevertheless it has very similar content and reaches remarkably similar conclusions. This may mean either that we have not progressed far in the intervening five years or that there is some consensus on the major themes which still remain the most salient. The latest symposium and these proceedings show that while good and occasionally exciting advances have been made in some areas since 1990, we have progressed little in others.

The origins of leaf surface microbiology were in plant pathology and this discipline remains the dominant influence today. There is the danger of parochialism and progress in phyllosphere microbiology tends to be equated with progress in biocontrol. If our science is to thrive in the coming decades it must move aggressively outward to embrace other disciplines and fresh perspectives.

ACKNOWLEDGMENTS

I thank Cindy Morris and Linda Kinkel for their major role in organising the discussion session from which this chapter emerges and for help in many ways. Assistance of the leaders of the individual discussion groups is acknowledged with appreciation: Gwyn Beattie, Richard Belanger, Greg Boland, James Buck, Tom Burr, Vern Elliott, Regine Samson, and Mark Wilson. I thank Susan Hirano and Steve Lindow for comments on a draft of this chapter. Finally, I acknowledge program support from the NSF, USDA, and EPA.

REFERENCES

Andrews, J. H. 1991, Future research directions in phyllosphere ecology. pp. 467-479, IN: Andrews, J. H. and Hirano, S. S. (eds.) *Microbial Ecology of Leaves* Springer-Verlag, NY.

Andrews, J. H. and Hirano, S. S. (eds). 1991, *Microbial Ecology of Leaves*. Springer-Verlag, NY.

Beattie, G.A. and Lindow, S.E. 1995, The secret life of foliar bacterial pathogens on leaves, *Annu. Rev. Phytopathol.* 33:145-172.

Blakeman, J. P. (ed). 1981, *Microbial Ecology of the Phylloplane*. Academic Press, NY.

Braam, J. and Davis, R. W. 1990, Rain-, wind-, and touch- induced expression of calmodulin and calmodulin-related genes in *Arabidopsis. Cell* 60:357-364.

Caldwell, D. E., Korber, D. R., and Lawrence, J. E. 1992, Confocal laser microscopy and digital image analysis in microbial ecology. *Adv. Microb. Ecol.* 12:1-67.

Diamond, J. 1986, Overview: laboratory experiments, field experiments, and natural experiments, pp. 3-22, IN: Diamond, J. and Case, T. J. (eds), *Community Ecology*. Harper and Row, NY.

Dickinson, C. H. and Preece, T. F. (eds). 1976, *Microbiology of Aerial Plant Surfaces*. Academic Press, NY.

Fisher, R. F. and Long, S. R. 1992, *Rhizobium*-plant signal exchange. *Nature* 357:655-660.

Fokkema, N. J. and van den Heuvel, J. (eds). 1986, *Microbiology of the Phylloplane*. Cambridge Univ Press, NY.

Hamdan, H., Weller, D. M. and Thomashow, L. S. 1991, Relative importance of fluorescent siderophores and other factors in biological control of *Gaeumannomyces graminis* var. *tritici* by *Pseudomonas fluorescens* 2-79 and M4-80R. *Appl. Envir. Microbiol.* 57:3270-3277.

Hirano, S. S. and Upper, C. D. 1989, Diel variation in population size and ice nucleation activity of *Pseudomonas syringae* on snap bean leaflets. *Appl. Environ. Microbiol.* 55:623-630.

Hwang, I., Li, P.-L., Zhang, L., Piper, K., Cook, D., Tate, M. and Farrand, S. K. 1994, TraI, a LuxI homologue, is responsible for production of conjugation factor, the Ti plasmid N-acylhomoserine lactone autoinducer. *Proc. Nat. Acad. Sci. USA* 91:4639-43.

Javois, L. C. (ed). 1994, Immunocytochemical Methods and Protocols. *Methods in Molecular Biology* vol. 34. Humana Press, Totowa, NJ.

Juniper, B.E. and Jeffree, C.E. 1983, *Plant Surfaces*, Edward Arnold, London.

Kinkel, L. L., Andrews, J. H., and Nordheim, E. V. 1989, Fungal immigration dynamics and community development on apple leaves. *Microb. Ecol.* 18:45-58.

Kombrink, E. and Somssich, I. E. 1995. Defense responses of plants to pathogens. *Adv. Bot. Res.* 21:1-34.

Li, S., Spear, R., and Andrews, J. H. 1994, Detection and quantification of fungal cells on leaves by *in situ* hybridization based on 18S rRNA-targeted fluorescent oligonucleotide probes. *Abstracts, Fifth Inter. Mycol. Congr.* Vancouver, B.C., August 1994. p. 125.

Lindow, S. E. 1991, Determinants of epiphytic fitness in bacteria, pp. 295-314. IN: Andrews, J. H. and Hirano, S. S. (eds). *Microbial Ecology of Leaves*, Springer-Verlag, NY.

Linton, A. H., Handley, B., and Osborne, A. D. 1978, Fluctuations in *Escherichia coli* O-serotypes in pigs throughout life in the presence and absence of antibiotic treatment. *J. Appl. Bact.* 44:285-298.

Mechaber, W. L., Marshall, D. B., Mechaber, R. A., Jobe, R. T., and Chew, F. S. 1996, Mapping leaf surface landscapes. *Proc. Natl. Acad. Sci.* USA 93: 4600-4603.

Meighen, E. A. 1991, Molecular biology of bacterial bioluminescence. *Microbiol. Rev.* 55:123-142.

Preece, T. F. and Dickinson, C. H. (eds). 1971, *Ecology of Leaf Surface Micro-organisms*. Academic Press, NY.

Reid, P. H., Pont-Lezica, R. F., and del Campillo, E. and Taylor, R. (eds.). 1992, *Tissue printing: Tools for the Study of Anatomy, Histochemistry, and Gene Expression*. Academic Press, NY.

Savka, M. A., and Farrand S. K. 1992, Mannityl opine accumulation and exudation by transgenic tobacco. *Plant Physiol.* 98:784-789.

Shotton, D. (ed). 1993, Electronic light microscopy. Wiley-Liss, NY.

Weising, K., Schell, J. and Kahl, G. 1988, Foreign genes in plants: transfer, structure, expression, and applications. *Annu. Rev. Genet.* 22:421-477.

Weller, D. M. and Thomashaw, L. S. 1990, Antibiotics: Evidence for their production and sites where they are produced, pp. 703-711, IN: Baker, R. R. and Dunn, P. E. (eds.) *New Directions in Biological Control: Alternatives for Suppressing Agricultural Pests and Diseases*, Liss, NY.

Wilson, M., Savka, M. A., Hwang, I., Farrand, S. K., and Lindow, S. E. 1995, Altered epiphytic colonization of mannityl opine-producing transgenic tobacco plants by a mannityl opine-catabolizing strain of *Pseudomonas syringae*. *Appl. Environ. Microbiol.* 61:2151-2158.

Wilson, M. and Lindow, S. E. 1994, Ecological similarity and coexistence of epiphytic ice-nucleating (Ice+) *Pseudomonas syringae* strains and a non-ice-nucleating (Ice-) biological control agent. *Appl. Environ. Microbiol.* 60:3128-3137.

Wilson, M. and Lindow, S. E. 1995, Enhanced epiphytic coexistence of near-isogenic salicylate-catabolizing and non-salicylate-catabolizing *Pseudomonas putida* strains after exogenous salicylate application. *Appl. Environ. Microbiol.* 61:1073-1076.

Winans, S. C. 1992, Two-way chemical signalling in *Agrobacterium*-plant interactions. *Microbiol. Rev.* 56:12-31.

CONTRIBUTORS

John H. Andrews
Department of Plant Pathology
University of Wisconsin
Madison, WI 53706-1598, USA
tel: (1)-608-262-9642
fax: (1)-608-263-2626
e-mail: jha@plantpath.wisc.edu

Mark J. Bailey
Institute of Virology & Environmental
  Microbiology
Mansfield Road
Oxford OX1 3SR, United Kingdom
tel: (44)-1-865-512361
fax: (44)-1-865-59962
e-mail: mbj@mail.nerc-oxford.ac.uk

Alain Baille
INRA - Station de Bioclimatologie
Site Agroparc
84914 Avignon cedex 9, France
tel: (33)-4-90-31-60-79
fax: (33)-4-90-89-98-10

Peter Baur
The Phytodematology Group
Institute of Fruit and Nursery Sciences
University of Hannover, Am Steinberg 3
D-31157 Sarstedt, Germany
tel: (49)-50-66-826110
fax: (49)-50-66-826111
e-mail: baur@mbox.iob.uni-hannover.de

Katarina Björklöf
Dept. of General Microbiology
University of Helsinki
Mannerheimintie 172
SF 00300 Helsinki, Finland
tel: (358)-0-708-59213
fax: (358)-0-708-59262
e-mail: katarina.bjorklof@helsinki.fi

Ulla Bonas
CNRS, Institut des Sciences Végétales
Avenue de la Terrasse
Bât. 23
91198 Gif sur Yvette cedex, France
tel: (33)-1-69-82-36-12
fax: (33)-1-69-82-36-95
e-mail: bonas@trefle.isv.cnrs-gif.fr

Edo Chalutz
Institute for Technology & Storage
of Agricultural Products
ARO, The Volcani Center, P.O. Box 6
Bet Dagan 50250, Israel
tel: (972)-3-9683615
fax: (972)-9683622
e-mail: vtfrst@volcani.agri.gov.il

Donald D. Clarke
Dept. of Botany
University of Glasgow
Glasgow G12 8QQ, United Kingdom
tel: (44)-141-330-5833
fax: (44)-141-330-5834

Sylvie Derridj
Laboratoire des médiateurs chimiques
INRA - Station de Phytopharmacie
Route de St. Cyr
78026 Versailles cédex, France
tel: (33)-1-30-83-31-64
fax: (33)-1-30-83-31-49

Julian P. Diaper
Institute of Virology & Environmental
  Microbiology
Mansfield Road
Oxford OX1 3SR, United Kingdom
tel: (44)-1-865-512361
fax: (44)-1-865-59962

Samir Droby
Institute for Technology & Storage
of Agricultural Products
ARO, The Volcani Center, P.O. Box 6
Bet Dagan 50250, Israel
tel: (972)-3-9683618
fax: (972)-9683622
e-mail: vtsdroby@volcani.agri.gov.il

Kielo Haahtela
Dept. of General Microbiology
University of Helsinki
Mannerheimintie 172
SF 00300 Helsinki, Finland
tel: (358)-0-708-59258
fax: (358)-0-708-59262
e-mail: kielo.haahtela@helsinki.fi

Susan S. Hirano
Dept. of Plant Pathology
1630 Linden Dr.
University of Wisconsin
Madison, WI 53706, USA
tel: (1)-608-263-2092
fax: (1)-608-263-2626
e-mail: ssh@plantpath.wisc.edu

Marie-Agnès Jacques
INRA - Station de Pathologie Végétale
BP 94
84143 Montfavet cedex, France
tel: (33)-4-90-31-63-84
fax: (33)-4-90-31-63-35
e-mail: jacques@avignon.inra.fr

Linda Kinkel
Dept. of Plant Pathology
University of Minnesota
1991 Upper Buford Ave., Room 495
St. Paul, MN 55108, USA
tel: (1)-612-625-0277
fax: (1)-612-625-9728
e-mail: lindak@puccini.crl.umn.edu

Kurt J. Leonard
Dept. of Plant Pathology
University of Minnesota
1991 Upper Buford Ave., Room 495
St. Paul, MN 55108, USA
tel: (1)-612-625-9728
fax: (1)-612-649-5054

Andrew K. Lilley
Institute of Virology & Environmental
  Microbiology
Mansfield Road
Oxford OX1 3SR, United Kingdom
tel: (44)-1-865-512361
fax: (44)-1-865-59962

Steven E. Lindow
Dept. Environmental Science, Policy and
  Management
Univ. of California, 108 Hilgard Hall
Berkeley, CA 94720-3110, USA
tel: (1)-510-642-4174
fax: (1)-510-643-5098
e-mail: icelab@violet.berekely.edu

Elisa Longo
Department of Microbiology, Faculty of
  Sciences
University of Vigo, Spain
tel: (34)-81-592490
fax: (34)-81-594631

Jacques L. Luisetti
INRA - CIRAD
Laboratoire de Phytopathologie
BP 180
97455 St. Pierre cedex, La Réunion
tel: (262)-35-76-34
fax: (262)-25-83-43

Peter G. Mantle
Dept. of Biochemistry
Imperial College of Science, Technology &
  Medicine
London SW7 2AY, United Kingdom
tel: (44)-171-594-5245
fax: (44)-171-225-0960
e-mail: p.mantle@ic.ac.uk

Cindy E. Morris
INRA - Station de Pathologie Végétale
B.P. 94
84143 Montfavet, France
tel: (33)-4-90-31-63-84
fax: (33)-4-90-31-63-35
e-mail: morris@avi-amp.avignon.inra.fr

Miriam Newton
Dept. of Plant Pathology
University of Minnesota
1991 Upper Buford Ave., Room 495
St. Paul, MN 55108, USA
tel: (1)-612-625-0277
fax: (1)-612-625-9728

Christophe Nguyen-the
INRA - Station de Technologie des
  Produits Végétaux
Site Agroparc
84914 Avignon cedex 9, France
tel: (33)-4-90-31-61-60
fax: (33)-4-90-31-62-98
e-mail: nguyenth@avignon.inra.fr

Philippe C. Nicot
INRA - Station de Pathologie Végétale
B.P. 94
84143 Montfavet, France
tel: (33)-4-90-31-63-66
fax: (33)-4-90-31-63-35
e-mail: nicot@avignon.inra.fr

Eeva-Liisa Nurmiaho-Lassila
Dept. of General Microbiology
University of Helsinki
Mannerheimintie 172
SF 00300 Helsinki, Finland
tel: (358)-0-708-59209
fax: (358)-0-708-59262

Tuula Ojanen
Dept. of General Microbiology
University of Helsinki
Mannerheimintie 172
SF 00300 Helsinki, Finland
tel: (358)-0-708-59205
fax: (358)-0-708-59262
e-mail: tuula.ojanen@helsinki.fi

William Pfender
Dept. of Plant Pathology
Kansas State University
Manhattan, Kansas 66506, USA
tel: (1)-913-532-6176
fax: (1)-913-532-5692
e-mail: pfender@plantpath.pp.ksu.edu

Alan D.M. Rayner
School of Biology & Biochemistry
University of Bath
Claverton Down
Bath BA2 7AY, United Kingdom
tel: (44)-0225-826826 ext. 5419
fax: (44)-0225-826779
e-mail: a.d.m.rayner@uk.ac.bath.ss1

Elina Roine
Dept. of General Microbiology
University of Helsinki
Mannerheimintie 172
SF 00300 Helsinki, Finland
tel: (358)-0-708-59214
fax: (358)-0-708-59262
e-mail: elina.roine@helsinki.fi

Martin Romantschuk
Dept. of General Microbiology
University of Helsinki
Mannerheimintie 172
SF 00300 Helsinki, Finland
For courier mail:
University of Helsinki, Biocenter
Division of General Microbiology
Viikinkaari 9
FIN-00710 Helsinki, Finland)
tel: (358)-0-708-59219
fax: (358)-0-708-59262
e-mail: martin.romantschuk@helsinki.fi

Jörg Schönherr
The Phytodematology Group
Institute of Fruit and Nursery Sciences
University of Hannover, Am Steinberg 3
D-31157 Sarstedt, Germany
tel: (49)-50-66-826110
fax: (49)-50-66-826111

Christen D. Upper
Dept. of Plant Pathology
1630 Linden Dr.
University of Wisconsin
Madison, WI 53706, USA
tel: (1)-608-263-2092
fax: (1)-608-263-2626
e-mail: cdu@plantpath.wisc.edu

Guido Van den Ackerveken
CNRS, Institut des Sciences Végétales
Avenue de la Terrasse
Bât. 23
91198 Gif sur Yvette cedex, France
tel: (33)-1-69-82-36-12
fax: (33)-1-69-82-36-95

Tomás G. Villa
Facultad de Farmacia
Universidad de Santiago de Compostela
15071 Santiago de Compostela, Spain
tel: (34)-81-592490
fax: (34)-81-594631
e-mail: mpvilla@usc.es

Charles L. Wilson
USDA, ARS
Appalachian Fruit Research Station
45 Wiltshire Road
Kearneysville, WV 25430, USA
tel: (1)-304-725-3451
fax: (1)-304-728-2340

Michael E. Wisniewski
USDA, ARS
Appalachian Fruit Research Station
45 Wiltshire Road
Kearneysville, WV 25430, USA
tel: (1)-304-725-3451
fax: (1)-304-728-2340

INDEX

DATE DUE

DEC 0 1 2016	
OCT 1 8 2017	